Max-Planck-Institut für ausländisches öffentliches Recht und Völkerrecht

Beiträge zum ausländischen öffentlichen Recht und Völkerrecht

Begründet von Viktor Bruns

Herausgegeben von
Armin von Bogdandy · Rüdiger Wolfrum

Band 235

Susanne Wasum-Rainer • Ingo Winkelmann
Katrin Tiroch (eds.)

Arctic Science, International Law and Climate Change

-Legal Aspects of Marine Science in the Arctic Ocean-

Papers from the International Conference at the German Federal Foreign Office in cooperation with the Ministry of Foreign Affairs of Finland

Berlin, 17/18 March 2011

Springer
the language of science

ISSN 0172-4770
ISBN 978-3-642-24202-1 ISBN 978-3-642-24203-8 (eBook)
DOI 10.1007/978-3-642-24203-8
Springer Heidelberg New York Dordrecht London

Die Deutsche Nationalbibliothek verzeichnet diese Publikation in der Deutschen Nationalbibliografie; detaillierte bibliografische Daten sind im Internet über http://dnb.d-nb.de abrufbar.

Gedruckt auf säurefreiem Papier

Springer ist Teil der Fachverlagsgruppe Springer Science+Business Media (www.springer.com)

Preface and Acknowledgements

This book is the product of the Second International Berlin Conference on Arctic issues. The Conference titled 'Arctic Science, International Law and Climate Change: Legal aspects of Marine Science in the Arctic Ocean' was organized in March 2011 by the German Federal Foreign Office together with the Finnish Foreign Ministry. It is a sequel to the first such Conference in 2009 organized jointly by Denmark and Norway on 'New Chances and New Responsibilities in the Arctic Region'.

The melting of the Arctic Ocean's ice masses is causing dramatic changes in the area's natural environment, especially for Arctic fauna. At the same time new opportunities are opening up for resource exploration and exploitation, easier or entirely new shipping routes and for fisheries. Since 2007 most of these issues have been discussed intensively in a large number of political and scientific forums. There have also been documents on the Arctic published by the EU Council, the European Commission and the European Parliament. Numerous Arctic countries have formulated Arctic strategies.

One aspect, however, which has been largely neglected in the international discussion to date are the parameters within which marine science research in the Arctic Ocean is conducted. Given the current challenges, this is a key concern. The Arctic Ocean is both a showcase for global climate change and a scientific site supplying crucial data for foundational research into climate change. So it is extremely important to focus not only on the economic, logistical and fisheries-related aspects of the Arctic but also on the role of polar scientists and the parameters within which they pursue their research.

What framework does international law currently provide for Arctic marine science? Is it likely to change in future? Will it remain in its present form? What are the future priorities for Arctic marine science? What can it tell us about climate change, what legal aspects are involved here? How can international cooperation on Arctic issues be strengthened? How far does the common (ecological) heritage of mankind principle shape the answers to these questions?

Bringing together the views of prominent experts in the field of international law, scientists, researchers and diplomats, this book will open up new horizons, we hope, for combating global climate change through unhampered marine science research in an Arctic Ocean accessible to all.

We would like to express our sincere thanks to all Conference moderators and contributors for their valuable input. Our colleagues from the Finnish Foreign Ministry did sterling work as co-organizers of the Conference. Special thanks go to Ambassador Hannu Halinen and Petri Hakkarainen for their close and excellent cooperation. Volker Rachold and Bernhardt Coakley contributed greatly to the conceptual design of the Conference.

The support we received from five renowned academic institutions in Bremerhaven, Fairbanks, St. Petersburg, Heidelberg and Rovaniemi and their respective directors was critical to the Conference's success. In this connection we would particularly like to thank Karin Lochte from the Alfred-Wegener-Institut für Polar- und Meeresforschung; Larry Hinzman from the International Arctic Research Center, Ivan Frolov from the Arctic and Antarctic Research Institute, Rüdiger Wolfrum from the Max Planck Institute for Comparative Public Law and International Law as well as Paula Kankaanpää from the Arctic Center of the University of Lapland.

The members of the Conference task force worked almost round the clock to make the event an organizational and academic success. Sincere thanks (in alphabetical order) go to Dagmar Amling, Helmut Aust, Johannes Buckow, Jan Flasche, Mario Hemmerling, Anja Jaekel, Franziska Scherff, Hannelene Schilar, Theresa Schönfeld, Hendrik Wassermann, Klaus Wendelberger and Nadine Zachariadis for their enthusiasm and commitment. Special thanks also go to the publishing staff at the Max Planck Institute, in particular Christiane Philipp, for their dedicated work.

Berlin, März 2012

Susanne Wasum-Rainer
Ingo Winkelmann
Katrin Tiroch

Table of Contents

Guido Westerwelle

Climate Change, International Law and Arctic Research –
Legal Aspects of Marine Research in the Arctic Ocean 1

Maria Damanaki

The Arctic: A Test Bench for International Dialogue 5

Jaakko Laajava

Arctic Science, International Law and Climate Change 9

Lars-Otto Reiersen & Simon Wilson

The Arctic – a Sentinel for Environmental Processes and
Effects 15

Michael Gahler

More Relevant Research for the EU Arctic Policy 37

Peter Lemke

Arctic Processes and the Global Climate 45

Keith Alverson

Developing and Sustaining an Arctic Ocean Observing
System 55

Igor Ashik

Recent Russian Marine Research Activities in the Arctic Ocean 59

Uwe Nixdorf

Arctic Research in Practice 67

Larry Mayer

Arctic Marine Research: The Perspective of a US Practitioner 83

Bernard Coakley

Environmental Permitting Constraints on Arctic Marine Scientific Research 97

Alexander S. Studenetsky

Marine Scientific Research in Waters under the Jurisdiction of the Russian Federation 101

Klemetti Näkkäläjärvi

Climate Change and Traditional Knowledge 105

Christian Marcussen

Extended Continental Shelf Issues in the Arctic Ocean: A Modern 'Land Grab' Or an Example of Cooperation between the Arctic Coastal States? 111

Ted L. McDorman

Setting the Stage: The Continental Shelf and Marine Science in the Arctic Ocean 119

Zhiguo Gao

Legal Issues of MSR in the Arctic: A Chinese Perspective 141

Paul Arthur Berkman

'Common Interests' as an Evolving Body of International Law: Applications to Arctic Ocean Stewardship 155

Michael W. Lodge

The International Seabed Authority and the Arctic............ 175

Alex G. Oude Elferink

The Regime for Marine Scientific Research in the Arctic: Implications of the Absence of Outer Limits of the Continental Shelf beyond 200 Nautical Miles............ 189

Betsy Baker

Common Precepts of Marine Scientific Research Access in the Arctic............ 209

Appendix I: White Paper: Marine Research Access in the Arctic Ocean: Background for Potential Guidelines in a Changing Arctic by *Betsy Baker and Hajo Eicken*............ 233

Vladimir Golitsyn

Climate Change, Marine Science and Delineation of the Continental Shelf............ 245

Jean-Claude Gascard

From the DAMOCLES to ACCESS Projects (Sixth & Seventh EU Framework Programmes 2005-2015) IAOOS – An Advanced Arctic Ocean Observing System (2011-2019)............ 261

Hajo Eicken

Internationally Coordinated, Cooperative Arctic Marine Science during the Fourth International Polar Year: Lessons for Future Arctic Ocean Science Agreements............ 285

Udloriak Hanson

Inuit, Circumpolar Law & Politics, Resource Development 299

Susanne Wasum-Rainer

Conclusions of the Chair 307

Annex 1

Second International Arctic Forum 'The Arctic – Territory of Dialogue' Arkhangelsk, 22 – 23 September 2010 *by Vladimir Putin* 311

Annex 2

Letter dated 18 July 2011 from the Permanent Representative of Germany to the United Nations addressed to the Secretary-General *by Peter Wittig* 317

Annex 3

Report on a sustainable EU policy for the High North (2009/2214(INI)) Committee on Foreign Affairs *by Michael Gahler* 323

Annex 4

Programme 353

Annex 5

List of Participants 361

List of Abbreviations

AARI	Arctic and Antarctic Research Institute
ABE-LOS	Advisory Body of Experts on the Law of the Sea
AC	Arctic Council
ACEX	Arctic Coring Expedition
ACIA	Arctic Climate Impact Assessment
ACOBAR	Acoustic Technology for Observing the interior of the Arctic Ocean
AD	*Anno domini*
AEPS	Arctic Environmental Protection Strategy
AERONET	Aerosol Robotic Network
AIRSS	Arctic Ice Regime Shipping System
AITP	Acoustic Ice Tethered Platforms
AMAP	Arctic Monitoring and Assessment Programme
AMEC	Arctic Military Environmental Cooperation
AMSR-E	Advanced Microwave Scanning Radiometer for the Earth Observing System
AOD	Atmospheric optical depth
AON	Arctic Observing Network
AOSB	Arctic Ocean Science Board
approx.	Approximately
ARM	Atmospheric Radiation Measurement

Art(s)	Article/Articles
ASPPR	Arctic Shipping Pollution Prevention Regulations
ATS	Antarctic Treaty System
AUVs	Autonomous underwater vehicles
AWI	Alfred Wegner Institute for Polar and Marine Research
BASC	Barrow Arctic Science Consortium
BEAC	Barents Euro-Arctic Council
BGR	Bundesanstalt für Geowissenschaften und Rohstoffe
C	Celsius
CAA	Chinese Arctic and Antarctic Administration
CAFF	Conservation of Arctic Flora and Fauna
CAO	Cold air outbreaks
CCOM	University of New Hampshire's Center for Coastal and Ocean Mapping
CHINARE	Chinese National Arctic Research Expedition
CliC	Climate and Cryosphere
CLCS	Commission on the Limits of the Continental Shelf
cm	Centimeter
CO	Carbon monoxide
Cs	Caesium
CTD	Conductivity, Temperature and Depth
DAMOCLES	Developing Arctic Modeling and Observing Capabilities for Long-term Environmental Studies
DDT	Dichlorodiphenyltrichloroethane
Declaration of Principles	Declaration of Principles Governing the Seabed and the Ocean Floor and the Subsoil Thereof

DF	Directorate of Fisheries
DOALOS	United Nations Division of Ocean Affairs and Law of the Sea
DOE	Department of Energy
E	East
EA	Environmental Assessment
ECASE	Eastern Canadian Arctic Seismic Experiment
ECMWF	European Centre for Medium-Range Weather Forecasts
EEZ(s)	Exclusive economic zone(s)
e.g.	for example
EIS	Environmental Impact Statement
EM	Electromagnetic sensor
eNGO	Environmental Non-Governmental Organization
Eos Trans. AGU	Eos, Transactions, American Geophysical Union
EPPR	Emergency preparedness, prevention and response
ERS	European Remote Sensing
ESA	Endangered Species Act
ESFRI	European Strategy Forum for Research Infrastructures
etc.	*et cetera*
EU	European Union
FAO	Food and Agriculture Organization
FARO	Forum of Arctic Research Operators
Fig.	Figure
FP 7	Seventh EU Framework Programme
FWS	Fish and Wildlife Service
FYI	First-year ice
GCHS	1958 Geneva Convention on the High Seas

GEUS	Geological Survey of Denmark and Greenland
GHG	Greenhouse gases
GOOS	Global Ocean Observing System
GPS	Global Positioning System
HCB	Hexachlorobenzene
HCH	Lindane
hPa	Hectopascals
IABP	International Arctic Buoy Program
iAOOS	Ice, Atmosphere, Ocean Observing System
IARC	International Arctic Research Center
IASC	International Arctic Science Committee
IASI	Infrared Atmospheric Sounding Interferometer
IASSA	International Arctic Social Sciences Association
Ibid.	*ibidem* (the same place)
IBRU	International Boundaries Research Unit
ICES	International Council for the Exploration of the Sea
ICJ	International Court of Justice
ICSU	International Council for Science
id.	*idem* (the same)
i.e.	*id est* (in other words)
IHA	Incidental Harassment Authorization
ILM	International Legal Materials
IMB	Ice Mass Balance
IMO	International Maritime Organization
INC	Intergovernmental Negotiating Committee
INSU/DT	Institut national des sciences de l'univers/Division technique

IOC	Intergovernmental Oceanographic Commission
IODP	Integrated Ocean Drilling Program
IPEV	Institut Polaire Français
IPSL	Institut Pierre Simon Laplace
IPY	International Polar Year
IR	Infrared
ISA	International Seabed Authority
ISAC	International Study of Arctic Change
ITK	Inuit Tapiriit Kanatami
ITP	Ice-Tethered Profilers
IUCN	International Union for the Conservation of Nature
JCOMM	Joint IOC-WMO Technical Commission on Oceanography and Marine Meteorology
JL	Jurists/Linguists
Jr.	Junior
kHz	Kilohertz
kg	Kilogram
km	Kilometer
lidar	Light detection and ranging
LRTAP	Long-range Transboundary Air Pollution
m	Meter
MARCOPOLI	Marine, Coastal and Polar Systems
MCS	Multi-channel seismic
mm	Millimeter
MMPA	Marine Mammal Protection Act
MoU(s)	Memorandum (Memoranda) of Understanding
MPAs	Marine protected areas
MSR	Marine scientific research
MYI	Multi-year ice
N	North

NABOS	Nansen and Amundsen Basins Observational System
NACG	North Atlantic Coast Guard Forum
NASA	National Aeronautics and Space Administration
NC	Nordic Council
NEA	Convention on Future Multilateral Cooperation in North-East Atlantic Fisheries
NEAFC	North-East Atlantic Fishery Commission
NEFCO	Nordic Environmental Finance Corporation
NEPA	National Environmental Policy Act
NERSC	Nansen Environmental and Remote Sensing Center
NF	Northern Forum
NGDC	National Geophysical Data Center
(e)NGO(s)	(environmental) Non-governmental Organization(s)
NH	Northern hemisphere
NIRB	Nunavut Impact Review Board
nm	Nautical miles
NMFS	National Marine Fisheries Service
No.	Number
NOAA	National Oceanic and Atmospheric Agency
NRI	Nunavut Research Institute
NSF	National Science Foundation
NSIDS	National Snow and Ice Data Center
NRCAN	Natural Resources Canada
NWP	Numerical Weather Prediction
OD/A	Open Data/Access
ODS	Optical Depth Sensor

OES	Bureau of Oceans and International Environmental and Scientific Affairs
OPA	Office of Ocean and Polar Affairs
OSPAR Convention	Convention for the Protection of the Marine Environment of the North-East Atlantic
PACES	Polar regions and Coasts in a changing Earth System
PAHs	Polycyclic aromatic hydrocarbons
PAME	Protection of the Arctic Marine Environment Working Group
PB	Agreement on the Conservation of Polar Bears
PBDEs	Polybrominated diphenyl ethers
PCBs	Polychlorinated biphenyls
PCNs	Polychlorinated naphthalenes
PFOA/PFOS	Perfluorooctanoic acid/Perfluorooctane sulfonate
PI	Principal Investigator
PICES	North Pacific Marine Science Organization
PICES Convention	Convention for a North Pacific Marine Science Organization
POPs	Persistent Organic Pollutants
POPS	Polar Ocean Profiling System
PSC	Polar Stratospheric Cloud
QA/QC	Quality assurance/quality control
QIA	Qikiqtani Inuit Association
San Diego Int'l L.J.	San Diego International Law Journal
SAAOs	Senior Arctic Affairs Officials
SAO(s)	Senior Arctic Official(s)
SAON	Sustaining Arctic Observing Networks
SAR	Agreement on Cooperation on Aeronautical and Maritime Search and Rescue in the Arctic

SAT	Surface air temperature
SCAR	Scientific Committee on Antarctic Research
SCPA	Standing Committee of the Conference of Parliamentarians of the Arctic Region
SDWG	Sustainable Development Working Group
SHEBA 1997-1998	Surface Heat Budget of the Arctic Ocean project
SEARCH	Study of Environmental Arctic Change
SIOS	Svalbard Integrated Arctic Earth Observing System
SPIT	Treaty Concerning the Archipelago of Spitsbergen, and Protocol
SSW	Sudden stratospheric warmings
SWIPA	Snow, Water, Ice and Permafrost in the Arctic
SYI	Second-year ice
S4D	SEARCH for DAMOCLES
Tc	Technetium
TIAS	Treaties and Other International Acts Series
TMT	Transfer of Marine Technology
UIC	Ukpeagvik Inupiat Corporation
ULS	Upward looking sonar
UN	United Nations
UNCLOS	United Nations Convention on the Law of the Sea
UNCLOS III	Third United Nations Conference on the Law of the Sea
UNDP	United Nations Development Programme
UNECE	United Nations Economic Commission for Europe

UNEP	United Nations Environment Programme
UNESCO	United Nations Educational, Scientific and Cultural Organization
UPMC	Université Pierre et Marie Curie
US	United States
USC	United States Code
USCG	United States Coast Guard
USGS	United States Geological Survey
USSR	Union of Soviet Socialist Republics
UV	Ultraviolet
Vol.	Volume
vs.	Versus
W	West
WCRP	World Climate Research Program
WHO	World Health Organization
WMO	World Meteorological Organization
ZaöRV	Zeitschrift für ausländisches öffentliches Recht und Völkerrecht
1994 Agreement	1994 Agreement Relating to the Implementation of Part XI of the United Nations Convention on the Law of the Sea
%	per cent
‰	per mil

Climate Change, International Law and Arctic Research – Legal Aspects of Marine Research in the Arctic Ocean

(Speech)

by Guido Westerwelle[*]

Commissioner,

Under-Secretary of State,

Excellencies,

Ladies and gentlemen,

Arctic issues are at present overshadowed by developments on the world stage, by the catastrophic events plaguing Japan and the globe, by the escalation of hostilities in Libya. But the Arctic is a subject that will be of crucial significance for the long-term survival of mankind. I would like to thank all our international guests who have travelled to Berlin to share their thoughts and opinions with us. And above all, I would like to thank the Finnish Ministry for Foreign Affairs, which is co-hosting this Conference with us.

Ladies and gentlemen,

Germany has for decades been involved in Arctic activities. We are proud of our successful research in the polar region. We are an active observer on the Arctic Council. But of course we respect the natural leadership of the Arctic Council's full members and permanent participants, be they coastal States, non-coastal States or indigenous representatives. With this in mind, I hope you will permit me to mention three goals pursued by Germany's Arctic policy.

---

[*] Member of the German Bundestag and Federal Minister for Foreign Affairs.

The first goal is to ensure the greatest possible freedom of research. For several years now, the economic exploitation of the Arctic has been a real possibility. It must not however impede research. Research must in principle be open to all, because the challenges of climate change affect us all. It is a sad fact that all States produce emissions and so contribute to climate change. They can no longer be allowed to be just part of the problem, but must also become part of the solution. Solutions require the firm basis in fact provided by research. That is why we have to encourage the efforts of the international community so that research continues. I therefore call on all those who exercise sovereignty in the High North, in the Arctic Ocean, to seek research-friendly solutions. That means as little bureaucracy as possible. Of course, a State should know what research is being conducted in its territorial waters and exclusive economic zone, by whom, when and why. But a requirement to simply notify the State concerned is much more conducive to research than having to apply for a permit. And less freedom of research surely cannot be in the long-term interests of the coastal States.

The second goal is ensuring that the strictest environmental standards are observed. The Arctic is a unique habitat, beautiful and fragile. The injuries inflicted by mankind will not heal without help. This context also reveals the limits of the traditional concept of sovereignty, which views sovereignty above all as a right. This concept of sovereignty has long since been complemented by the idea of sovereignty as a duty. Perhaps there has been too much talk in the past of 'Arctic rights' and not enough of 'Arctic duties'. It is not enough to set environmental standards and fishing quotas. You also have to observe them. Only then are partners willing to accept them. Effectively enforcing environmental standards is a task for the Arctic coastal States. But the Arctic Council, too, could become a guardian of the environment. The better able both the coastal States and the Arctic Council are to monitor and enforce existing standards, the more their authority as natural leaders will be accepted by other interested parties.

The third goal of German Arctic policy is to ensure that responsibility is taken for any environmental damage that occurs. This means there must be clear rules on liability which are effectively enforced. Sensibly limiting the risk posed by economic activities in the Arctic and imposing a clear liability on polluters is a serious policy challenge. The risk of being held liable for damage is the best incentive there is to stop people causing damage – it is in the would-be polluter's best interests to observe the highest standards. The coastal States' agencies are responsible for enforcing standards in the areas under their exclusive sovereignty.

But the highest standards must also apply on the high seas. That is where the Arctic Council must assume responsibility.

Ladies and gentlemen,

Preserving the common heritage of mankind must be a paramount goal of any policy concerning the Arctic Ocean. Looking beyond the legal definition of this common heritage, a heritage we fully support, we should all draw inspiration from the underlying idea of a common heritage of mankind. A policy that builds on this idea is in the interests of the coastal States and serves the legitimate interest of all States in the preservation of this heritage and the protection of this environment. We are counting on cooperation with the Arctic Council in this regard. Because the Arctic is so important for all mankind, and by no means just for the members of the Arctic Council, it is vital that the Council does not close its doors, but remains open to the world. The Arctic States and, above all, the indigenous peoples of the Arctic, have a natural leadership role to play. For they have known life in the Arctic for centuries. We have confidence in their knowledge. It is up to them to protect not just their national interests, but also the interests of the international community as a whole.

Wherever our country can help, we will be glad to do so.

The Arctic: A Test Bench for International Dialogue

(Speech)

by Maria Damanaki[*]

Mr Chairman, distinguished speakers, ladies and gentlemen,

If the rest of the world is hot from climate change, the Arctic region is burning. In the Arctic surface temperatures are rising twice as fast as in the rest of the world. Over the past decades, sea ice has been thinning and retreating to record lows of ice surface. The melting of the ice poses unprecedented ecological risks, but at the same time new opportunities open up for mankind in transport, trade, fisheries, oil and gas drilling. These opportunities make the Arctic the 'new frontier' in economic and political terms.

This dichotomy between economic opportunities and environmental risks is the challenge the international community is facing today. This dichotomy overwhelms all actions of mankind. From the exploitation of resources in North Africa to the disaster in Japan there is clearly the same dilemma that ancient Greek tragedies faced: man against nature. The winner we already know.

So the Arctic will test our ability to work together and our willingness to put environmental protection, sustainability and public safety first. It will show whether we have understood – or not – that all we have is one planet after all. The message I bring to you today is that it must be possible to reap the economic benefits opening up in the Arctic while at the same time preserving the environment from further damage. Let us not forget that the people living in the Arctic must benefit from the process too: their way of life, their heritage and their livelihoods are at stake.

---

* European Commissioner for Maritime Affairs and Fisheries.

Since 2008 we have been formulating a coherent and comprehensive policy to tackle the ever growing challenges of the Arctic region. Our purpose is threefold: Firstly, we want to bring a decisive contribution to preserving the Arctic region, in unison with the people of the Arctic. Secondly, we want to make sure that the emerging industrialization and exploitation of Arctic resources follow the highest environmental and safety standards, with fair access and treatment to European Union ('EU') citizens and businesses. Thirdly, and following logically from the other two, we want closer and enhanced international cooperation in the Arctic region.

Our priorities and our approach are very similar to those advocated by the Arctic States, which undeniably have a primary responsibility. The competences of the Arctic coastal States, as laid down by the United Nations Convention on the Law of the Sea,[1] are undisputed. The Arctic Council which is the main platform to exchange knowledge and best practices on the region's challenges has a fundamental role to play in Arctic cooperation. We already attend its working groups on a case-by-case basis and we hope to do so on a permanent basis in the future. The EU presence under the status of permanent observer can advantage all the parties. Let me show how you, by referring to our work, till now.

Ladies and gentlemen,

The EU is a world leader in the fight against climate change and promotes environmental sustainability through reform, research and concrete initiatives. A successful policy to mitigate climate change will naturally also benefit the Arctic area and its people. A report published in January 2011 and funded by the EU studies the consequences of European activities in nine areas, such as biodiversity, transport, energy and climate change. It shows that our impact is significant: for example, 24% of mercury depositions and 42% of sulphur dioxide emissions to the Arctic come from the Union! So, we are aware of our share of responsibility. We intend to work for environmental protection and sustainable development – for example to come to good safety standards for polar vessels or for oil and gas exploitation.

We want concrete projects delivering concrete results. This is why in the last decade the European Commission has spent € 200 million on Arctic research projects which focus on key areas such as sea ice retreat-

---

1 United Nations Convention on the Law of the Sea (concluded 10 December 1982, entered into force 16 November 1994) 1833 UNTS 397.

ing and thinning, rising sea levels and Arctic pollution. Let me give you a few examples:

Among the projects strengthening international efforts to mitigate climate change or to adapt to the effects of climate change there are:

- The 'Arctic Tipping Points project'. This project is about identifying the tipping points of marine ecosystems and understanding what happens when those lines are crossed.
- The 'ice2sea project' seeks to inform the international debate on climate change mitigation and the European debate on coastal adaptation and sea-level defense.
- The 'Hermione project' studies how deep-sea ecosystems work and how they can be used to produce goods and services.

This year, we will publish a call for a large-scale project to improve our knowledge of the dynamic processes affecting permafrost and of the implications for global warming. There are also a number of recent EU projects that help indigenous Arctic people protect their lifestyles. For instance, the projects ArcRisk and CLEAR investigate the links between climate change, contaminants, and human health and are supported by the EU with € 6 million. We have also started to study the impact of man's activities on Arctic ecosystems; in the coming years we will have to look into the socio-economic consequences of human activities. The European Commission is also funding observation and research infrastructure in the Arctic. We support long-term measurements and reporting of marine data in the context of the European Marine Observation and Data Network; and we support the establishment of the Arctic component of Global Earth Observation System of Systems.

We have funded feasibility studies for the Aurora Borealis project: an innovative research ice-breaker that could work all year round on Arctic ice. I would like to thank the German Federal Ministry for Education and Research for its strong support to this project. If realized, this major project will be emblematic of the good cooperation between the Arctic partners.

The most recent building block in this domain emerged from the joint call 'Ocean of Tomorrow' and is worth € 11 million. It is called ACCESS, which stands for Arctic Climate Change, Economy and Society. It will assess and quantify the impact of climate change on the Arctic's main economic sectors in the next 20 years: maritime transport, fisheries, tourism and resource extraction. The output of this project will underpin the strategic choices that policy makers will have to make in the years to come.

A status of permanent observer for the EU to the Arctic Council would have several advantages; it would foster international cooperation, which is one of the EU's primary objectives; it would make our work for the region more effective; and it would bring benefits for the Arctic States, because we would bring the EU experience, knowledge and resources into the Council's work. Most importantly: admitting the EU as permanent observer to the Arctic Council would send a clear signal that the Arctic States welcome investors and researchers from the EU; that our partners, particularly Canada and Russia, know that foreign investments and concerted research efforts are needed to let the region develop sustainably and in the interest of the Arctic people. It is the only way to appreciate the region's intricate links to the rest of the world and tackle the very acute threats posed to our common heritage in modern times.

Ladies and gentlemen,

As members of a global community sharing a collective responsibility, we must find every opportunity to work together. This conference is one such opportunity and I wish it every success.

Thank you.

Arctic Science, International Law and Climate Change

(Speech)

by Jaakko Laajava[*]

Excellencies, ladies and gentlemen,

It gives me a great pleasure to address the Arctic Conference organized jointly by the Foreign Ministries of Germany and Finland. Finland is an Arctic country, with one third of our territory above the Polar Circle. And Germany is an important Arctic stakeholder, with a longstanding participation and expertise in the region – and more precisely, expertise in marine science research, the main topic of our joint conference today and tomorrow.

I. Arctic Strategy of Finland

In the Arctic the need for a fundamental change in our approach has been recognized during the last few years by both the circumpolar governments as well as researchers. What used to be viewed as a periphery is now becoming a center of global attention. Many nations assess and reassess their approaches to the region. Finland has adopted a new national Arctic Strategy[1] in order to face the new challenges and to seize the new opportunities. Our basic view is that Arctic issues should be dealt with within a rules-based multilateral framework with an emphasis on comprehensive security and environmental sustainability. All

* Undersecretary of State, Finland.

1 Finland, Prime Minister's Office, Finland's Strategy for the Arctic (2010), see also <http://www.geopoliticsnorth.org/images/stories/attachments/Finland.pdf> (30 June 2011).

Arctic and non-Arctic actors should be committed to an approach based on constructive cooperation, not confrontation.

The Finnish Arctic Strategy from June 2010 is an effort to put the various aspects regarding the Arctic in one comprehensive package and to provide an assessment of the challenges and the potential of the region from a Finnish perspective. The Strategy defines our goals in the Arctic region as well as the means to achieve them; it deals *inter alia* with the utilization of Finland's Arctic know-how and research, institutional issues as well as questions of regional cooperation. It also emphasizes the importance of environmental matters and questions related to the indigenous peoples.

The opening of the Arctic Sea offers new perspectives for exploitation of natural resources in energy, mining and fish stocks. New sea routes attract both tourists and commercial transport. Finland has wide Arctic expertise and know-how to offer in this context. A key issue for Finland – and I believe for all stakeholders in the Arctic – is how to organize economic activities in the Arctic while fully taking into account environmental concerns and keeping the need for sustainable development in the Arctic as the fundamental perspective.

The utilization of the region's natural resources requires know-how, caution and a sense of responsibility. Due to the fragile Arctic environment the principle of sustainable development must be respected. We for our part believe that education, research and application of our Arctic expertise are the key to a responsible exploitation of the Arctic. Finland has strong traditions in winter shipping and technology, shipbuilding as well in offshore industries such as oil and gas rigs and vessels needed for Arctic circumstances.

II. The Arctic Council

Finland seeks close cooperation with all partners in the Arctic region. Today, when all our harbors are covered with ice, our attention is drawn more and more to the North and particularly to our neighbors Norway and Russia. We are engaged in an active dialogue with both countries – bilaterally and trilaterally – in order to share our expertise with them. As an example, let me mention that just a few weeks ago we launched an Arctic Partnership between Russia and Finland in St. Petersburg.

But the primary intergovernmental forum to deal with Arctic policies is the Arctic Council. Last summer Mr. Alexander Stubb, the Foreign Minister of Finland, presented some concrete proposals on the strengthening of the Arctic Council such as the establishment of a permanent secretariat for the Council; better burden sharing with a joint budget; the extension of the Council's mandate by enhancing its political and legal role and, finally, an improvement of its working methods including the role of observers.

Interaction between Arctic and non-Arctic stakeholders and players is of key importance – an integrated approach requires engagement from all with legitimate interest in the Arctic. The eight Member States have concluded that the Council is the proper platform for Arctic considerations. This includes the bilateral as well as multilateral cooperation between the five coastal States, on one hand, and indigenous peoples and observer countries, institutions and organizations on the other. The Arctic Council Foreign Ministers will meet in Nuuk in Greenland in less than two months time. In our view a forward-looking decision on observers at that meeting is indispensable for the future of the Council.

Finland has also proposed the idea of a meeting at the top level to discuss the Arctic issues. Such a first Arctic Summit, under the auspices of the Arctic Council, would give new direction to the Arctic cooperation and become a milestone in the development of the Council itself. The high profile of such a meeting and the attention given by the Heads of States and Governments of the Arctic countries could substantially contribute to the reaffirmation of the multilateral and rules-based approach we are witnessing in the Arctic today. The idea of an Arctic Summit is not new; it has been raised by researchers for many years. A serious consideration of the initiative gives in itself added value and content to this emerging region with global reach.

III. The European Union

The Arctic policy of the European Union ('EU') is evolving but to some extent is still a work in progress.[2] I am very pleased to welcome

---

2 See e.g.: European Parliament, 'Resolution on Arctic Governance', P6_TA(2008)0474 (9 October 2008); Commission of the European Communities 'Communication from the Commission to the European Parliament and the Council – The European Union and the Arctic Region', COM (2008) 763 final (20 November 2008); see also European Commission, Maritime Affairs, 'The

Commissioner Damanaki to our Conference and thank her for her leadership in these issues. She just outlined the actions taken by the Commission so far. During 2008 and 2009 we have seen indeed the European Commission and the European Council publish Arctic Communications and Conclusions that have laid the foundation for Arctic thinking within the Union. Finland will continue to assist in shaping the Union's Arctic policies for the years to come. A new Communication is currently under preparation in the Commission. This will be, we are confident, a step again in the right direction.

The European Parliament has consistently contributed to the formulation of the EU's Arctic policy with resolutions, statements and conferences. The Parliament recently adopted a much awaited '*Report on a Sustainable EU Policy for the High North*' by Michael Gahler.[3] Finnish members of the European Parliament took an active part in the preparations of the Report. I am confident that the Report will be duly noted in discussions within the EU institutions, including in the Commission when preparing its Communication.

Based on the rapidly increasing importance of the Arctic to the EU and the growing need to reach out and communicate on Arctic issues both internally and externally, Finland has proposed the establishment of an EU Arctic Information Centre as a network undertaking by European Arctic institutes. The Centre would, on one hand, support the formation of coherent Arctic policy for the EU, and, on the other, provide a channel for dissemination of accurate Arctic information within and outside the EU. The Arctic Center at the University of Lapland in Rovaniemi would, in our view, be the best location for the EU Arctic Information Centre for a number of reasons, the most important being the strong and internationally acknowledged cross-disciplinary Arctic scientific research conducted in Rovaniemi. The University of Lapland already coordinates the activities of the existing network of Arctic Universities, known as UArctic. Furthermore, the Sami, as the only indigenous people in the EU, would have a best possible access to this Arctic location on the Polar Circle. Since the initiative is gaining ever wider

---

EU and the Arctic Region', see <http://ec.europa.eu/maritimeaffairs/arctic_overview_en.html> (30 June 2011).

3 European Parliament, 'Report on a Sustainable EU Policy for the High North', A7-0377/2010 (16 December 2010); see also European Parliament, 'Resolution on a Sustainable EU Policy for the High North', P7_TA(2011)0024 (20 January 2011).

support within the EU, we believe that the time for the Commission to move on the modalities is now.

Another dimension of EU's Arctic policy is the concept of the so-called Arctic window of the Northern Dimension policy of the EU.[4] Geographically, the region covered by the Northern Dimension closely coincides with the Barents Euro Arctic Council. Synergies can be found also with activities of other regional councils and cooperation structures. The newest Partnership on Transport and Logistics is particularly relevant in dealing with the development of transport corridors in the North, covering the Arctic maritime routes and rail and road connections in the Barents region. This could be the platform to extend the cooperation more broadly to the Arctic.

With the emerging climate change – twice as fast as anywhere else in the globe – the Arctic is rapidly reaching a tipping point. Accurate information on the situation as well as assessment on measures to be taken is now of paramount importance. A conference like this can provide additional scientific evidence and legal advice for governments to make the right decisions.

I would like to thank Minister Westerwelle and the Auswärtiges Amt for the excellent cooperation in the preparation of this important conference. I would like to extend my warmest thanks to all presenters and participants. Thank you for joining us. I look forward to interesting and rewarding discussions here in Berlin regarding Arctic issues.

Thank you.

---

[4] See European Union, External Action, 'Regional Policies: Northern Dimension', see <http://ec.europa.eu/external_relations/north_dim/index_en.htm> (30 June 2011).

The Arctic – a Sentinel for Environmental Processes and Effects

*by Lars-Otto Reiersen & Simon Wilson**

I. Introduction

In a speech in Murmansk, in October 1987, the then president of the USSR, Mikhail Gorbachev, called for international cooperation to address pollution in the northern territories of the USSR and the bordering Arctic region. This environmental initiative was one of the first signals that the Cold War was coming to an end, and one of the first steps in a process that ultimately led to the establishment of the Arctic Council. The first significant result of the call from the Russian president was a Finnish initiative in January 1989 to promote international cooperation in the Arctic. This led to the arrangement of a consultative meeting on the protection of the Arctic environment in Rovaniemi in September 1989. At this meeting, possibilities for environmental cooperation among the Arctic States were discussed and a follow-up work plan approved. Among other things, it was agreed that Canada would prepare a background report on persistent organic pollutants ('POPs') in the Arctic, and Norway together with the USSR would review existing national and international monitoring programs operating in the Arctic and present proposals for future action in this field.[1] Between Septem-

---

* Arctic Monitoring and Assessment Programme ('AMAP'), Secretariat, Oslo, Norway. This contribution is mainly based on the assessments presented by AMAP for the period 1993-2011. For more detailed information the readers are advised to read these reports, see <www.amap.no> (15 September 2011).

[1] L-O Reiersen, S Wilson and V Kimstach, 'Circumpolar Perspectives on POPs: the Case of the Arctic Monitoring and Assessment Programme', in DL

ber 1989 and June 1991, several expert meetings were held to develop these reports.

In 1991, Ministers of the eight Arctic countries (Canada, Denmark, Iceland, Finland, Norway, Russia, Sweden and the USA), adopted the Arctic Environmental Protection Strategy ('AEPS').[2] To implement part of this strategy, the Arctic Monitoring and Assessment Programme ('AMAP') was established and requested by Ministers to 'examine the levels of anthropogenic pollutants [from any sources] and to assess their effects in all relevant compartments of the Arctic environment'.[3] In 1993, AMAP was asked by the Arctic Ministers of Environment to include assessment of climate change and UV-radiation/stratospheric ozone depletion. In 1996 the AEPS, including all its working groups, was reorganized under the newly formed Arctic Council. The geographical area covered by the AMAP monitoring and assessment programs is shown in Figure 1.

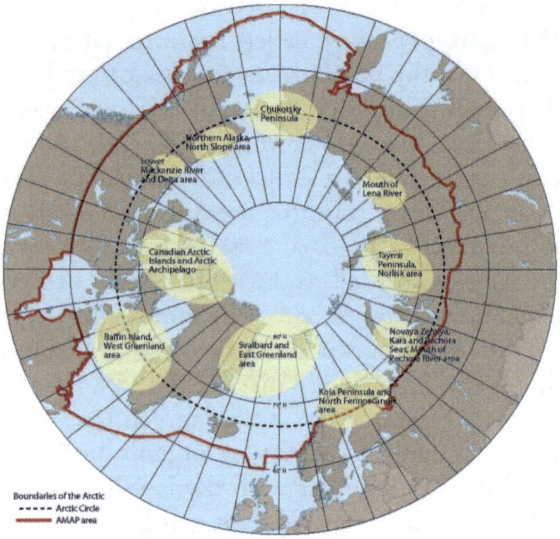

Fig. 1. Source: Arctic Monitoring and Assessment Programme

Downie and T Fenge (eds), *Northern Lights against POPs: Combatting Toxic Threats in the Arctic* (2003) 60 et seq.

[2] Arctic Environmental Protection Strategy, 'Declaration on the Protection of the Arctic Environment' (14 June 1991), see <http://arctic-council.org/filearchive/artic_environment.pdf> (15 September 2011).

[3] *Ibid.*

II. The AMAP Monitoring Programme

Between 1990 and 1992, experts from the eight Arctic countries and some of the Arctic indigenous organizations designed the first detailed AMAP Monitoring Programme.[4] The program was divided into five sub-programs (covering the atmospheric, terrestrial, freshwater and marine environments and human populations), and designed to provide the data necessary to perform integrated scientific assessments of the pollution status of the Arctic, to explain the mechanisms behind the observed levels, and to propose actions that could reduce Arctic pollution and its effects. Therefore, AMAP designed a monitoring program that could follow contaminants from any sources at any latitudes, via transport pathways (atmosphere, rivers and ocean) into all compartments of the Arctic environment and its ecosystems, and ultimately into top predators including humans. The program has subsequently been updated to also address effects due to climate change and UV/ozone.[5]

A key feature of the implementation of AMAP was that the program could be initiated in a step-by-step manner. This gave the eight Arctic countries the freedom to adapt or develop their national programs based on AMAP's recommendations, and adjust them according to their own priorities and financial and scientific possibilities. As a part of its general strategy, the AMAP Monitoring Programme has always tried to build on ongoing (national and international) research and monitoring activities. Thus, from the outset, AMAP recognized that research (in addition to national monitoring work) would provide much of the relevant information necessary for assessing levels of contaminants and their effects in the Arctic, a region where logistical constraints impose major limitations on routine monitoring activities. In many other regional monitoring programs, research is largely ignored. In order to address quality assurance issues, AMAP encouraged all participating laboratories to join appropriate international QA/QC programs and, where relevant, adopt existing international recommendations for methodology and parameters to be analyzed. By doing this AMAP was able to compare the levels of contaminants observed in the Arctic with levels published from other monitoring and research programs being conducted at lower latitudes.

---

4 AMAP, The Monitoring Programme for Arctic Monitoring and Assessment Programme, AMAP Report 93:3 (1993) ('AMAP [1993a]').

5 AMAP, AMAP Trends and Effects Programme 1998-2003, AMAP Report 1999:7 (1999).

At the beginning of the 1990s, few international monitoring and assessment programs included more than one ecological system, e.g. the marine or terrestrial environments, and possibilities of following contaminants from their sources through the environment to their ultimate fate were very limited. AMAP was thus one of the first international monitoring programs to design and implement a monitoring program covering all major ecological systems (atmospheric, marine, freshwater and terrestrial – and humans), and all major contaminant groups in one program, and at the same time fully integrating its monitoring and assessment activities.[6]

III. Transport Mechanisms for Contaminants and Energy

The Arctic plays a key role in the global energy budget. Both energy and contaminants are transported to and redistributed within the Arctic by the three main environmental pathways mentioned above; ocean currents, atmospheric winds, and rivers. The AMAP assessments have documented that snow and ice also play a significant role in the transport and fate of contaminants in some areas of the Arctic. Snow and ice are also key factors in affecting the energy balance through changes in albedo. As ice and snow melt, the amount of solar radiation reflected back to space decreases.[7]

Atmospheric transport is by far the fastest transport mechanism both for contaminants and energy and can bring contaminants from sources at southern latitudes into the Arctic within days or weeks, depending on wind speed and direction. Major sources of atmospheric transported contaminants are point sources such as power generating stations, smelters, incinerators and dump sites and diffuse sources such as agriculture and transportation.

Some of the world's largest rivers (Ob, Yenisey, Lena and Mackenzie) empty into the Arctic Ocean, Figure 1. More than 70% of the freshwater entering the Arctic Seas originates south of the Arctic region as defined in Figure 1. These rivers drain vast areas of land. In addition to direct discharges into the rivers from municipal sewage works, mines, metal processing facilities, factories, oil and gas exploration, exploitation and transportation, waste dumps, etc., the rivers pick-up runoff of

---

6 Reiersen, Wilson and Kimstach, see note 1.

7 ACIA, Arctic Climate Impact Assessment (2005) ('ACIA [2005]').

pesticides from agriculture further south together with long-range transported contaminants deposited in their catchment areas.

The flow in Arctic rivers is highly seasonal with most transport of water and sediments occurring during the spring snowmelt period. Large amounts of nutrients are also transported with the spring flood, so that in both the rivers themselves and their estuaries an increased level of contaminants (some soluble in the water, others adsorbed to particles) become available for organisms, at the same time as biological productivity is at its peak.

Contaminants and energy transport via ocean currents is much slower. Radionuclides such as ^{137}Cs have been used as ocean tracers, and releases from re-processing plants entering the Irish Sea and the North Sea have been found to take two to four years to reach the Arctic. However, due to the volumes of water involved, and their capacity to store and transport heat, ocean currents are the main driving force in the redistribution of energy around the globe. Arctic freshwater and marine systems also receive contaminants from secondary sources, such as glaciers and snow and ice melt.

A hundred years ago, Fritjof Nansen observed that sea ice in the polar basin transported sediments deposited onto or incorporated in the ice. Recent studies have shown that bottom sediments from Russian estuaries and the continental shelf can be frozen into sea ice during the autumn when the ice is formed. Contaminated sediments can thus be transported across the Arctic Ocean as the sea ice drifts. During its drift over the Arctic, sea ice will receive additional contaminants from atmospheric deposition.

Normally most of the multi-year sea ice will melt when it passes/has passed through the Fram Strait, while one-year ice tends to melt on the continental shelf, e.g., in the Barents Sea, Kara Seas, Beaufort Sea and the Chukchi Sea. However, during the last decade a significant part of the multi-year sea ice has melted before it entered the Fram Strait.[8] The release of contaminants from sea ice to the marine environment during the short Arctic spring/summer coincides with the maximum in primary production along the ice edge. At this time of the year there is therefore large biological activity along the ice edge involving plankton, invertebrates (ice-fauna), seabirds, marine mammals, etc., and the contaminants may easily be accumulated into food chains. The sea ice may

---

8 AMAP, *Snow Water Ice and Permafrost in the Arctic (SWIPA)* (forthcoming 2011) ('AMAP [2011a]').

therefore play an important role both in the transport of contaminants within the Arctic Ocean, and also in making them available to organisms living in the Arctic.[9]

Sea ice also plays a significant role for the Arctic climate. The white surface reflects most (ca. 90%) of incoming solar radiation – the albedo effect.[10] With the increased melting of sea ice, more energy from the sun is absorbed by the dark ocean waters thereby increasing the temperature of surface waters. The same mechanism occurs over land areas where snow and ice melt exposes darker land surfaces that can absorb the energy. The consequence of this increased melt for the Arctic cryosphere has been further assessed by AMAP in a report released in May 2011 – the Snow, Water, Ice and Permafrost in the Arctic ('SWIPA') assessment.[11]

IV. Results

Over the past 20 years, the priority 'issues of concern' for the Arctic environment and its people with respect to pollution have been associated with the following contaminant groups: radionuclides, persistent organic pollutants, certain heavy metals – especially mercury –, acidifying substances, radiatively important trace species (climate forcers), and petroleum hydrocarbons. Some of these contaminants are of circumpolar concern, while others are of concern at more regional and local levels.

AMAP's first reports concerning 'Proposals for Environmentally Sound Investment Projects in the Russian Part of the Barents Region'[12] were used by the Nordic Environmental Finance Corporation

---

[9] AMAP, *AMAP Assessment Report: Arctic Pollution Issues* (1998) ('AMAP [1998]').

[10] ACIA [2005], see note 7.

[11] AMAP [2011a], see note 8.

[12] AMAP/NEFCO, *Barents Region Environmental Programme: Proposals for Environmentally Sound Investment Projects in the Russian Part of the Barents Region, Volume I: Non-Radioactive Contamination* (1995) ('AMAP/NEFCO [1995a]') and AMAP/NEFCO, Barents Region Environmental Programme: Proposals for Environmentally Sound Investment Projects in the Russian Part of the Barents Region, Volume II: Radioactive Contamination (1995) ('AMAP/NEFCO [1995b]').

('NEFCO') as a basis for funding regional environmental improvement projects.

AMAP's first comprehensive assessment was delivered in the 'Arctic Pollution Issues: A State of the Arctic Environment' report,[13] together with its detailed scientific background report 'AMAP Assessment Report: Arctic Pollution Issues'.[14] These reports documented the state of knowledge regarding pollution threats to Arctic ecosystems and humans and established a baseline against which future developments could be compared. Subsequent AMAP 'State of the Arctic Environment' reports and their scientific background documentation – the AMAP assessment reports – have updated the information on all of these pollution issues.

The 1998 AMAP report[15] also identified gaps in knowledge related to the effects due to climate change on the Arctic. This triggered the work that led to the 'Arctic Climate Impact Assessment',[16] produced by AMAP in association with the International Arctic Science Committee ('IASC') and the Conservation of Arctic Flora and Fauna ('CAFF') group.

1. Radioactivity

In the early 1990's, the general view among 'experts' was that radioactive contamination from the former USSR constituted a major pollution threat to the Arctic. The AMAP (1995b) report[17] documented several significant sources of radioactive pollution on Russian territories and in adjacent Northern Seas related to dumped and stored nuclear wastes, and sources connected with the operation of nuclear power plants and nuclear submarines, and nuclear explosions used for civilian engineering purposes. This report, presented to the Barents Council in December 1995, together with the AMAP 1998 assessment report,[18] formed a key

---

[13] AMAP, *Arctic Pollution Issues: A State of the Arctic Environment Report* (1997) ('AMAP [1997]').

[14] AMAP [1998], see note 9.

[15] *Ibid.*

[16] ACIA [2005], see note 7.

[17] AMAP/NEFCO [1995b], see note 12.

[18] AMAP [1998], see note 9.

part of the documentation that led to a number of international initiatives to clean up radioactive sources in northwestern Russia. These efforts have focused on decommissioning nuclear submarines, improving the safety culture at nuclear power plants and other related facilities, and improved handling and safe storage of nuclear waste and fuel. As of September 2010, 118 out of 122 nuclear submarines from the Russian Northern Fleet had been decommissioned and two were in the process of being decommissioned. The resulting nuclear waste is shipped to the reprocessing plant in Majak. This cleanup work (2002 – 2012) has a price tag of approximately US$ 20 billion.

More surprising to some experts was the information contained in the AMAP (1997[19], 1998[20] and 2002[21]) reports about the contribution to the radioactive contamination of the Arctic Seas that originated from European nuclear re-processing plants at Sellafield (UK) and Cap de la Hague (France). This documentation supported international pressure on the UK to reduce the releases of Cs and later Tc from Sellafield, which has proved effective in reducing the levels of these contaminants entering the Arctic Seas.

The main source of radioactive contamination in the Arctic, as elsewhere in the world, is associated with fallout from (atmospheric) nuclear weapons tests conducted in the early 1960s. Since 1995, AMAP has prepared three major assessment reports focusing on radioactivity, the latest in 2009.[22] All have documented declining levels of radionuclides in the Arctic derived from atmospheric fallout following introduction of the ban on atmospheric nuclear testing; however, the reports also document the increase and subsequent decline in contamination following the Chernobyl accident in 1986.

Other sources of radioactive contamination in the Arctic described in the AMAP assessment reports include the dumping of reactors from nuclear submarines at Novaya Zemlya, the accident involving a B52 carrying nuclear weapons at the Thule airbase in 1968, the radioactive debris from 'Cosmos-954' spread over part of the North West Territories of Canada in 1978 and the *Komsomolets* nuclear submarine accident in 1989. Contamination in these cases is largely limited to the marine

---

19 AMAP [1997], see note 13.

20 AMAP [1998], see note 9.

21 AMAP, *AMAP Assessment 2002: Radioactivity in the Arctic* (2004).

22 AMAP, *AMAP Assessment 2009: Radioactivity in the Arctic* (2009) ('AMAP [2009a]').

environment close to the dumping/accident sites. The situation around the Thule airbase is frequently monitored; however, the dumping sites at Novaya Zemlya are in need of improved knowledge regarding corrosion of storage containers and potential leakages. Monitoring is currently underway to determine whether releases following the accident at the nuclear power plant in Japan (spring 2011) can be detected in the Arctic.

Recent AMAP assessments have focused attention on technological (and industrial) activities, such as oil and gas extraction, that may lead to enhanced releases of naturally occurring radionuclides.[23]

2. Persistent Organic Pollutants ('POPs')

AMAP assessments of POPs date from 1992 and initially[24] focused mainly on 'legacy' POPs including pesticides such as lindane ('HCH'), toxaphene, and DDT and their metabolites, certain industrial chemicals (e.g. PCBs), and anthropogenic and natural combustion products [e.g. dioxins/furans and polycyclic aromatic hydrocarbons ('PAHs')]. While the assessments show that levels in Arctic air, water, soil and most animals are generally lower than those found in temperate areas, this is not the case for higher trophic level Arctic marine species. Due to bioaccumulation, and more importantly biomagnification of POPs in Arctic food webs, and the fact that many of these legacy POPs are lipophilic (concentrating in fatty tissues), the levels in some animals – especially marine mammals and some seabirds – are highly elevated. Furthermore, this has resulted in high dietary POPs exposures for some Arctic indigenous peoples that consume these species as part of their traditional diet.

POPs in certain Arctic animals and human populations are therefore at or approaching levels which are of concern for biological effects. POPs such as DDT and PCBs have been shown (largely through laboratory experiments) to have the potential to cause biological effects in animals and humans including neurological, reproductive, immunosuppressive and carcinogenic effects. Increasingly the laboratory evidence is being

---

23 *Ibid.*

24 AMAP [1998], see note 9; AMAP, *Arctic Pollution 2002: Persistent Organic Pollutants, Heavy Metals, Radioactivity, Human Health, Changing Pathways* (2002).

confirmed through field studies. Studies on polar bears in the Hudson Bay area indicate that mothers with the highest levels of some POPs are more prone to losing their cubs. Studies from Svalbard have found a clear correlation between reduced immuno-suppression and increased levels of PCB.

POPs contamination observed in most of the Arctic cannot be explained by local use or releases, and is therefore the result of long-range transport from lower latitudes. Based on the AMAP documentation, AMAP recommended already in 1993 that

> [g]iven the increasing substantiation of reasons for concern related to persistent organics in the Arctic, the eight Arctic countries [should] agree to support activities that will lead to the development of a protocol to control the emissions of these substances under the UNECE LRTAP Convention.[25]

In response to this recommendation, (AEPS) Ministers agreed

> to support the development of appropriate protocols under LRTAP auspices, and to consult with non-ECE nations whose emissions and discharges may affect the Arctic, to achieve their participation in the protocols [and] to continue to take measures to reduce and/or control the use of a number of persistent organic pollutants.[26]

The Arctic Environmental Ministers further agreed, in 1997,

> to take [the AMAP] findings and recommendations into consideration in [their] policies and programmes [and] to increase [...] efforts to limit and reduce emissions of contaminants into the environment and to promote international co-operation in order to address the serious pollution risks reported by AMAP [and to] draw the attention of the global community to the content of the AMAP reports in all relevant international fora [and] make a determined effort to secure support for international actions which will reduce the Arctic contamination.[27]

---

[25] AMAP, *Report to Ministers: Update on Issues of Concern to the Arctic Environment, including Recommendations for Actions*, AMAP Report 93:4 ('AMAP [1993b]').

[26] AEPS, *The Nuuk Ministerial Declaration on Environment and Development in the Arctic & Report, Arctic Environmental Protection Strategy: Second Ministerial Conference* (1993).

[27] Arctic Council, *The Alta Declaration on the Arctic Environmental Protections Strategy & Senior Arctic Affairs Officials (SAAOs) Report to Ministers* (1993) ('Arctic Council [1997]').

These messages were repeated when the Arctic Council was established in 1998 and AMAP and the AEPS were subsumed under the Arctic Council. The (Arctic Council) Ministers reaffirmed their agreement

> to work vigorously for the early ratification and implementation of the Protocols on the elimination or reduction of discharges, emissions and losses of Persistent Organic Pollutants (POPs) [...] under the framework of the United Nations Economic Commission for Europe Convention on Long-Range Transboundary Air Pollution, [to] encourage other states to do the same, with the aim to bring the Protocols into force as early as possible.[28]

The establishment firstly of the POPs Protocol under the United Nations Economic Commission for Europe ('UN ECE') (adopted in 1998 and entering into force in 2003)[29] and thereafter the United Nations Environmental Programme ('UNEP') Stockholm Convention on Persistent Organic Pollutants (adopted in 2001 and entering into force in 2004)[30] are probably the clearest examples of the way that AMAP information has been effectively used in substantiating the need for the development of international agreements to reduce environmental pollution. In these processes, the Permanent Participants to the Arctic Council – the Arctic Indigenous Peoples' organizations – were empowered by the information available from AMAP to take an active part in the negotiating process.[31] AMAP expertise was used in the development of the Stockholm Convention's 'Global Monitoring Programme' for POPs, and AMAP results have been fed into the 'effectiveness and sufficiency' assessments conducted as part of the implementation of both the UNECE and UNEP agreements.

Levels of 'legacy' POPs have generally decreased in Arctic air and biota over the last years,[32] however there is evidence that some of these decreasing trends are now flatting out – DDT which continues to be used

---

[28] Arctic Council, *The Iqaluit Declaration. The First Ministerial Meeting of the Arctic Council, Iqaluit* (1998) ('Arctic Council [1998]').

[29] *Protocol to the 1979 Convention on Long-Range Transboundary Air Pollution on Persistent Organic Pollutants* (24 June 1998, entered into force 23 October 2003) 2230 UNTS 82.

[30] Stockholm Convention on Persistent Organic Pollutants (done 22 May 2001, entered into force 17 May 2004) (2001) 40 ILM 532.

[31] Reiersen, Wilson and Kimstach, see note 1.

[32] AMAP, *AMAP Assessment 2009: Persistent Organic Pollutants (POPs) in the Arctic (2009)* ('AMAP [2009b]').

against malaria is an example. In other cases there are indications that levels of some 'legacy POPs' may even be increasing again. As these substances are now subject to widespread bans, the likelihood is that climate change is playing a role. Warmer temperatures are likely to lead to increased re-volatilization of some POPs, such as hexachlorobenzene ('HCB'), from the environmental 'reservoirs' (soils and ocean waters) where they have accumulated in past decades. In the Arctic, loss of sea ice is increasing the connectivity between the ocean and atmosphere, and resulting exchange of contaminants between these two compartments. These observations, first addressed by AMAP in 2002[33] have led to an increasing focus in AMAP on the combined effects of climate change and pollution.[34] AMAP has recently undertaken work together with the UNEP Stockholm Convention to produce a technical report on this subject.[35]

In addition to the 'legacy POPs', some groups of persistent organic substances that are only now starting to be regulated or considered for regulation. Many of these have been detected in the Arctic for a decade or more, and some are produced in large quantities. Examples include polybrominated diphenyl ethers ('PBDEs') and polychlorinated naphthalenes ('PCNs') used, e.g., as flame retardants; chlorinated paraffins, used in cutting oils; and compounds such as PFOA/PFOS (perfluorooctanoic acid/perfluorooctane sulfonate) used as surfactants and flame suppressants. PCNs are chemically similar to PCBs and some PCNs have toxic properties similar to those of chlorinated dioxins, furans, and dioxin-like PCBs. In addition to these POPs that sometimes are called the 'new POPs', AMAP has reported in 2009[36] that several currently used pesticides (the real 'new' ones) can be detected in a variety of matrices throughout the Arctic. The fact that chemicals (not or little used in the Arctic) are persistent enough to transport to and accumulate in the Arctic environment has become an important criteria in the definition of 'persistence' used in classifying new substances for inclusion under existing international agreements such as the Stockholm

---

[33] AMAP, *AMAP Assessment 2002: The Influence of Global Change on Contaminant Pathways to, within, and from the Arctic* (2003).

[34] AMAP, *Arctic Pollution 2009* (2009) ('AMAP [2009c]'); AMAP, *Update on Selected Climate Issues of Concern* (2009) ('AMAP [2009d]').

[35] UNEP/AMAP, *Climate Change and POPs: Predicting the Impacts* (2011).

[36] AMAP [2009b], see note 32.

Convention. Thus again the Arctic acts as a 'sentinel'. Recent AMAP POPs assessments[37] have focused on these 'new' POPs, and on following the temporal trends in 'legacy' POPs.

3. Mercury

Mercury is a naturally occurring environmental contaminant and a metal that has been used by man for centuries. However environmental levels throughout the world have increased dramatically over the past 150 or so years as a result of human activities in the post-industrial period – most notably the burning of fossil fuels (especially coal). The Arctic is no exception. Recent analyses of hard tissues (hair, feathers and teeth) taken from museum samples dating as far back as 1100 AD provide evidence that the levels observed in the Arctic today are 10 times higher than they were in pre-industrial times.[38] Although mercury emissions to the atmosphere have decreased over the past two decades in Europe and North America, following the introduction of improved emission control devices at power stations and shifts in fuel used for energy, this has been offset by increasing emissions from South East Asia - mainly China, Figure 2.

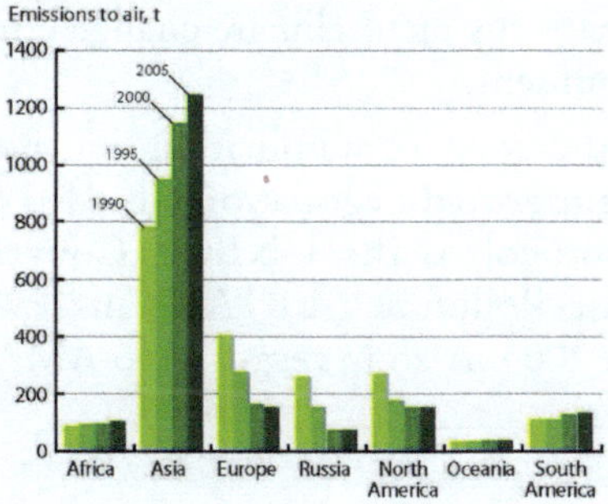

Fig. 2. Source: Arctic Monitoring and Assessment Programme

[37] *Ibid.*

[38] AMAP, *Arctic Pollution 2011* (2011) ('AMAP [2011b]').

The net result is that anthropogenic emissions to air have remained relatively stable since 1990. There is however some evidence that emissions may be increasing again, and projections show that if controls are not introduced this is certainly likely to be the picture in the coming years.

As with POPs, the highest levels of mercury have been detected in Arctic marine mammals, and like POPs this has resulted in high dietary exposure to mercury for some indigenous peoples who eat large amounts of marine food.[39] Daily intakes of mercury by some Inuit in Greenland and northeastern Canada are up to eight times the limit set by the World Health Organization ('WHO'), and high percentages of mothers and women of reproductive age in some communities in West Greenland and northeastern Canada exceed guidelines for mercury in blood. Human health effects of mercury exposure such as effects on neurological development in young children and associations with cardiovascular disease have been documented in some Arctic populations.[40] Blood monitoring has shown that levels of mercury in the most affected populations are declining – in some areas due to dietary advice but in others likely due to both changing levels in traditional food items and shifts from consumption of traditional foods to a more western diet (with a greater proportion of store-bought foods).

Recent observations of increasing trends in mercury levels in a range of species in the North American/western Greenland Arctic segment reported in the most recent AMAP assessment[41] are therefore a cause for concern, however it is not yet clear whether this is related to changes in global emissions patterns or to climate change effects on mercury pathways in the environment.

Information on mercury contamination in the Arctic reported by AMAP in 1997[42] supported negotiations that led to the adoption of the Heavy Metals Protocol to the UNECE Convention on Long-range Transboundary Air Pollution ('LRTAP') in 1998.[43] The Protocol entered into force in 2003. Also in response to AMAP's findings, the Arc-

---

39 *Ibid.*

40 AMAP, *AMAP Assessment 2009: Human Health in the Arctic* (2009) ('AMAP [2009e]').

41 AMAP [2011b], see note 38.

42 AMAP [1997], see note 13.

43 Protocol to the 1979 Convention on Long-Range Transboundary Air Pollution on Heavy Metals (adopted 24 June 1998, entered into force 29 December 2003) 2237 UNTS 7.

tic Council Ministers (in 2000) called on UNEP to initiate a global assessment of mercury that could form the basis for appropriate international action, and in February 2001 UNEP's Governing Council decided to initiate the UNEP Global Mercury Assessment. In 2003, UNEP agreed that there was sufficient evidence of significant global adverse impacts from mercury and its compounds to warrant further international action to reduce the risks to human health and the environment from the release of mercury and its compounds to the environment. In 2009, UNEP began a process aimed at negotiating, by 2013, a legally-binding international agreement to limit emissions of mercury. If implemented, this agreement has the potential to significantly reduce Arctic mercury contamination. Since 2005, AMAP has worked closely with UNEP to support the UNEP mercury process through its Intergovernmental Negotiating Committee ('INC') process, and parts of the 2011 AMAP assessment on Mercury in the Arctic[44] have been specifically developed to support the ongoing negotiations.[45]

4. POPs, Mercury and Human Health

The information presented above has highlighted the potential negative impacts of POPs and mercury contamination on the health of some Arctic populations. Notwithstanding this, AMAP assessments have always endeavored to present the risks associated with contamination of Arctic traditional foods in a responsible and balanced manner.[46]

Traditional foods are an important part of the cultural identity of northern communities. They are also important sources of healthy nutrients and vitamins. The evidence presented by AMAP confirms that traditional food is more nutritious than market food, reduces risk factors for several disease conditions such as heart disease, obesity, and diabetes, and can bind communities together in ways that market food does not.

AMAP human health experts have concluded that, despite the presence of contaminants in some marine animals and human milk, consumption of traditional food should be continued; and that breastfeeding is the

---

[44] AMAP [2011b], see note 38.

[45] AMAP/UNEP, *Technical Background Report to the Global Atmospheric Mercury Assessment* (2008).

[46] AMAP [2009e], see note 40.

best and safest form of infant nutrition, critical for proper development of the infant immune system.[47] It is also clear that in some areas of the Arctic there is a need for some groups, such as pregnant women and women of reproductive age, to restrict their intake of the most contaminated food items by substituting it with less contaminated but similarly nutritious items in order to minimize the risks for their babies.

The results of AMAP human health studies are fed into the 'effectiveness and sufficiency' assessments conducted as part of the implementation of both the UNECE and UNEP agreements. Together with air sampling, human blood and breast milk sampling are the core components of the Stockholm Conventions' 'Global Monitoring Programme' for POPs. AMAP is responsible for arranging an international program to QA/QC laboratory analyses of contaminants in human blood samples.

5. Petroleum Hydrocarbons

One of the (sub-regional) priority issues for AMAP concerns pollution of the Arctic by petroleum hydrocarbons and oil. AMAP has produced two major assessments[48] addressing this subject and documenting oil pollution in the Arctic from natural seepages and spills during oil and gas resource exploitation, including transportation (by both pipelines and ships). Contamination also occurs as a result of operational discharges from ships and runoff from land. Within the Arctic, atmospheric PAH contamination is mainly linked to incomplete combustion of oil and coal, especially in Russian Arctic cities such as Murmansk and Arkhangelsk.[49] PAH levels in marine and freshwater sediments reflect various combustion sources and natural fossil fuel deposits, such as those occurring in the Canadian Beaufort Sea, and natural seepages in the basin of the Mackenzie River and some Alaskan rivers.

An oil spill in the Arctic waters, especially in ice- or partially ice-covered seas, may remain in the environment for a long period due to low degradation rates and difficulties in cleaning-up spills under dark and cold conditions. The ice edge is an important Arctic habitat for

---

47 *Ibid.*

48 AMAP [1998], see note 9; AMAP, *AMAP Assessment 2007: Oil and Gas Activities in the Arctic: Effects and Potential Effects* (2010).

49 AMAP/NEFCO [1995a], see note 12.

both primary production, and seabirds and marine mammals that aggregate at leads to feed during certain parts of the year. An oil spill in the Arctic at the wrong time of the year and place might therefore have serious potential consequences for vulnerable Arctic ecosystems.

With increasing Arctic sea ice melt, it is not only the oil and gas drilling activities that are likely to increase. Sea transportation, including transportation of oil and gas within and across the Arctic is set to increase, taking advantage of new sea routes between the North Atlantic and the Pacific. Stricter regulations on the type of ships and their operations in Arctic waters are needed to reduce the risk of accidents and spills.[50]

The AMAP Oil and Gas assessment concludes that although technologies used for exploration and resource development have been improved significantly since oil and gas exploration in the Arctic began in the 1920's, the existing stores of equipment to combat spills in Arctic areas are inadequate to deal with accidents that will inevitably occur.[51] If oil and gas activity is to be extended in ice-prone marine areas there is a need for both better technologies and better operational practices, including compliance with regulations.

A regulation of the International Maritime Organization ('IMO') to define part of the Arctic as a particularly sensitive area is one option that has been proposed for increasing the protection of Arctic marine and coastal ecosystems. Under the Arctic Council work involving several of its working groups [AMAP, Protection of the Marine Environment Working Group ('PAME'), CAFF and Sustainable Development Working Group ('SDWG')] is ongoing to further address these issues.

6. Climate Change

AMAP's first scientific assessment[52] addressed climate change in the Arctic mainly from the perspective of evaluating the adequacy of available information. It concluded that there were important gaps in the knowledge and data required to assess this subject. At the same time, it provided some of the first circumpolar reviews of changes taking place in the Arctic as a result of climate change and depletion of stratospheric

---

50 ACIA, *Impact of a Warming Arctic: Arctic Climate Impact Assessment* (2004).

51 AMAP, *Arctic Oil and Gas 2007* (2008).

52 AMAP [1998], see note 9.

ozone. Improved monitoring and modeling of effects due to climate change in the Arctic were therefore identified as priority areas for further work.

As a direct follow-up, AMAP expanded its climate monitoring efforts and initiated a new assessment – the Arctic Climate Impact Assessment ('ACIA'). ACIA was implemented jointly with the IASC and CAFF, and delivered the first comprehensive assessment of climate change and its effects in the Arctic in 2004.[53] The ACIA provided some of the first indications that the climate change ongoing in the Arctic might lead to a situation where the Arctic Ocean could be ice free in summer by the end of the 21st century, opening the region for increased shipping, and new access to resources such as fish, oil and gas, and minerals. It further examined the possible (significant) impacts that such development could have on Arctic human populations, including impacts on the traditional lifestyles of Arctic indigenous peoples. The ACIA also highlighted the potential feedbacks from the Arctic to the global climate system, in particular the possible role of the Arctic in raising global sea levels.

The ACIA[54] had a significant impact on the public awareness of (Arctic) climate change, and this was reflected in the assessment reports prepared by the IPCC in 2007.[55] The ACIA put Arctic climate change higher on the political agenda. Since 2005, an increasing number of countries have recognized that the Arctic Council is an important forum for addressing international Arctic affairs, and many new countries have applied for Arctic Council observership. This increased interest is an important driver behind the ongoing reshaping of the structure of the Arctic Council.

In May 2011, AMAP in collaboration with IASC and the World Meteorological Organization ('WMO')/Climate and Cryosphere ('CliC') Project presented a major work following-up on the ACIA. The SWIPA assessment focuses on the changing state of the Arctic cryosphere and the implications at local, regional and global scales.[56] The assessment based on new scientific results from national programs and research radically updates projections made as recently as 2007 by the IPCC.

---

53 ACIA [2005], see note 7.

54 *Ibid.*

55 IPCC, *Climate Change 2007* (2007).

56 AMAP [2011a], see note 8.

The last five years have been the warmest five-year period recorded in the Arctic since 1880. The SWIPA assessment documents that the melting of snow and ice in the Arctic is proceeding at rates faster than those reported in previous (ACIA and IPCC) assessments, and in particular that the mathematic models applied to project sea ice decline are too conservative (Figure 3).

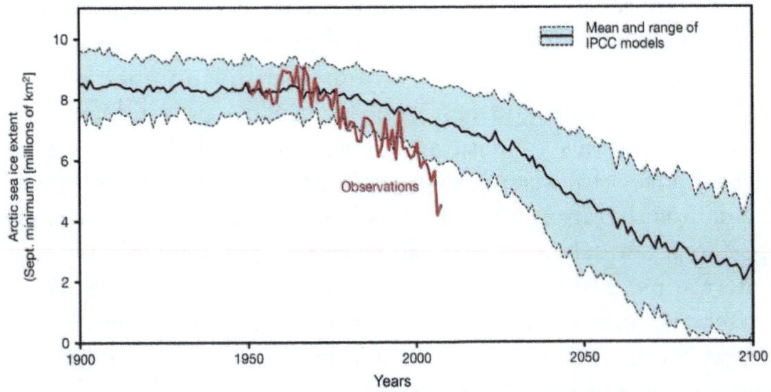

Fig. 3. Source: Potsdam Institute for Climate Impact Research

Melting and run-off from the Greenland ice sheet are also shown to be increasing at previously unsuspected rates, with projections that global average sea level may rise up to 0.9-1.6 m by 2100 as a consequence of this and the melting of other terrestrial ice in Arctic glaciers and ice caps.[57] These changes may have huge implications not only at a local Arctic level, but also on global level especially for countries, societies and peoples living in areas that are already close to sea level.

AMAP is currently implementing two further new assessments related to Arctic climate change: the first addresses acidification of the Arctic Ocean and will be presented in 2013. The second concerns the role played by short lived climate forces – especially black carbon, on the warming of the Arctic. This report was presented in May 2011.[58]

---

57 *Ibid.*

58 AMAP, *The Impact of Black Carbon on Arctic Climate* (forthcoming 2011) ('AMAP [2011c]').

V. Challenges for the Arctic

The Arctic territory is vast, remote and difficult to access for long periods of the year due to climate and darkness. This limits the possibilities of performing monitoring and research across the entire geographical area. Despite increasing availability of information from satellite-based remote sensing, there is still a lack of sensors or automated monitoring equipment that can be deployed on e.g. drifting buoys to monitor many key parameters. Field observations are needed not only to fill these gaps but also to 'ground-truth' satellite data. One of the main challenges for the scientific community is to improve the number of platforms for Arctic monitoring and research – in space, on land, in the ocean. Increasing volumes of data are being collected, but access to many of these datasets for science remains a problem. Improving (free) access to the data is therefore a priority, as is securing long-term data storage and appropriate quality control, so that data are available for use by future generations. These issues are central elements in the Sustaining Arctic Observing Network ('SAON') initiative that is being developed in cooperation between the Arctic Council and relevant international organizations and non-Arctic countries.

The Arctic contains vast natural resources – renewable (fisheries, forests, etc.) and non-renewable (oil and gas, minerals, etc.). These are of increasing interest to global societies. The melting of the sea ice will open access to some of these resources from the sea, but at the same time the thawing of the permafrost and reduced ice on rivers may reduce access on land. Transpolar shipping will increase and the region will also become more accessible for other activities such as tourism.

With a warming Arctic, new species – both plants and animals – will migrate northwards and come into contact with species currently resident. The same will apply to human populations. During the Cold War, huge military infrastructures existed in the North. Even if some of these installations have been decommissioned, the Arctic will remain an important military-strategic area.

As described in this article, the Arctic is under pressure not only from climate change and pollution, but increasingly from interest in exploiting its resources. Most countries and local northern societies are not prepared for the changing situation and therefore there is a need to conduct new studies of the combined effects of these multiple drivers of change in order to better inform about options for adaptation to change, both on shorter and longer timescales.

A key question for the Arctic is: Do we have the necessary instruments in place, i.e. international laws and organizations, to manage this 'new Arctic'? Are the UN Convention on the Law of the Sea[59] and the Arctic Council sufficient? And are they prepared to complete the necessary tasks?

VI. Summing Up

In this article we have described a small part of the very interesting developments that have occurred within the Arctic since Gorbachev's speech in 1987. The AEPS initiatives to 'monitor and assess' the Arctic environment have been met with huge enthusiasm by the scientific community, but on the other hand have been greeted with less enthusiasm (and a large degree of skepticism) by others, including the military establishment, elements of industry, and some working within the business of 'foreign affairs'.

However, the material presented above shows how important this scientific activity to 'monitor and assess' the Arctic environment has been, not only for the Arctic but for the globe.

Actions resulting (in part at least) from the work of AMAP, such as the clean-up of radioactive sources within the Arctic and the establishment of international agreements to reduce global pollution by POPs and mercury, and actions to follow-up on AMAP health assessments have directly benefited Arctic communities by reducing the potential risk of accidents and reducing their exposure to contaminants. There remains, however, much to do in this respect.

AMAP's documentation of the rather dramatic developments in the Arctic due to climate change and the connections and feedback from the Arctic to areas further south will hopefully strengthen the work to reduce emissions of greenhouse gases and black carbon to the atmosphere and thereby reduce the resulting impacts of these emissions on global climate and the melting of snow and ice with its secondary effects around the globe.

The Arctic environment is a sentinel for change and a barometer for the health of the global environment. It is extremely important that systems to adequately and effectively monitor and assess the Arctic environ-

[59] United Nations Convention on the Law of the Sea (concluded 10 December 1982, entered into force 16 November 1994) 1833 UNTS 397.

ment are continued and improved so that trends and effects can be documented over the long term, so that new concerns can be identified in a timely fashion, and so that the effectiveness of mitigation actions can be evaluated – and most importantly – so that the Arctic can be prepared for the changes that are coming.

More Relevant Research for the EU Arctic Policy

(Speech)

by Michael Gahler[*]

Herr Staatssekretär,
liebe Susanne Wasum-Rainer,
sehr geehrte arktisch-wissenschaftliche Gemeinde,
meine Damen und Herren,

zunächst möchte ich mich herzlich für die Einladung bedanken und gleichzeitig zum Ausdruck bringen, dass ich es für eine gute Entscheidung halte, einen Vertreter des Europäischen Parlaments zu einer solchen Konferenz einzuladen. Dabei geht es nicht um mich persönlich, sondern um die Institution, ohne die es in Europa nicht mehr geht bei der Formulierung von gemeinschaftlichen Politiken einerseits und ihrer Finanzierung durch den EU-Haushalt andererseits.

The Arctic is changing rapidly and attracts more and more attention. Not only the planting of someone's flag raised awareness, but in particular the challenges of climate change and the prospects of resources extraction and economic development are filling the headlines.

Most obvious changes become visible to the wider public. They might be illustrated by tour operators offering luxurious expeditions on icebreakers: 'Choose the seven-day "Polar Bear Experience" ' or 'Take the "High Arctic & Northwest Passage" – two weeks on a comfortable ship', calling at places on our planet that not so long ago were accessible only to courageous and well-equipped researchers like you.

* Member of the European Parliament.

The Northwest Passage is now open in summertime, and, thanks to icebreakers, people in sunglasses can watch polar bears from their deck chairs. More than any political document or research paper these images portray how fundamentally the Arctic is changing.

Europe not only bears a responsibility as one of the main contributors to pollution and greenhouse gas emissions, but also has a particular economic and scientific interest in the Arctic. We will have to deal with the consequences of the changes happening in the region – from environmental and climate change issues to the geopolitics of shipping routes and security of supply of resources.

Already the European Union ('EU') is involved and actively engaged in the Arctic arena. Of the eight members of the Arctic Council three are EU Member States: Denmark, Sweden and Finland. As an EU accession country Iceland will become number four – perhaps. And Norway is a member of the European Economic Area, almost number five. The three remaining countries are strategic EU partners: Canada, Russia and the US.

Even without a share of the Arctic coastline the EU is an Arctic player in a number of relevant fields. Several of the EU competences and policies potentially touching upon the region are either shared with the Member States or complementary to their policies: others, like fisheries, are exclusive EU competences. In areas such as research, environment, climate change, energy, transport and fisheries, EU action or non-action can have a direct bearing on the Arctic.

It is worth noting that the Lisbon Treaty[1] changed the internal procedures of the EU towards a stronger involvement of the European Parliament as a co-legislator. Since 2008, the European Union has been formulating an Arctic Policy in several steps.[2] In January this year the

[1] Treaty of Lisbon amending the Treaty on European Union and the Treaty Establishing the European Community (signed 13 December 2007, entered into force 1 December 2009) [2007] OJ C306.

[2] See e.g.: European Parliament, 'Resolution on Arctic Governance', P6_TA(2008)0474 (9 October 2008); Commission of the European Communities 'Communication from the Commission to the European Parliament and the Council – The European Union and the Arctic Region', COM (2008) 763 final (20 November 2008).

European Parliament adopted my Report on the High North with an overwhelming majority.[3]

The Report clearly refers to the existing set of international rules and networks of cooperation thereby rejecting the idea of an Arctic Treaty on the model of the Antarctic Treaty,[4] because of the fundamentally different legal and political situation. As regards Arctic governance, the European Union has repeatedly underlined the applicability of the UN Convention on the Law of the Sea (UNCLOS),[5] and of other relevant international agreements.

The vast areas of the Arctic region mostly belong to sovereign countries which are either EU members or closest partners. To enhance cooperation the European Union finds it useful to be a permanent observer in the Arctic Council. Mutual trust and international coordination are crucial. We favor a further development of the body of international law. Furthermore the Report clearly establishes the interest of the EU in the Arctic region and highlights EU contributions.

During the preparation of the Report we founded the EU-ARCTIC-Forum in Brussels, which has become a platform for exchange and debate between politicians, business and civil society, stakeholders and experts, as well as researchers from the Arctic or those who deal with it. Developing a common understanding of the facts was crucial for the formulation of our policy document. The EU-ARCTIC-Forum can also serve as a platform for the presentation of your work as researchers to the political sphere in the EU.

Although the formulation of a comprehensive EU Arctic Policy is relatively new, the EU's involvement in Arctic issues is not. We have been involved through different sectoral policies, especially in the field of Arctic research. Over the years, the EU itself has contributed more than € 200 million in a wide range of areas through the research framework programs. Substantial support for Arctic observation during the International Polar Year 2007-2008 is one example. The EU's contribution to

---

[3] European Parliament, 'Report on a Sustainable EU Policy for the High North', A7-0377/2010 (16 December 2010); European Parliament, 'Resolution on a Sustainable EU Policy for the High North', P7_TA(2011)0024 (20 January 2011).

[4] Antarctic Treaty (signed 1 December 1959, entered into force 23 June 1961) 402 UNTS 71.

[5] United Nations Convention on the Law of the Sea (concluded 10 December 1982, entered into force 16 November 1994) 1833 UNTS 397.

projects on 'the impact of climate change on human health in the Arctic' is another.

The Integrated Project for Developing Arctic Modelling and Observing Capabilities for Long-term Environmental Studies ('DAMOCLES') monitored the Arctic environment and detected significant changes occurring in the sea-ice surface area and thickness. Sea-level rise is at the heart of the ongoing discussions, and one of the main issues of this debate focuses on how much the change in ice-mass balance contributes to this phenomenon. This is also a key objective of another flagship project: Ice2sea – estimating the future contribution of continental ice to sea-level rise. It will provide improved sea-level estimates based on the most up-to-date climate projections. In addition we are supporting research projects which investigate the environmental pollution in the Arctic and its impact on human health.

The European Commission is also supporting observations and research infrastructure in the Arctic. The aim is to satisfy the scientific community's needs for high-quality data and to establish long-lasting research capabilities. The European Commission is supporting long-term measurements and reporting of marine data in the context of the European Marine Observation and Data Network and of the Global Earth Observation System of Systems. As regards infrastructures, the European Strategy Forum for Research Infrastructures ('ESFRI') plays an important science-policy interface role. ESFRI brings together representatives of EU Member and Associated States and the European Commission to support a coherent approach to policy-making on research infrastructure in Europe. In the ESFRI strategic road map a novel European research icebreaker with drilling capability has been identified as one of the most important needs for research infrastructures of pan-European relevance.

In this context, the EU is supporting the preparatory phase of the Aurora Borealis icebreaker through the project ERICON with € 4.5 million. The research icebreaker Aurora Borealis is now one of the most ambitious and prestigious European projects. It will support multidisciplinary research in areas such as climate change, biodiversity assessments and long-term monitoring. The ship and the organizations that will be created around it will provide an excellent platform for coordinated European polar research.

European research projects have contributed significantly to the success of the International Polar Year. While the Polar Year is now over, we have continued to contribute to the IPY legacy through the 'Ocean of

Tomorrow' joint call in the context of the Seventh EU Framework Programme ('FP 7'), launched in 2010.

The preparatory phase of a multidisciplinary research infrastructure, the Svalbard Integrated Arctic Earth Observing System ('SIOS'), has been funded through the FP 7 Capacities Programme. SIOS was conceived in 2007 during the International Polar Year and is designed to be an integral part of the Sustained Arctic Observing Network. Moreover, the current 'Environment theme call' includes a topic on permafrost thawing, which basically completes the picture of the ongoing EC research activities.

For many years, EU Member States such as Germany, the Netherlands, the United Kingdom, France, Poland, Spain or Italy have dedicated substantial human and financial resources to Arctic research.

Alongside growing economic demands, climate change is the main driver of change in the Arctic more than elsewhere. It is commonly agreed that the Arctic is a region that is affected earlier and more heavily by climate change and pollution originating from other parts of the world. Whilst fighting climate change, the EU must acknowledge the need to adapt to the unavoidable changes and have a rational assessment of the risks, threats, challenges and opportunities arising from them. All our actions need to be science-based and this is a major task for research and science – to provide decision makers with better, more precise and reliable data and advice!

A spirit of economic and political competition would not serve the future of the Arctic. We actually need the spirit of dialogue and cooperation: a scientific dialogue among experts examining the changing situation on the ground; a political dialogue among all the stakeholders, i.e. the States, communities and political authorities that play a role in the region. The European Union is fully willing to assume its role in this regard.

What are the next steps for future research? The orientation should include long-term monitoring of the Arctic environment in a coordinated international effort. This is essential for a better understanding of change in the Arctic and provision of data for deeper scientific analyses. Furthermore, the development of climate and socio-economic scenarios must be considered a priority to help policy-makers build a sustainable future for the Arctic region.

Let me be clear on this. We need to learn and understand better how the various systems of the Arctic work, but we also need applied science that helps in managing the adaptation and finding ways of balancing environmental protection and sustainable development of the region for the good of the inhabitants of the Arctic!

In the EU context, the Framework Programme research outcomes have a key role to play in shaping the EU's policy development. But more effort is needed to ensure that project results inform policy-making in a meaningful way.

In this, we need to keep in mind that policy has a different perspective and time frame than research. The short time line in which policy operates requires very fast mechanisms to answer scientific and technical issues. Research projects by nature have a much longer perspective and, therefore, science should be able to foresee future policy needs. At the same time, the policy-makers need to be informed of the existing scientific results. Early engagement with scientists by policy-makers is also essential to get policy definition and strategic direction understood by both sides.

So let us establish such a closer relationship between science and politics to ensure that research agendas also tackle issues of relevance to policy-makers. Scientific findings should be conveyed on a regular basis to decision-makers in a manner that emphasizes the implications for making and implementing policies. Such meaningful dialogue should then promote enhanced interactions between scientists and policy-makers to pursue sustainable development in the Arctic.

The Arctic region provides a unique model to explore the interdisciplinary nexus of science and policy that is essential for the sustainable development of our society from local to global levels. Due to the global importance of the Arctic region, we also need to establish an active dialogue between Arctic and non-Arctic States in order to increase awareness among the general public and governments. The European Union is inextricably linked to the Arctic region by a unique combination of history, geography, economy and scientific achievements. In this context, the Commission is interested in broadening its dialogue and international cooperation on Arctic issues with other Arctic players, especially in the research field.

The Arctic can have different prospects. The future will not only be determined by the blind fortune of natural forces, such as climate change. The future of the Arctic is shaped by political and economic decisions of today and tomorrow. The model we would like to see develop is a model of sustainable development and adaptation of the Arctic – a region heavily affected by climate change which turns into a laboratory of how to manage the adaptation in a way that balances nature conservation and necessary economic and social development. If we are to succeed with this we need a sound and broad foundation in sciences!

Thank you.

Arctic Processes and the Global Climate

by Peter Lemke[*]

I. Polar Processes

High latitudes have received attention recently because of significant changes in the atmosphere, sea ice, and ocean, and on land, especially in the Arctic. The surface air temperature in the Arctic has increased about twice as fast as the global air temperature. The Arctic sea-ice extent in summer has decreased by 35 % since 1979, and the sea ice thickness during late summer has declined in the Central Arctic by about 40 % since 1958. A warming has also been observed in the Arctic Ocean at intermediate depths. Permafrost is warming, glaciers, ice caps and the Greenland ice sheet are melting, and the sea level is rising. These observed amplified trends are in agreement with warming scenarios performed with coupled climate models, which indicate an amplified response in high latitudes to increased greenhouse gas concentrations. But details of the complex interaction between atmosphere, sea ice and ocean, and the impacts on the ecosystem and the human society are still only marginally known.

The Arctic and Antarctic are quite different as regards their geography. The Arctic is a mediterranean sea surrounded by continents, while the Antarctic is a continent surrounded by the sea. In the Arctic the smaller ice sheet (Greenland) is located at the margin, while in the Antarctic the huge ice sheet is in the center covering nearly the entire continent. The seasonal cycles of sea ice extent are also quite different. Whereas a substantial amount of sea ice survives the Arctic summer, the Antarctic sea ice being at lower latitudes nearly disappears in summer. Winter sea ice

---

* Alfred Wegener Institute for Polar and Marine Research, Bremerhaven, Germany.

extent is substantially larger in the Antarctic as compared to the land-locked Arctic. In contrast to the southern hemisphere, there are vast regions on the northern continents characterized by permafrost. These differences between the Arctic and Antarctic have significant consequences for their role in the climate system and for their response to current global warming.

Generally, polar regions are zones of energy loss, whereas equatorial regions receive a surplus of solar energy. The deficit at the poles is balanced by winds and ocean currents which transport huge amounts of energy from the tropics to the poles. The associated atmospheric and oceanic circulations are a response to the temperature contrast between the poles and the equatorial regions, which is largely dependent on the areal extent of snow and ice surfaces at the poles. The strengths and patterns of these circulations are influenced by various interactions and feedbacks in the climate system (Fig.1).

Because snow and ice surfaces reflect up to 90% of the incoming solar energy, cold air is produced in these regions. With the receding snow and ice surfaces the equator-to-pole temperature contrast and subsequently the atmospheric and oceanic circulations are likely to be changing. At the same time, polar regions are very sensitive to global climate change. An initial retreat of snow and ice unveils a dark surface (ocean, land), which strongly absorbs solar radiation with subsequent warming and a further retreat of snow and ice. This positive feedback is partly responsible for the observed polar amplification of climate signals.

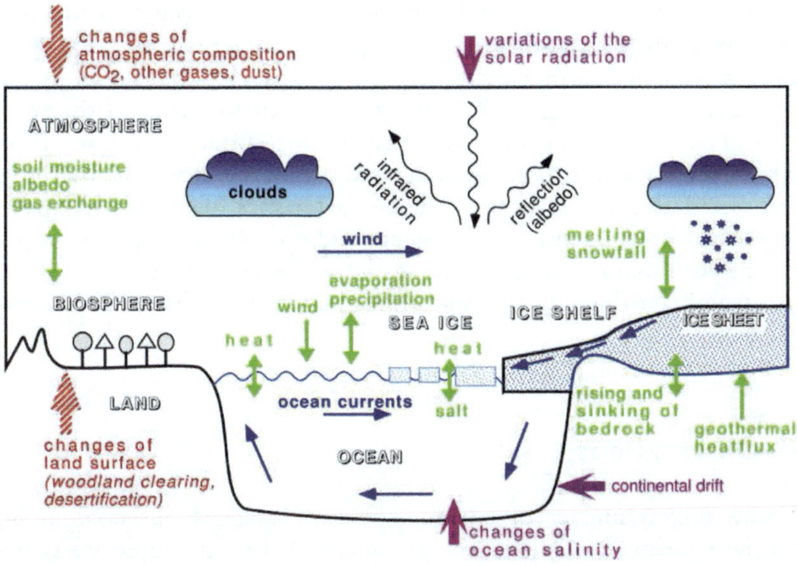

Fig.1. Schematic depiction of the climate system. Thin black arrows denote radiative processes; blue arrows show advection processes, green arrows indicate interactions within the climate system, bold magenta arrows represent changes of the boundary conditions, and the hatched red arrows show impacts of human activities on the climate system

Other feedbacks involve the atmospheric boundary layer – including the clouds – and its interaction with the sea ice cover, which is a mixture of sea ice floes and open water. The surface temperature of the open water is year-round typically at -2° C, the freezing point of sea water, whereas the temperature at the sea ice surface ranges from 0° C in summer to about -40° C in winter. Even though the open water represents only a few percent of the Arctic Ocean area in winter, it dominates the heat exchange between ocean and atmosphere.

Sea ice plays an important role in the climate system since it modifies the surface radiation balance due to its high albedo, and it influences the exchange of momentum, heat and matter between atmosphere and ocean because of its insulating behavior. During cooling periods the freezing of sea ice initiates brine expulsion and subsequent convection with a deepening of the surface mixed layer and the formation of deep and bottom water. During the melting period relatively fresh water stratifies the oceanic surface layers, i.e. the mixed layer retreats to shallower depths. In contrast to low latitudes the mixed layer evolution in polar regions is dominated by surface fluxes of salt or freshwater (positive or negative freezing rates).

Because of the sea ice motion, melting generally occurs in places far away from the formation area. Averaged over one seasonal cycle the net freezing rate in a certain area is therefore rarely zero, but rather positive or negative depending on the divergence or convergence of the sea ice flow. The net freezing rates represent strong surface buoyancy fluxes, which heavily affect the density structure and, therefore, the thermohaline circulation in the ocean. The characteristics of the sea ice motion (especially convergence or divergence and the net freezing rate) depend on the rheological aspects of the sea ice as a plastic material, on the geometry of coastlines and on the atmospheric forcing fields, in particular the wind.

II. Observations

The strongest impact of current climate change is observed in high northern latitudes. An extensive summary of current conditions is given in the Arctic Report Card, which is updated annually.[1] The positive temperature trend during the past 50 years recorded at weather stations north of 65 N is about twice the global temperature increase shown in Fig. 2.

---

[1] J Richter-Menge and JE Overland (eds), 'Arctic Report Card 2010' (2010), see <www.arctic.noaa.gov/reportcard/> (12 August 2011).

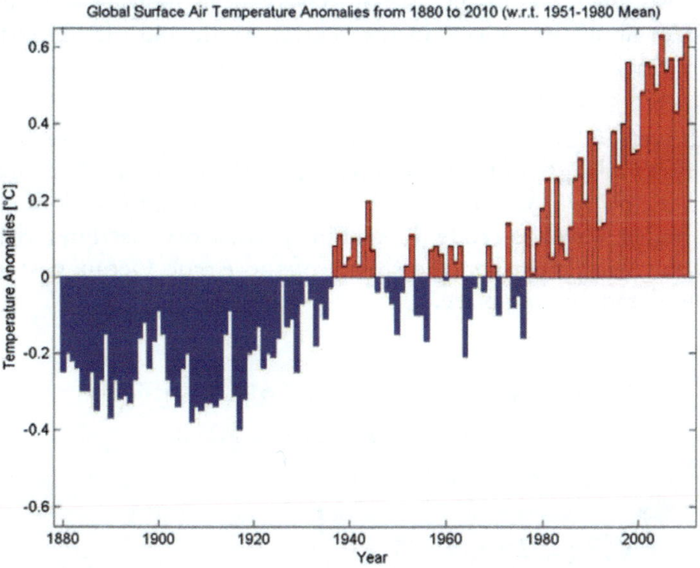

Fig. 2. Global surface temperature anomalies from 1880 to 2010 as compared to the 30-year mean temperature (1951-1980)[2]

The warming air is accompanied by a warming ocean. Observations from moorings in Fram Strait deployed since 1997 indicate that the temperature of the Atlantic inflow into the Arctic Ocean through the West Spitsbergen Current has followed a positive warming trend during the entire period.

Over the period 1979-2010 sea ice has receded in winter (March) at a rate of 2.6% and in summer (September) at a rate of 11.5% per decade (Fig. 3). During the last four years (2007-2010) the sea ice minimum was below that of all other years since 1979. The absolute minimum so far was observed in 2007, when persistent low pressure over Siberia and high pressure over the Beaufort Sea forced warm air from the North Pacific across the North Pole into the North Atlantic. An extrapolation of the sea ice trend shown in Fig. 3 indicates that the Arctic Ocean might be ice free in summer in the early 2070s.

The stronger retreat in summer as compared to winter hints at a significant reduction of sea ice thickness. There are only a few sea ice thick-

---

2 Data source: NASA, Goddard Institute for Space Studies, see <http://data.giss.nasa.gov/gistemp/> (12 August 2011).

ness measurements available, but all observations indicate a strong reduction of about one meter in the central Arctic and off the coast of North America during the past three decades, which represents a thinning of more than 25%.[3]

With the retreat of the sea ice, a rich ecosystem adapted to this porous medium is in danger of being eliminated, i.e. the entire food chain from algae to the polar bear may be destroyed. In addition, the lack of sea ice allows higher waves to erode the melting permafrost coastlines, thereby threatening settlements on the shores of the Arctic Ocean; a scenario which is already taking place.

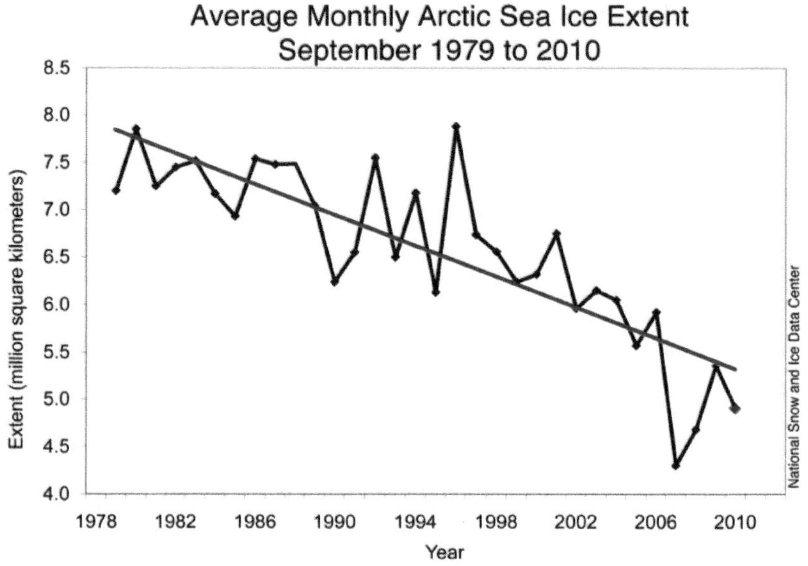

Fig. 3. Average Arctic sea ice minimum extent since 1979. The linear trend shown amounts to 11.5% per decade[4]

---

[3] DA Rothrock and J Zhang, 'Arctic Ocean Sea Ice Volume: What Explains Its Recent Depletion?', *Journal of Geophysical Research* 110 (2005) C01002; C Haas et al., 'Synoptic Airborne Thickness Surveys Reveal State of Arctic Sea Ice Cover', *Geophysical Research Letters* 37 (2010) L09501.

[4] National Snow and Ice Data Center, 'Arctic Sea Ice News & Analysis' (4 October 2010), see <http://nsidc.org/arcticseaicenews/2010/100410.html> (12 August 2011).

The Greenland ice sheet is losing mass, approximately in equal parts from melting and iceberg discharge.[5] In the last five years the mass loss has doubled, currently contributing about 0.4 mm per year to sea level rise. Currently, global sea level is rising by about 3 mm per year, with approximately equal contributions from melt water and ocean warming. Arctic glaciers and ice caps are shrinking at a rate that has been increasing since 1987.

The snow cover has decreased in most regions in the northern hemisphere ('NH'), especially in spring and summer. The snow cover observed by satellite over the 1966 to 2005 period decreased every month except November and December, with a stepwise drop of 5% in the annual mean in the late 1980s.

The freeze-up and break-up dates for river and lake ice show considerable spatial variability. Averaged over the available data for the NH spanning the past 150 years, the freeze-up date has occurred later at a rate of 6 days per century, while the break-up date has occurred earlier at a rate of 6 days per century.

The temperature at the top of the permafrost layer has increased in the Arctic by up to 3° C since the 1980s. The permafrost base has been thawing at a rate ranging up to 0.04 m per year in Alaska since 1992. The extent of seasonally frozen ground has decreased on average by 7% and in spring by 15% since 1900. Permafrost degradation is leading to changes in land surface characteristics and drainage systems. Melting permafrost may also be a source of increasing methane fluxes into the atmosphere, with a subsequent strengthening of the greenhouse effect, which will lead to further warming. This positive feedback effect is currently one of the most pressing scientific questions associated with permafrost degradation.

The Arctic-wide warming trend is accompanied by considerable temporal (see Fig. 3) and regional variability. There are strong impacts on the terrestrial and marine ecosystems, including the highly specialized ecosystem associated with the sea ice. Due to the warming, an increase in greening of up to 15% was observed between 1982 and 2008 in the High Arctic of Canada, West Greenland and Northern Alaska.

---

[5] P Lemke et al., 'Observations: Changes in Snow, Ice and Frozen Ground', in S Solomon et al. (eds), *Climate Change 2007: The Physical Science Basis. Contribution of Working Group I to the Fourth Assessment Report of the Intergovernmental Panel on Climate Change (2007)* 337 et seq.

The Arctic Ocean is characterized by decreasing sea ice cover and increasing open water, where wind systems can easily excite high surface waves, which have a strong capability to erode the melting permafrost coasts. In some places erosion has already forced villages to move further inland.

III. Scenarios

Projections for the last decade of this century with sophisticated coupled climate models indicate significant warming, depending on the level of anthropogenic CO_2 emissions. With a low emission scenario (B1), climate models suggest a global warming of about 1.8° C (1.1–2.9° C). With a high emission scenario (A2, business as usual), global temperature will increase on average by 4.0° C (2.4–6.4° C).

The greatest warming will occur in high northern latitudes. As a result, snow, permafrost, sea ice, glaciers, and the Greenland ice sheet will recede considerably. The Arctic Ocean will most likely be ice-free in summer from the 2080s on. Projected sea level rise for the end of this century will be up to 38 cm for the low emission scenario and 59 cm for the high emission scenario. There is considerable uncertainty concerning ice loss by the two ice sheets in Greenland and Antarctica, especially with respect to dynamic loss through glacier discharge. There are indications from recent studies that the sea level rise may reach up to 1 m or more by 2100. Furthermore, model studies suggest that the Greenland ice sheet will not recover if the warming exceeds 3 to 4° C. Then Greenland will melt continuously, resulting in a sea level rise of 7 m. But this process will most likely last several centuries or a thousand years.

In some projections Arctic late-summer sea ice disappears almost entirely around 2080. An extrapolation of the current trend of summer sea ice loss suggests that this may be the case even earlier.

In response to these strong climatic changes Arctic species that have adapted to the extreme environment are expected to be displaced towards the Central Arctic by the approaching Sub-Arctic species, which will invade the margins.

IV. Conclusion

Currently, the strongest warming is observed in high northern latitudes. As a consequence, snow, permafrost, sea ice, glaciers, ice caps and the Greenland ice sheet are melting. These changes have a strong impact on Arctic ecosystems and on human settlements on Arctic coastlines. Projections suggest an even stronger warming with more drastic impacts by the end of this century.

With the warming and the receding sea ice, use of Arctic resources and shipping through the North-East and North-West passages are becoming increasingly attractive. This will have a significant, and in most cases negative, impact on the Arctic environment.

Comparing the various coupled climate models, it is obvious that most models agree in their simulation of the decay of the sea ice in winter, but there are great differences in the simulation of summer decay. This is a consequence of our lack of understanding of polar processes in the summer, which are associated with a wet sea ice surface, including melt ponds and a thin summer stratus in the atmospheric boundary layer. The optimal parameterization of these processes is currently one of the grand challenges in polar research. Another grand challenge is understanding the detailed mechanisms which determine the impact of Arctic processes on the global climate system.

Developing and Sustaining an Arctic Ocean Observing System

(Abstract)

by Keith Alverson[*]

The Global Ocean Observing System ('GOOS') has been in existence for over a decade. During this first decade, GOOS was primarily a planning exercise, developing observational strategies and the international governance structures required to facilitate multi-national ownership and development of the system. This work is mostly done and the greatest challenge now facing GOOS is to complete and sustain an integrated, global system with clear user benefits. Substantial progress has been made with more than half of the *in situ* open ocean observing system for climate already in the water, including buoys, moorings, floats, tide gauges and repeat hydrographic lines. In addition to the *in situ* system, GOOS has also benefitted from the continuous satellite record of essential climate variables from space including, for example, sea surface temperature, height, surface vector winds, color (an indicator of primary productivity), and sea ice extent. The system is now operational, and serving as the observational backbone for near- and real-time coastal inundation warnings, weather services and even nascent climate services.

At the same time, substantial challenges remain. One of the most challenging of these is to develop a sustainable regional system in the Arctic Ocean. The Arctic presents unique logistical and technical challenges due to its remote and challenging environment and sea ice cover as well as governance challenges, due to the evolving nature of national research interests and territorial claims. The state of the existing near real-time Arctic Ocean observing system is shown in figure 1. Key future

* United Nations Environment Programme.

challenges will include *in situ* measurements from sea ice based platforms, particularly with decreasing areas of multiyear ice and increasingly large seasonal extent changes, measurements of changing biogeochemistry, for example air-sea CO2 fluxes and related acidification in Arctic waters, as well as ecosystem monitoring. Despite polar amplification of climate change and ongoing high profile rapid changes such as the record reduction in summer sea ice extent, the Arctic remains one of the only substantial ocean areas with ecosystems thought to have experienced 'very low' levels of human impact to date (figure 2). This seems highly unlikely to remain true in the near future, thus the importance of filling the Arctic gap in GOOS is given additional impetus by the need to document and understand the impacts of ongoing rapid changes, be they climatic, biological or within the social and economic realms. For example, increased ship traffic in the Arctic Ocean and development along its coasts, as well as the livelihood of indigenous residents, all depend on delivery of accurate synoptic weather forecasts and marine safety and distress systems that depend on sustained ocean observations.

References:

K Alverson, 'Sustained Arctic Observations: A Legacy of the Polar Year', *EOS Trans. AGU* 89(39) (2008).

UNESCO, Climate Change and Arctic Sustainable Development: Scientific, Social, Cultural and Educational Challenges (2009).

IOC/UNESCO, Why Monitor the Arctic Ocean? Services to Society from a Sustained Ocean Observing System (2010).

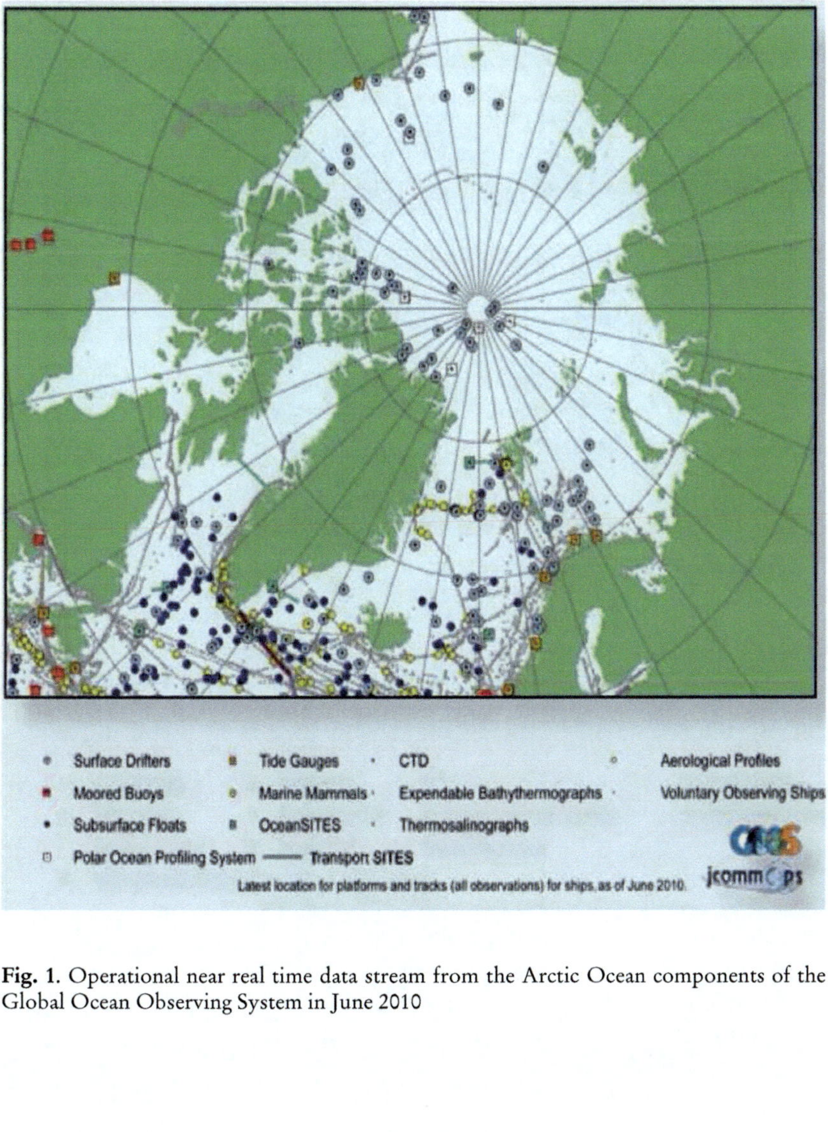

Fig. 1. Operational near real time data stream from the Arctic Ocean components of the Global Ocean Observing System in June 2010

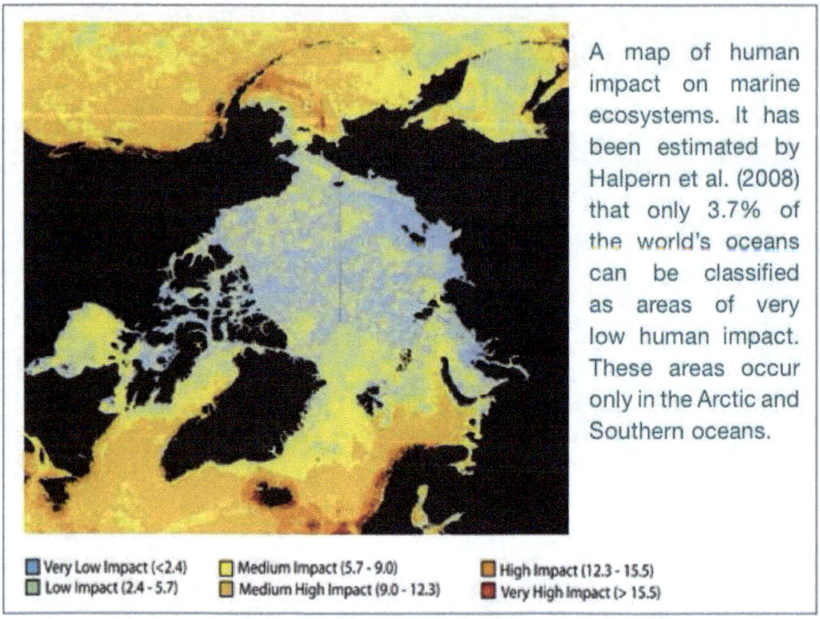

Fig. 2. Human Impacts on Marine Ecosystems. Adapted from BS Halpern et al., 'A Global Map of Human Impact on Marine Ecosystems', *Science* 319 (2008) 948 et seq.

Recent Russian Marine Research Activities in the Arctic Ocean

*by Igor Ashik**

Russia sees the Arctic region as an area of special interest in terms of economic, geopolitical and social development. Organizing and conducting complex real-time monitoring of meteorological, glacial, hydrophysical, geochemical, geophysical, biological and other parameters in the whole water column of Arctic Ocean is becoming the top priority in view of the active exploration and development of the Arctic region and the need to keep track of the ecological state of the ocean and especially its coastal areas. The environmental, economic, social and geopolitical problems of the modern age suggest three main lines of research:

1. Tracking climate change of the natural environment, explaining and forecasting climate change in the Arctic region; evaluating the impact of climate variations on economic infrastructures, the economy, the environment and the living conditions in the Arctic region;

2. Studying changes in environmental conditions related to natural resource development on the Arctic shelf, including exploration and development of gas and oil, their extraction, the construction of waterworks, developing maritime traffic in the Arctic region, the stimulation of various kinds of industries and improving living conditions there;

3. Studying hydrometeorological and glacial processes, providing current and prognostic information on natural conditions and processes to people, organizations, companies, and governmental bodies under the new economic management conditions found in the Arctic region.

* Arctic and Antarctic Research Institute, St. Petersburg, Russia.

The condition of the Arctic Ocean was constantly monitored during the Soviet period. From 1948 to 1993, annual oceanographic surveys were performed by marine and airborne expeditions and daily oceanographic observations were carried out on Severny Polyus drifting stations. The results of these observations afforded Russia a leading position in Arctic Ocean research.

In the 1990s, economic recession and reduced governmental funding undermined the Russian system of obtaining data in the northern polar region. Field work suffered the most, being the most costly element of the monitoring system. The changes also affected the network of coastal observation posts, which in case of the maritime environment are limited to measuring the sea level, the water temperature and salinity: their number decreased notably in the period from 1992 to 1997.

In the early and mid-2000s, field work in the high latitudes of the Arctic region became more frequent. In 2000, the research vessel 'Akademik Fyodorov' carried out a complex marine expedition to the Arctic Ocean in order to perform geological and geophysical field observations on the Mendeleev Rise so as to determine the borders of the Russian continental shelf.

In summer 2004, research vessel 'Akademik Fyodorov', accompanied by the 'Arktika' nuclear ice-breaker, reached the 85° N latitude where a floe suitable for a SP-33 floating station setup was found. A hydrological testing area was established near the site where the station was set up. After the disembarkation and construction of the drifting station were complete, complex oceanological observations of the sections connecting Severnaya Zemlya, the Franz Josef Land archipelago and the northern part of Novaya Zemlya were performed. Throughout the expedition, the natural environment of the Arctic region was examined thoroughly.

The 'Arktika-2005' expedition included two stages. The first stage involved defining continental shelf borders near the Mendeleev Rise on board the research vessel 'Akademik Fyodorov', and the second stage mainly encompassed the evacuation of the SP-33 drifting station and the setup of a new SP-34 station. On 29 August 2005, at 18:50, research vessel 'Akademik Fyodorov' arrived at the geographical North Pole. For the first time in the history of seafaring, a non-icebreaker vessel reached the northernmost point of the earth without escort. During the entire expedition, a research team performed complex studies on board of the research vessel under the following sub-programs: physical oceanography and water dynamics; interactive processes within the system 'atmosphere-sea ice-upper layers of the sea'; vessel performance in ice; marine geology.

During International Polar Year 2007-2008 ('IPY'),[1] Russian scientific organizations led by the Arctic and Antarctic Research Institute ('AARI') took part in large-scale oceanological observations in many foreign and domestic marine expeditions. One of the most prominent Arctic expeditions to the high latitudes of the Arctic region in the same IPY was the trip of research vessel 'Akademik Fyodorov' in 2007, during which the manned deep submergence vehicles Mir-1 and Mir-2 performed the first descent to the seabed beneath the geographic North Pole in the history of polar expeditions and planted a Russian flag there.

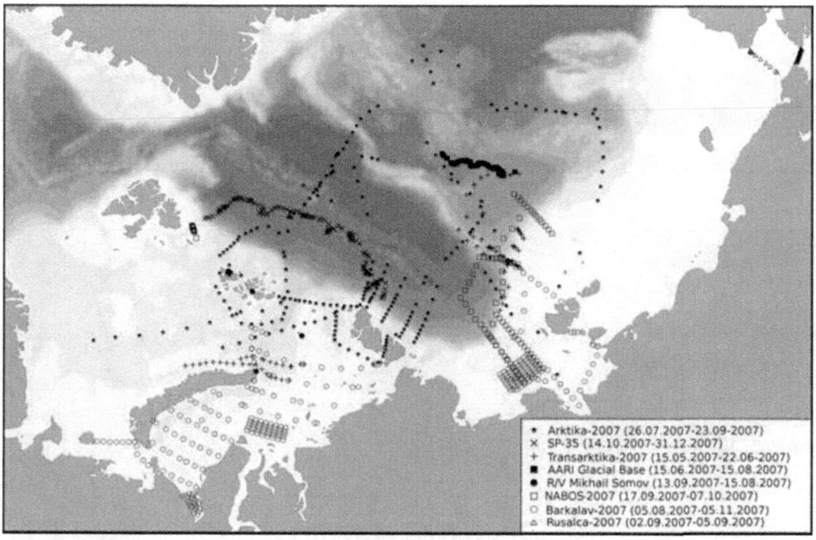

Fig. 1. Expeditions of the AARI in 2007 in the waters of Arctic Ocean and Arctic Seas

[1] IE Frolov, VT Sokolov and IM Ashik, 'Russian Marine Research of AARI during the IPY 2007/08', in *AARI,* Proceedings of the International Conference: *Marine Research in the Polar Areas of the Earth in the International Polar Year 2007/08* (2010) 25 et seq.

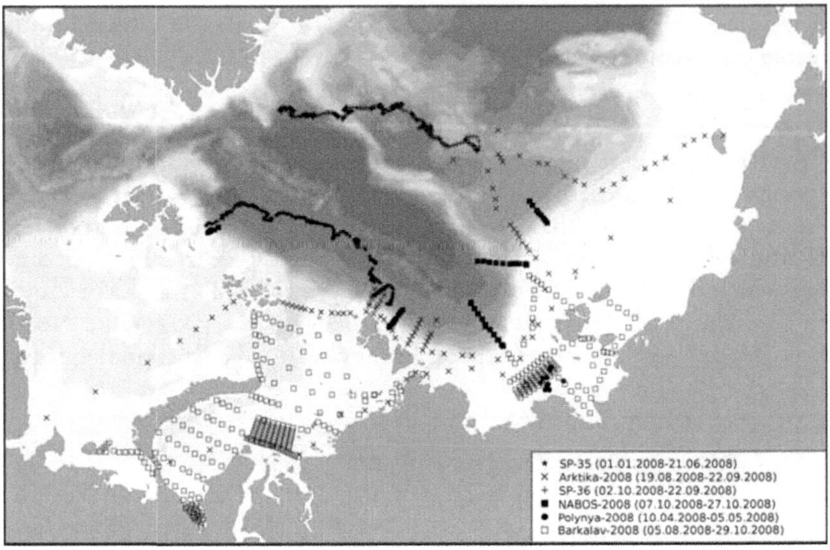

Fig. 2. Expedition of the AARI in 2008 in the waters of Arctic Ocean and Arctic Seas

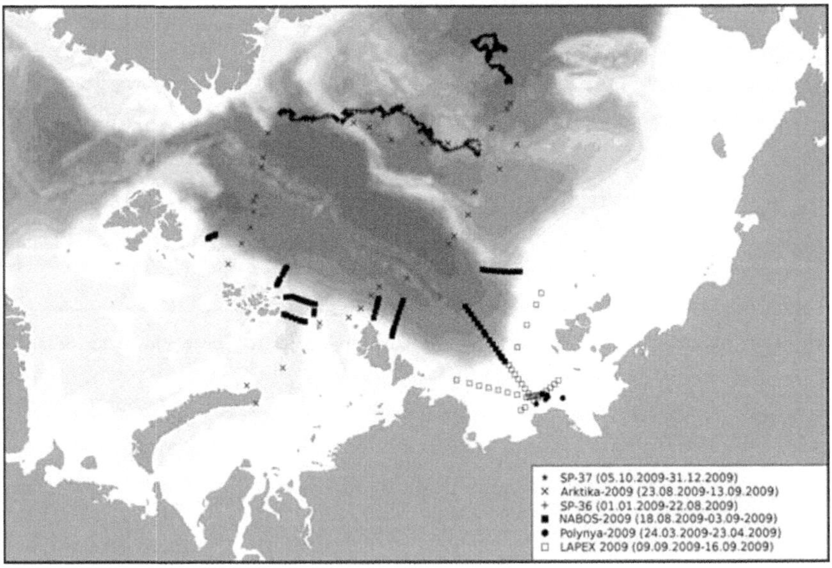

Fig. 3. Expeditions of the AARI in 2009 in the waters of Arctic Ocean and Arctic Seas

Throughout the International Polar Year as well as in 2009, the experts of the AARI participated in a total of more than 20 scientific expeditions, most of which were organized by the AARI. Among others, there are such ambitious projects as the continuation of hydrophysical monitoring of the waters in the Arctic basin using drifting stations SP-36, SP-37 and SP-38. Furthermore, the Russian National Arctic Expeditions Program involved complex research in the vast waters of Arctic basin under the 'Arktika' research program and in the Barents, Kara, Laptev and the East Siberian Seas under the BARKALAV program in 2007-2009 (fig. 1 – fig. 3).

Participation of AARI specialists in international projects allowed to perform a number of quite expensive expeditions in collaboration with German (Laptev Sea System project, LAPEX expeditions) and US research institutes (Nansen and Amundsen Basins Observational System, NABOS expeditions; Russian American Long-term Census of the Arctic, RUSALCA expeditions) during the International Polar Year.

The total number of oceanographic stations established by AARI specialists or involving their participation during the International Polar Year amounted to around 2500. The data collected by these stations provided new insights into the nature of thermohaline changes in the marine environment of the northern polar region and its relation to global climatic changes. The IPY coincided with a period of dramatic warming in the Arctic region, which makes the data obtained during this period extremely valuable. Getting an overall picture of the changes currently underway in the Arctic marine environment, that are linked to this fact, became possible to a great extent due to AARI's efforts aimed at studying the Eurasian part of the Arctic Ocean.

The Russian and foreign data obtained during the IPY brought some new results of great importance, which primarily concern the formation of vast zones of abnormal salinity in the surface layer of the Arctic Ocean. In general, these anomalies reveal a dipole structure, in which negative anomalies of salinity (reaching 2-4‰ from long-time average annual figures) were found in the waters of the Amerasian Basin. At the same time, positive anomalies of salinity (up to 2‰) were found in the Eurasian Basin, the division between positive and negative anomalies between the two basins being located along the Lomonosov Ridge. The results of the observations in the following years showed that the contrast between the anomalies of surface-water salinity in the Amerasian and Eurasian basins has somewhat decreased, even though in terms of structure the situation of the anomaly zones has not changed.

Recent years have also been abnormally warm in the entire history of observations of the Arctic Ocean surface layer throughout the 20th and early 21st centuries. In 2007, the deviations of water temperature established in the Amerasian basin reached +5° C. The following years were also abnormally warm, although temperatures deviated from the long-time annual average less than previously, reaching +2° C in the Beaufort Sea, the southern part of the Podvodnikov Basin and the western part of the East-Siberian Sea.

Changes in the thermohaline structure were found not only in the surface layer, but also in deeper layers. The current stage of warming of the Atlantic water layer in the Arctic Ocean, which started in the early 21st century, notably differs from previously observed warming stages in terms of both temperature deviations in the layer and area affected. In 2007, the temperature anomalies of the Atlantic waters in some regions of the Arctic Ocean reached an unprecedented +1.5° C. In 2008-2009, there was a tendency towards a slight decrease of the positive anomaly of the maximum temperature of the Atlantic waters compared to the climatic data from all over the Eurasian basin, which suggests that the deep-water thermohaline structure has started to return to its average climatic state. At the same time, in 2009, the inflow of warm waters of the North Atlantic current through the Fram Strait intensified in the North Atlantic region which may cause another upturn in temperature of deep Atlantic waters in the Arctic region in the years to come.

The final stage of events carried out during the International Polar Year 2007-2008 was the International Science Conference 'Marine Research in Polar Areas of the Earth in IPY 2007/08' held by the AARI on 21-23 April 2010. The conference was dedicated to the 100th anniversary of the birth of EK Fyodorov.[2] 41 lectures out of the total number of conference reports dealt with problems of polar oceanology.[3] In general, Arctic Ocean marine researchers have been showing an increasing interest in new issues related to current objectives and methods of gaining insight into natural phenomena over the past two decades.

The first such issue is marked by approaching questions of fine structures of the ocean, which allows to expand human knowledge of the ocean's nature; in particular, this includes calculating the so-called fresh water reserves of ocean ice and water as well as using surface fluctuation

---

2 See AARI, *Marine Research in* the *Polar Areas of* the *Earth in the International Polar Year 2007/08* (2010).

3 *Ibid.*

data as a figure indicative of complex processes going on in the ocean in order to define ocean water zones.

The second new issue has been the creation of specialized databases on temperature and salinity of water, sea surface level (based on onshore observations) and ocean currents over the previous period of research, mainly from 1945 on. The specialized database does not only store natural data, but also organizes it more properly, compensating for some gaps in water zone observations which are inevitable when studying such complex systems as seas and oceans. Actually, it were the specialized databases on water temperature and salinity combined with modern computing technologies that enabled us to carry out research on fine structures of water thermohaline characteristics of the Arctic Basin and the Siberian Shelf Seas, from the Kara Sea in the West to the Chukchi Sea in the East.

The third new issue of the past two decades has been researchers' desire to base their knowledge on certain regularities not only on such a self-evident source as natural data, but also to make use of numerical computing (mathematical modeling) as much as possible which is the only way of completing the research results.

A large-scale expedition aimed at defining and providing scientific ground to the definition of the outer boundary of the continental shelf of the Russian Federation in the Arctic Ocean was carried out on board the research vessel 'Akademik Fyodorov' in the summer of 2010. The main goals of the expedition were as follows: collecting additional hydrographic data to determine the area of jurisdiction of the Russian Federation in accordance with the 1982 UN Convention on the Law of the Sea;[4] creating an open digital bathymetric database based on seabed configuration surveys to be submitted to the United Nations Commission on the Limits of the Continental Shelf.

During the expedition, 17,079 linear kilometers of bottom topography survey were carried out and ocean depth was measured at 822,802,528 points, including 386,450,640 points on the working site. As a result of primary data processing, grids of 100×100 m were set up, comprising 4,749,618 depth points. We obtained an array of bathymetric data by using multi-beam and single-beam echo sounders and a profiling sonar with a single-beam option for bottom topography survey.

---

[4] United Nations Convention on the Law of the Sea (concluded 10 December 1982, entered into force 16 November 1994) 1833 UNTS 397.

We also set up a data array based on the results of hydrological and associated meteorological work.

Considering the scale of works carried out in just one season in the high latitudes of the Arctic Ocean and considering the amount of obtained information meeting world quality standards, the results of this expedition considerably exceed those achieved by other Arctic States during the Arctic basin research. For the first time in history, a sea bottom relief survey was done in complicated ice conditions according to pre-set straight bathymetric profiles with a deviation from the profile axis of no more than ¼ of multi-beam echo sounder swath.

The current need for marine expeditions in the Arctic basin is caused by the fact that resumed national field research in Russian national seas and adjacent water zones of the world's oceans and seas is the key to stabilization and further development of Russia's activities in the World Ocean. Such research constitutes the basis for studying, monitoring and exploring the ocean. Full-scale exploration of the Russian Arctic Seas is particularly important due to the growing use of natural resources in the Arctic basin and the need for its environmental protection. An important part of the studies on the use of natural resources in Arctic Seas is dedicated to the interaction of means of transport and mining activities with the ice cover and the analysis of equipment operation at low temperatures.

The fact that the Arctic basin and its waters possess an abundant biological resource potential while being, at the same time, highly susceptible to anthropogenic influence (pollutants carried by the Atlantic waters, runoffs of the Pechora, Ob, Yenisei, Lena and other major rivers, dumping of radioactive wastes on the shelf) has also heightened the awareness of the Arctic region, the Arctic Ocean and the Arctic Seas. To this regard, it is highly urgent to make reliable prognostic evaluations of the ecological conditions in the region amid an expected increase of the anthropogenic influence. Such estimates can be made only based on detailed field data of the current condition of the main biota components in the Arctic basin, the Arctic Ocean and the seas of the Western sector of the Arctic region.

The Arctic region has been getting more and more attention from Arctic States and international organizations. Many national and international programs are being implemented here, and Russia, as the largest Arctic State, needs to take part in these activities. Russia's own studies in this area best of all meet its national economic and defense interests.

Arctic Research in Practice

*by Uwe Nixdorf**

I. Introduction

The Alfred Wegener Institute for Polar and Marine Research, Bremerhaven, Germany ('AWI'), conducts research in the Arctic and Antarctic as well as in the high and mid latitude oceans. The AWI coordinates polar research in Germany and provides the necessary equipment and key infrastructure for polar expeditions.

Climate change and geopolitical developments are putting the polar regions, especially the Arctic, progressively in the focus of scientific, political and economical attention. This is one consequence of the declining sea ice cover in the Arctic. This decline is a process that is occurring even faster than predicted by most climate models. Understanding the reasons and consequences of the changes is highly challenging as the central Arctic is still one of the least explored regions on earth and requires special research platforms. The AWI is therefore continuing with enhanced efforts its Arctic research program with the goal of identifying climate-relevant processes in the polar regions and their effects on middle and lower latitudes. At the Alfred Wegener Institute, scientists from various disciplines and nations are using state-of-the-art technology for cooperative and interdisciplinary investigations of the global atmospheric, biospheric, cryospheric and oceanic systems. Currently, the framework for ongoing scientific projects at the Alfred Wegener Institute is provided by the research program PACES (Polar regions and Coasts in a changing Earth System), following the program MARCOPOLI (2004-2008) (Marine, Coastal and Polar Systems).

---

* Alfred-Wegener-Institut für Polar- und Meeresforschung in der Helmholtzgemeinschaft, 27570 Bremerhaven, Am Handelshaven 12.

In order to effectively support this research the AWI operates the ice-breaker and research vessel *Polarstern*, the ski-equipped research aircraft *Polar 5* and the year-round research bases AWIPEV in the Arctic and Neumayer Station III in the Antarctic (see Fig. 1). AWIPEV is situated in Ny Alesund, Svalbard, and is operated in collaboration with the Institut Polaire Français Paul Emile Victor ('IPEV'), Brest, France.

Fig. 1. AWI research platforms in the Arctic and Antarctic

In addition to the year round bases we operate summer bases Samoylov in Siberia, in collaboration with Russian institutions; the Dallmann laboratory at the Argentine base Jubany on the Antarctic Peninsula, in collaboration with Argentina and the Netherlands; and the Kohnen Station on the Antarctic inland ice.

The AWI has considerable experience with scientific research expeditions in Arctic waters using *RV Polarstern* (see Fig. 2). On 7 September 1991 the *RV Polarstern* was the first vessel with a conventional propulsion system to reach the North Pole in a coordinated two-vessel operation in conjunction with the Swedish *RV Oden*.[1] Thus it was shown

---

[1] DK Fütterer, 'ARCTIC '91: die Expedition ARK-VIII/3 mit FS "POLARSTERN" 1991 – ARCTIC '91: the expedition ARK-VIII/3 of RV "PO-

that, by using modern icebreakers, most regions in the Arctic could be reached at will. In addition the *Polarstern* was able to acquire the first seismic reflection profile above the Lomonosov Ridge, showing that seismic data of a high quality can be obtained from the Arctic Ocean when two vessels operate, one breaking the ice, the second towing the instruments.[2] It must be noted, however, that such an operation is logistically very complex. After processing the seismic profiles above the Lomonosov Ridge it was obvious that this region qualifies for deep drilling in the framework of the Integrated Ocean Drilling Program ('IODP').The first deep drilling in the Arctic took place in 2004 during the ACEX expedition (Arctic Coring Expedition).[3] The *Polarstern* also reached the North Pole again in 2001 and 2011. All in all, AWI has coordinated and operated 25 Arctic and 27 Antarctic expeditions using the *Polarstern,* mainly during the respective summer seasons. The different number of cruises in the Arctic and Antarctic stems from the fact that the *Polarstern* stayed in Antarctic waters for two austral winter experiments and did not sail in the northern hemisphere during those years.

LARSTERN"', in DK Fütterer (ed.), *Berichte zur Polarforschung = Reports on Polar Research*, No. 107 (1992).

[2] W Jokat et al., 'Lomonosov Ridge – A Double-sided Continental Margin', *Geology* 20 (10) (1992) 887 et seq.

[3] HM Stoll, 'Climate Change: The Arctic Tells its Story', *Nature* 441 (2006) 579 et seq.

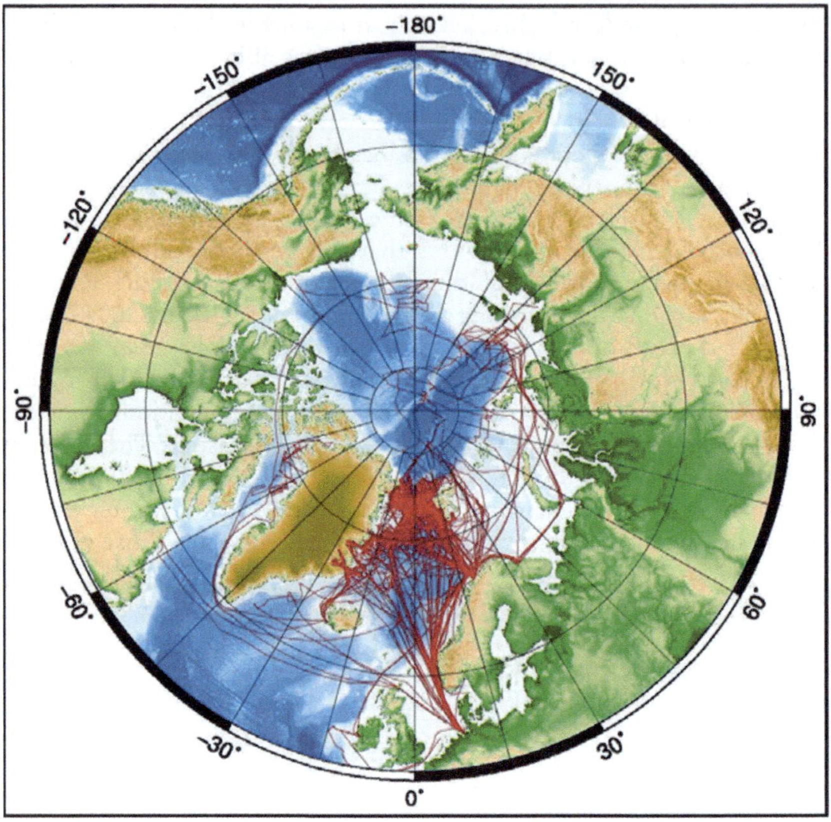

Fig. 2. *Polarstern* cruise tracks on the northern hemisphere between December 1982 and February 2011

II. The Legal Framework

The *Polarstern* is operated by a shipping company under contract to the AWI, where science and logistics are coordinated and managed. Thus the shipping operator and the AWI are jointly responsible for ensuring that the *Polarstern* complies with the all rules and regulations mandatory for research vessels. Table 1 shows the main rules and regulations of relevance in this context.

Table 1. Rules and regulations

UNCLOS – United Nations Convention on the Law	SOLAS – Safety of Life at Sea of the Sea
MARPOL – Prevention of Marine Pollution from Ships	ISM Code – International Safety Management Code
ISPS Code – International Ship and Port Facility Security Code	Code for the Safe Operation of Ships and for Pollution Prevention
STCW – Standards of Training, Certification and Watchkeeping for Seafarers	MDG Code – International Maritime Code for Dangerous Goods
Ballast Water Management – Regulations on Ballast Water Treatment	COLREG – Regulations for Preventing Collisions at Sea
International Convention on the Control of Harmful Anti-fouling Systems on Ships	Garbage Management – Regulations on Garbage Management
Seemannsgesetz, German	ILO – Maritime Labour Convention
Various BG-Verkehr Rules (ex. See-BG), German	Schiffsbesetzungsverordnung / Regulations on Safe Manning, German
Schiffsicherheitsverordnung, German	Verordnung über die Seediensttauglichkeit, German
Rules of Classification Societies (e.g. GL), German	GGVSee – Gefahrgutverordnung See, German
Polar Code (in progress)	Various national rules

Logistic coordination includes responsibility for all correspondence related to obtaining permits for marine scientific research ('MSR') cruises in foreign waters via diplomatic channels. The permit application process is a consequence of the United Nations Convention on the Law of the Sea ('UNCLOS'),[4] and will be outlined briefly in the following chapter.

The right to conduct marine scientific research is outlined in Art. 238 in Part III, Section 1 – General provisions. In detail it states:

> All States, irrespective of their geographical location, and competent international organizations have the right to conduct marine scientific research subject to the rights and duties of other States as provided for in this Convention.

The rights and duties of other States relevant in this context are enshrined in Art. 56 UNCLOS.

> Article 56.
>
> Rights, jurisdiction and duties of the coastal State in the exclusive economic zone
>
> 1. *In the exclusive economic zone, the coastal State has:*
>
> (a) sovereign rights for the purpose of exploring and exploiting, conserving and managing the natural resources, whether living or non-living, of the waters superjacent to the sea-bed and of the sea-bed and its subsoil, and with regard to other activities for the economic exploitation and exploration of the zone, such as the production of energy from the water, currents and winds;
>
> (b) *jurisdiction* as provided for in the relevant provisions of this Convention *with regard to*:
>
> (i) the establishment and use of artificial islands, installations and structures;
>
> (ii) *marine scientific research*;
>
> (iii) the protection and preservation of the marine environment;
>
> (c) other rights and duties provided for in this Convention.
>
> 2. In exercising its rights and performing its duties under this Convention in the exclusive economic zone, the coastal State shall have due regard to the rights and duties of other States and shall act in a manner compatible with the provisions of this Convention.
>
> 3. The rights set out in this article with respect to the sea-bed and subsoil shall be exercised in accordance with Part VI (*emphasis added*).

---

[4] United Nations Convention on the Law of the Sea (concluded 10 December 1982, entered into force 16 November 1994) 1833 UNTS 397.

That means in consequence – as emphasized above – that each operator has to seek the consent of the coastal State for marine scientific research cruises that shall be conducted in the exclusive economic zone ('EEZ') of that coastal State: 'Marine scientific research in the exclusive economic zone and on the continental shelf shall be conducted with the consent of the coastal State'.[5]

The operator can expect permission to be granted due to Art. 246(3) UNCLOS, according to which coastal States shall, in normal circumstances, grant their consent which shall not be delayed or denied unreasonably. Finally, four months after an operator's application for an MSR permit has been received via diplomatic channels by the coastal State and without response from this coastal State, consent is implied.[6] So far the AWI has not undertaken any MSR in foreign waters without the written permission of the responsible coastal State.

However, marine scientific research is not defined in UNCLOS, a fact which is highlighted in the white paper by Baker and Eicken:

> Several states take the view that, at a minimum, hydrographic surveys are excluded from the permission requirements of Part XIII (though other regulatory regimes may apply). This view is based on the distinction between "marine scientific research and hydrographic surveys" in Article 21 of the Convention, under which the coastal state may adopt laws and regulations regarding innocent passage through its territorial sea.[7]

This leads to uncertainties regarding – for instance – whether or not an MSR application needs to be made for the collection of marine meteorological data and other routine ocean observations, for example under the voluntary ocean observation programs of the Joint Intergovernmental Oceanographic Commission ('IOC') – World Meteorological Organization ('WMO') Technical Commission on Oceanography and Marine Meteorology ('JCOMM') and the Argo program. Moreover, scientific ambitions to investigate the state of all departments of the Arctic Ocean combined with technical developments have given rise to in-

---

5 *Ibid.*, Art. 246(2).

6 *Ibid.*, Art. 252.

7 B Baker and H Eicken, 'Marine Research Access in the Arctic Ocean: Background for Potential Guidelines in a Changing Arctic', unpublished White Paper (10 March 2010), see <http://www.iarc.uaf.edu/workshops/2009/4/> (8 June 2011) (click on 'download whitepaper'); the paper is also attached as an Appendix to the contribution of B Baker, in this volume.

struments and methods such as ice-tethered moorings and seagliders that operate autonomously. These obviously differ from classical ship-based expeditions and are thus not easily addressed in an MSR application.

During the recent past the rights of the indigenous peoples of the coastal States in the Arctic have been strengthened. There are independent administrations in Canada and Greenland, which have a special role in the permission process. In this context we have experienced additional expenditure as well as delays in the permission process, which we can only presume are due to new procedures that are still being improved. Furthermore, it seems that the indigenous peoples' new rights are accompanied by a new assertiveness on their part that sometimes results in an inner-State conflict of interests. Therefore, a foreign applicant seeking an MSR permit cannot – in the sense assumed by UNCLOS – refer to THE coastal State, whose consent needs to be sought, as long as these conflicts of interests are unresolved.

III. Practical Experience: *Polarstern* Cruise ARKIII/3

Due to the research activities performed on the research platform *RV Polarstern,* the AWI has experience of the permission processes of all Arctic coastal States with the exception of the USA. For *Polarstern* cruise ARKIII/3, which started from Reykjavik, Iceland, and during which not only the North-East passage but also the North-West passage was traversed, the consent of the following States had to be obtained, named in the order of passage through coastal waters: Iceland, Denmark/Greenland, Canada, Russia and Norway (see Fig. 3). To our knowledge this cruise was the first circum-Arctic voyage (circumnavigation of North Pole) by a research vessel. On the basis of these experiences I would like to outline in the following chapter the specific procedures adopted by each coastal State in its national legislation implementing UNCLOS.

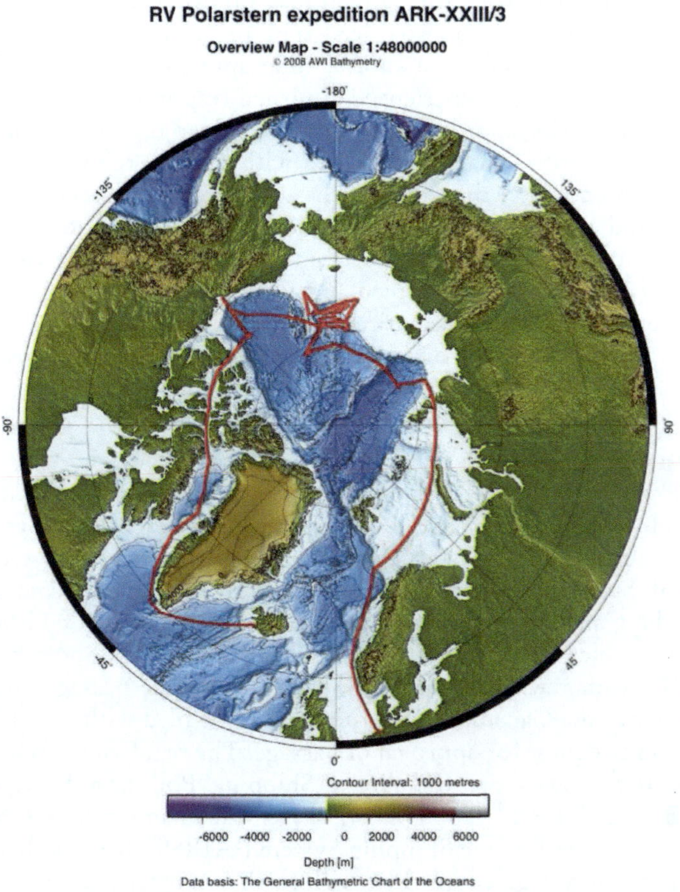

Fig. 3. Cruise track of *RV Polarstern* expedition ARK-XXIII/3

Norwegian legislation closely follows the provisions of Part XIII UNCLOS. Norway's procedures[8] use application forms which are based on the Draft Standard Forms developed by the United Nations Division of

[8] Norway, Directorate of Fisheries, Marine Scientific Research, 'Application Form and Regulations', see <http://www.fiskeridir.no/english/fisheries/marine-scientific-research>(15 June 2011).

Ocean Affairs and Law of the Sea pursuant to Part XIII UNCLOS.[9] Consent is required from the Directorate of Fisheries ('DF') which grants the permits. The requested lead time for an MSR application is six months unless DF allows a shorter time limit for an individual application. In practice we experienced a normal response time of as little as two months. The approvals are granted under the condition that Norwegian authorities shall be provided with a cruise report to be delivered to them not later than six months after the cruise has ended. Occasionally and on request Norwegian authorities are to be provided with access to all data and samples from the project and likewise furnished with data that may be copied and samples which may be divided without detriment to their scientific value, and given assistance in the assessment of such data. In our experience many other nations use the forms related to marine scientific research developed by Norway, with similar response and lead times of six months (Iceland) or as little as 30 days in case of Denmark.

Canada's official lead time for the approval of an MSR application is 45 days. However, in our experience one should allow for one year at least. The authorities involved in our application processes were the Geological Survey Canada, National Resources Canada, Transport Canada, the Canadian Coast Guard, and the Department of Fisheries and Oceans which issued the Marine Environmental Handbook. Before entering Canadian waters, especially the North-West passage, an announcement to the Canadian authorities had to be filed with vessel data, ice class and a request for approval of passage. The regulations to be followed in this respect are the Arctic Shipping Pollution Prevention Regulations ('ASPPR').[10] The *Polarstern* has Canadian ice class 4. Since 2003 the Arctic Ice Regime Shipping System ('AIRSS') has also been in effect. Approval was given quickly, with further contact dates provided for daily announcements and the exchange of ice reports, and from the initial announcement on we had to deliver weekly ice reports with NordREG being the only contact point. During the above-mentioned cruise ARK XXIII-3 a cruise participant had to be evacuated for medi-

---

9 United Nations, Division for Ocean Affairs and the Law of the Sea, *The Law of the Sea: Marine Scientific Research, A Guide to the Implementation of the Relevant Provisions of the United Nations Convention on the Law of the Sea* (1991).

10 Canada, Arctic Shipping Pollution Prevention Regulations, Consolidated Regulations of Canada (current to 14 September 2011) chapter 353.

cal reasons, and we would like to gratefully acknowledge the assistance we received from all institutions involved.

Polarstern cruise ARK XXV/3 (31 July – 9 October 2010) was planned under the lead of the Federal Institute for Geosciences and Natural Resources (Bundesanstalt für Geowissenschaften und Rohstoffe – 'BGR'), Hanover, Germany, and in close collaboration with Natural Resources Canada ('NRCAN') as part of the Eastern Canadian Arctic Seismic Experiment ('ECASE') continuing the land and sea-based geoscientific investigations aimed at understanding the geological structures and processes that led to the development of the Arctic Ocean and its bordering landmasses. The planned legs in Canadian waters are shown in Fig. 4.

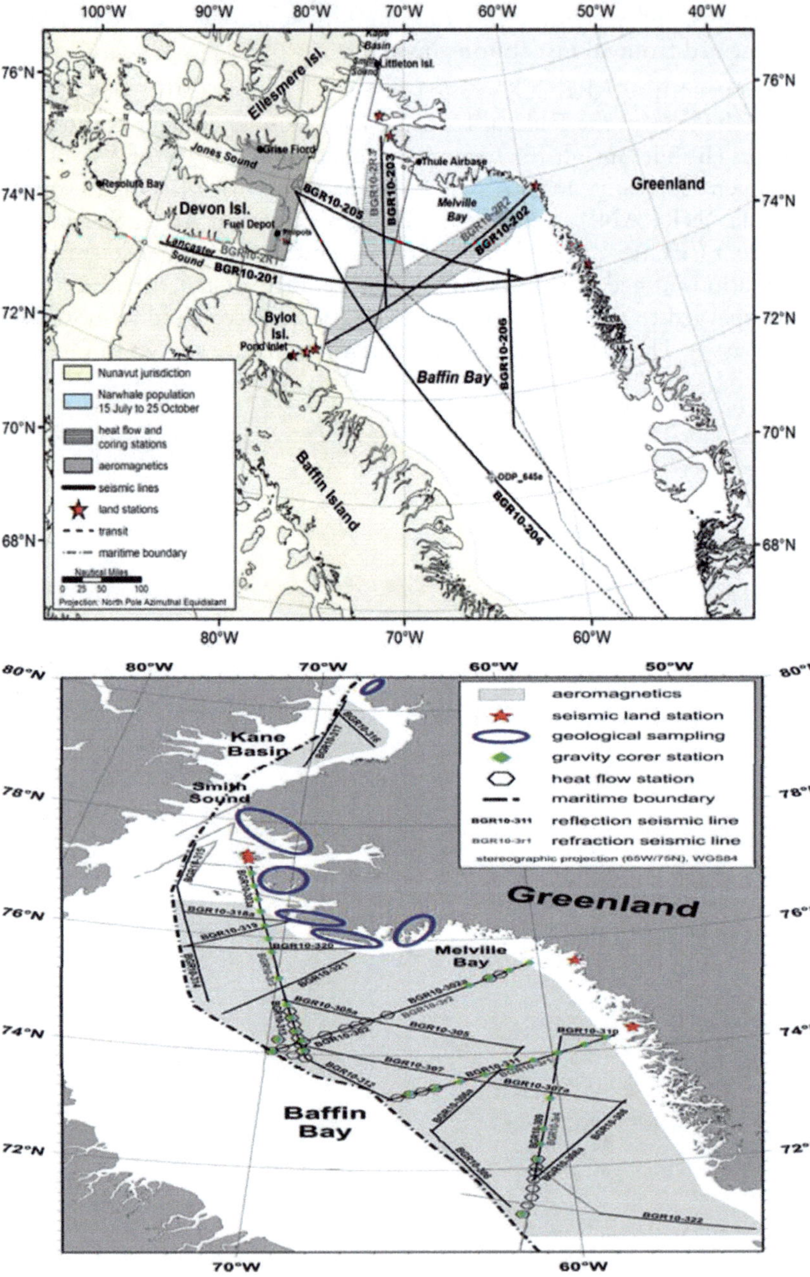

Fig. 4. Planned cruise tracks (on the right side) and changed program

By the end of July the relevant grants had been received from the Canadian authorities, including responses from the Nunavut Research Institute ('NRI') and the Nunavut Impact Review Board ('NIRB'). However, shortly before the actual scientific work was due to start, we were informed that an interlocutory order had been granted restraining the Canadian project partner Natural Resources Canada from proceeding with seismic testing pursuant to ECASE. The injunction had been applied for by the Qikiqtani Inuit Association ('QIA'). Shortly thereafter the German parties involved decided that no MSR should be undertaken in Canadian waters at all. Instead the experiments in Danish/Greenlandic waters were extended and the Danish/Greenlandic authorities approved the extension of the MSR in as little as 14 days. Subsequently, all permissions previously received were withdrawn by the Canadian authorities. Analyzing all activities undertaken in the permission process by German and Canadian project partners from November 2009 onwards, we concluded that this *Polarstern* cruise had been the subject-matter of an inner-Canadian juridical conflict.

Applications for MSR in Russian waters must comply with decree No. 391 of the Government of the Russian Federation dated 30 June 2004.[11] Applications have to be submitted via diplomatic channels to the Russian authorities at least six months before the intended start date of the expedition in the Russian language as well as in the applicant State's language. The passage through the Northern Sea Route requires approval by the Northern Sea Route Administration as well as a contract with the Russian State Atomic Energy Corporation ROSATOM, which operates icebreakers. In this context the ice class of the vessel is evaluated on the basis of the vessel documentation and at least once by visual inspection; a passage fee is charged in line with the tonnage of the vessel. This fee is reduced by 50% in the case of a successful application for an MSR. It takes time to obtain the approval of the Northern Sea Route Administration and to conclude the contract with ROSATOM; considerably more than six months lead time should be allowed for. The date and coordinates of the vessel's entry to the Northern Sea Route must be reported to the Northern Sea Route Administration 30 days prior to the intended date. Five days before entering the EEZ the vessel must announce itself to the Marine Operation Headquarters of ROSATOM. Having entered the EEZ daily reports have to be filed

---

[11] Decision of the Russian Federation Government No. 391 (30 July 2004) promulgated on August 2004 in the Russian Federation Code of Laws for 2004 (part 32, Art. 3338).

to different contact points at the above-mentioned institutions as well as at the military. The participation of a Russian observer during the cruise is mandatory, which is –among other things – very helpful for this reporting-related communication. During the International Polar Year the application process and collaboration during the cruise was excellent. However, permission for marine geophysical research was not granted, because this was seen to be a matter of national responsibility. Likewise, for the ongoing *Polarstern* cruise ARK XXVI-3 (5 August to 7 October 2011) the permission granted contained extensive limitations on marine research work, as no samples of seafloor, water or ice may be collected. Marine geophysical research in the Laptev Sea region was carried out by BGR in the early 1990s in a joint Russian-German expedition. This was only possible by chartering a Russian research vessel.

IV. Summary

In summary our experience agrees with both anecdotal reports that indicate a growing concern about the declining reliability of MSR access to the Arctic Ocean as well as with the findings of the Intergovernmental Oceanographic Commission of the United Nations Educational, Scientific and Cultural Organization ('UNESCO') Advisory Body of Experts on the Law of the Sea ('ABE-LOS') that published survey results regarding MSR permissions requested and granted by all coastal States from 1998-2002.[12]

In addition we would like to note that sometimes we have difficulty understanding regulations and that local authorities' responsibilities change over time, which makes applying for permits cumbersome. Our suggested solution would be to ask the foreign embassies in the respec-

---

12 E Tirpak, 'Excel File Analysis of Response to IOC Questionnaire No. 3' (13 April 2009), see <http://ioc-unesco.org/index.php?option=com_oe&task=viewDocumentRecord&docID=3571> (15 June 2011); E Tirpak, 'IOC Questionnaire No. 3: The Practices of States in the Field of Marine Scientific Research (MSR) and Transfer of Marine Technology (TMT)' (13 April 2009), see <http://ioc-unesco.org/index.php?option=com_oe&task=viewDocumentRecord&docID=3570> (15 June 2011); E Tirpak, 'Practice of IOC Member States in the Fields of Marine Scientific Research and Transfer of Marine Technology – An Analysis of Responses to ABE-LOS Questionnaire No. 3' (1 April 2009), see <http://www.ioc-unesco.org/index.php?option=com_oe&task=viewDocumentRecord&docID=3515> (15 June 2011).

tive countries to inform their national operators of applicable regulations.

Finally, the efforts of the Arctic coastal States to enlarge their respective EEZ give cause for concern about the accessibility of Arctic waters. If successful these efforts would leave only two comparatively small 'high sea areas' in the Arctic Ocean (see Fig. 5).

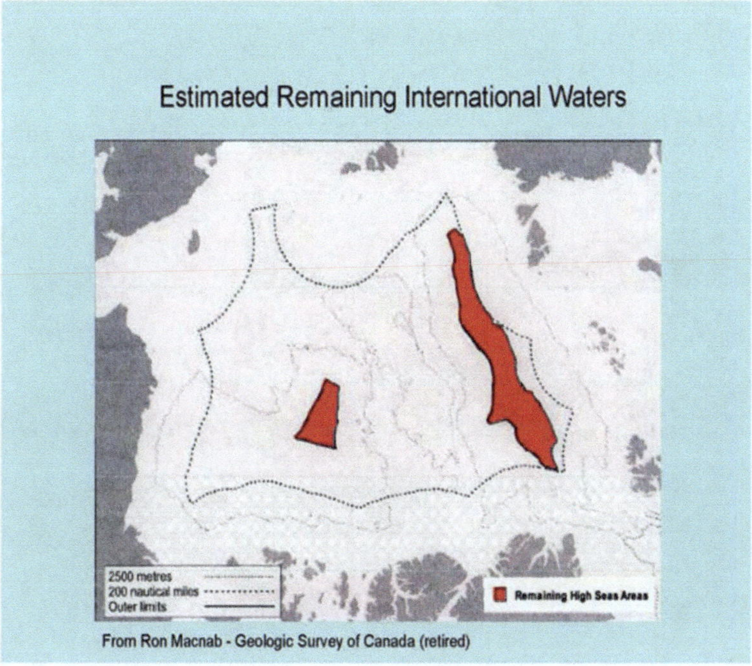

Fig. 5. Potentially 'permit free' high sea areas should all requests for EEZ extensions be honored

Only in these areas would it be possible to conduct marine scientific research without the permission of coastal States, which might not be readily granted. Hence, ecological, sea ice, oceanographical, marine geophysical and global tectonic studies could be impeded, with severe consequences for scientists' ability to predict future change.

Arctic Marine Research: The Perspective of a US Practitioner

by Larry Mayer[*]

I. Introduction

I have been asked to present the perspective of a US 'practitioner' of Arctic marine science – something that I am reasonably comfortable doing as I have had the privilege over the past few years to lead five research expeditions in the high Arctic on the *USCG Cutter Healy*. While I am comfortable offering my perspective, it is important to note that my perspective is limited to my experiences which are those of a sea-going academic US scientist collecting mapping and other data in support of a potential submission of the US for an extended continental shelf under Art. 76 United Nations Convention on Law of the Sea ('UNCLOS').[1] I emphasize the fact that I am an *academic* and that the views expressed are my own and not those of the US government. Despite this limited perspective, I believe that my experiences are reasonably representative of those of much of the US Arctic research community.

Through at least 15 federal agencies, the US invests more than US$ 400 million per year in Arctic research. The major supporters of US Arctic research are the National Science Foundation ('NSF'), the National Oceanic and Atmospheric Agency ('NOAA'), the National Aeronautics and Space Administration ('NASA'), the US Geological Survey ('USGS') and the US Navy though many other agencies also provide

* Professor and Director, Center for Coastal and Ocean Mapping, University of New Hampshire, Durham NH USA 03824.

1 United Nations Convention on the Law of the Sea (concluded 10 December 1982, entered into force 16 November 1994) 1833 UNTS 397.

support and play active roles in Arctic marine research. These agencies support individuals and teams of researchers and a substantial infrastructure that includes ice camps, aircraft, submarines, icebreakers, and several Arctic-based research facilities including the Barrow Arctic Science Consortium ('BASC'). BASC has been designated by the Ukpeagvik Inupiat Corporation ('UIC'), to manage UIC's research facilities at the former Naval Arctic Research Laboratory. BASC provides lab space and logistical support for researchers and their projects, including liaison with the local community. Support for researchers working on NSF projects is also provided through a contract with CH2MHill Polar Services. Both of these support organizations can provide food, housing, and many other services to Arctic researchers.[2] The National Oceanic and Atmospheric Administration and the Department of Energy ('DOE') also maintain research labs in Barrow focused on climate monitoring and global climate change studies including the recently opened, NOAA supported, 6000 square meter Global Climate Change Research Facility.

It is impossible for me to review the breadth of Arctic marine research being conducted by US scientists and so instead, I will focus on my personal experiences and, in doing so, try to address the questions of access, permitting, data collection, collaboration, and data availability posed by the session conveners. While these will touch on only a small segment of the broad reach of Arctic research, the experiences related are likely to be representative of the challenges faced by most practitioners of Arctic marine science.

II. US Mapping Efforts in Support of Art. 76 UNCLOS

Although the US Senate has not yet given advice and consent to the UN Convention on the Law of the Sea, the United States has recognized much of the Convention as representing existing international law and thus has, since 2003, been collecting bathymetric and other data to support delimitation of its juridical continental shelf. For simplicity and to avoid confusion, I will refer to the area of the juridical continental shelf beyond the 200 nm exclusive economic zone ('EEZ') as the extended continental shelf. Bathymetric data to support a potential sub-

---

[2] For further information see P Schlosser et al. (eds), *Logistics Recommendations for an Improved U.S. Arctic Research Capability* (1997).

mission under Art. 76 UNCLOS and the establishment of a US extended continental shelf have been collected (and are continuing to be collected) by the University of New Hampshire's Center for Coastal and Ocean Mapping ('CCOM') with NOAA funding under the auspices of an inter-agency Extended Continental Shelf Task Force that is led by the US State Department and co-chaired by NOAA and the USGS. To date, more than 1.648 million square km of high-resolution seafloor bathymetry data have been collected in numerous regions around the world where the US has the potential for an extended continental shelf under Art. 76 UNCLOS (Figure 1). It is the policy of the US to make the bathymetric data public soon after data collection and these data can be found at University of New Hampshire's Center for Coastal and Ocean Mapping web site[3] or at the National Geophysical Data Center's ('NGDC') web site.[4]

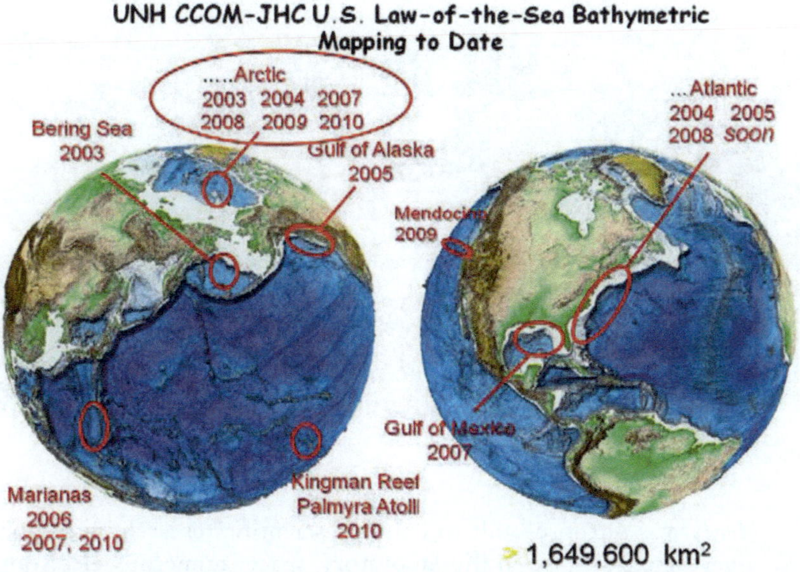

Fig. 1. Areas mapped in support of US Law of the Sea efforts

---

[3] University of New Hampshire, Center for Coastal & Ocean Mapping/Joint Hydrographic Center, see <http://www.ccom.unh.edu> (25 May 2011).

[4] National Oceanic and Atmospheric Agency, National Geophysical Data Center, see <http://www.ngdc.noaa.gov> (26 May 2011).

Of all the areas where the US has the potential for an extended continental shelf, it is in the Arctic region of the Chukchi Cap, north of Alaska where the US has the potential for the largest area of extension. For this reason, along with the fact that the Arctic Ocean is the least mapped and least studied of all of our ocean basins, the US has undertaken six expeditions in the Arctic regions north of Alaska on the *USCG Cutter Healy*. The *Healy* is a 128 m long, 25 m wide icebreaker that displaces 16,000 linear tons and can generate 30,000 horsepower through four diesel/electric engines (Figure 2). *Healy* has a maximum speed of 17 knots, a nominal cruising speed of 12 knots, and can break 1.4 m of ice continuously and 2.44 m of ice while backing and ramming. *Healy* carries a compliment of 19 officers, 12 petty officers, 54 enlisted personnel, and a scientific contingent of about 35 (though this can surge to 51 if necessary).

Fig. 2. USCG Cutter *Healy*

The *Healy* was purpose-built to conduct scientific research with more 390 square meters of scientific laboratory space, numerous electronic sensor systems, oceanographic winches, and the ability to operate in temperatures as low as -46° C. Most critically for our research, the *Healy* is also equipped with a multibeam bathymetric mapping system as well as a chirp subbottom profiler. When originally commissioned in 1999, the *Healy* had a Seabeam 2112, 12-kHz multibeam echosounder on board. Last year, the mapping system was upgraded to a Kongsberg EM-122 multibeam echosounder, representing the state-of-the-art in modern deep-water swath mapping systems.

A desktop analysis of data relevant to a potential US submission under Art. 76 UNCLOS[5] conducted in 2002 suggested that the region north of the Chukchi Cap had great potential for an extended continental shelf and that the critical bathymetric data needed would be the 2500 m depth isobath and the location of the foot of the slope (based on the reasonable assumption of very thick sediment accumulations in the Canada Basin). This region is typically ice-covered year-round and it was not clear whether it would be feasible to break ice, collect useful mapping data, and follow a depth contour all at once. Thus in 2003 we embarked on a feasibility study and demonstrated that it was indeed possible to map the 2500 m isobath while breaking through 8/9 tenths of the ice cover of first year ice. Since 2003, the *Healy* has returned to the Chukchi region for five additional mapping expeditions collecting more than 262,000 square km of high-resolution multibeam sonar bathymetry data. In the course of these surveys we have clearly demonstrated that our existing knowledge of the morphology of the region was inadequate; several new seamounts have been discovered and the mapping has radically changed our view of the location of both the 2500 m depth isobath and the foot of the slope with important ramifications for the extent of the juridical continental shelf. This mapping has also revealed morphologic features (grooves) indicative of a large grounded ice-sheet on the Chukchi Cap that has forced us to re-think the glacial history of the Arctic, and fields of pockmarks indicative of the resource potential of the region.[6] In 2008 and 2009 the *Healy* also collected dredge samples from outcrops on the Chukchi Cap and Alpha/Mendeleev Ridge complex. These rare samples, while still being analyzed, are already casting doubt on existing models for the geologic origins of the Amerasian Basin.[7] Ancillary programs conducted on these cruises have included ice monitoring, meteorological and oceano-

[5] LA Mayer, M Jakobsson and A Armstrong, *The Compilation and Analysis of Data Relevant to U.S. Claim Under United Nations Law of the Sea Article 76: A Preliminary Report* (2002), see also <http://www.ccom.unh.edu> (25 May 2011).

[6] LA Mayer et al., 'Seafloor Mapping in the Arctic in Support of a Potential U.S. Extended Continental Shelf', *International Hydrographic Review* 3 (2010) 14 et seq.

[7] K Brumley et al., 'Dredged Rock Samples from the Alpha Ridge, Arctic Ocean: Implications for the Tectonic History and Origin of the Amerasian Basin', *Eos Trans. AGU* 89(53) (2008) Fall Meeting Supplement, Abstract T43B-2013.

graphic measurements, marine mammal and bird observations, and ocean acidification studies.

In addition to bathymetric data, sediment thickness from seismic data is required to establish an extended continental shelf in the Arctic. Recognizing the logistical difficulties of collecting high-quality seismic data in ice-covered regions of the Arctic, and the proximity of regions of interest, the United States and Canada began a series of joint expeditions to collect seismic data relevant to both nations. These cruises have already collected more than 12,000 line km of high-quality multichannel seismic data in the Canada Basin and the surrounding regions;[8] another joint expedition is scheduled for this coming summer (2011).

III. The Cruise Scheduling Process

The *Healy* is operated by the US Coast Guard with funding for her operations (from NSF or other agencies) channeled through the National Science Foundation (though this is in flux right now). The *Healy* is typically the only US ice-breaker available to US scientists for Arctic research and thus competition for time on the *Healy* is great. Scientists funded through the NSF apply for research and ship-time funds through a peer-reviewed competitive process; scientists from other agencies may also go through a competitive procedure or programs are put forth based on agency priorities. Final selection of the scientific programs on the *Healy* for any given year is made collectively by NSF and the Coast Guard, balancing scientific and national priorities. The time available is very limited. For example for the 2011 season, the *Healy* has three programs scheduled, a NASA-funded program looking at ecosystem and chemistry changes in the Arctic in light of global climate change, our NOAA-funded program mapping in support of extended continental shelf efforts, and an NSF-funded program looking at western Arctic boundary currents. The total sea time allocated for these three programs is approximately 105 days. The demand for HEALY ship time is usually much greater than the time made available causing many scientific programs to be delayed – sometimes several years after funding is received.

---

8 DC Mosher et al., 'Sedimentation in Canada Basin, Western Arctic', *AGU* (2010) Fall Meeting, Abstract T31A-2126.

IV. The Authorization Process

Once a cruise is scheduled, the chief scientist of each cruise leg begins to work with the Coast Guard and the agency sponsoring the program to obtain appropriate permissions and permits for the proposed work. If work is to be conducted in the EEZ of another State, the US Department of State's Office of Ocean and Polar Affairs ('OPA') which is part of the Department's Bureau of Oceans and International Environmental and Scientific Affairs ('OES') is contacted as it serves as the 'appropriate official channel' for public- or privately-funded researchers seeking foreign coastal State consent (as well as for foreign researchers seeking US consent). The OPA uses the applications and reporting forms recommended by the UN's Division of Ocean Affairs and Law of the Sea ('DOALOS') for marine research and has established an online data management system designed to improve transparency and efficiency. It is US policy not to unreasonably deny or delay consent for marine scientific research though the United States reserves the right to participate in research activities conducted in the US territorial sea and/or EEZ, and on the US continental shelf.[9] Past practice has demonstrated that this is not the view of all Arctic States as many requests for scientific research in Russian waters have often been denied or not responded to (Figure 3). This is cause for great concern about access for researchers to large portions of the Arctic that may be deemed extended or juridical continental shelf under Art. 76 UNCLOS.

[9] See Ocean and Polar Affairs ('OPA') within the Department's Bureau of Oceans and International Environmental and Scientific Affairs ('OES'), 'Marine Scientific Research Authorizations', see <http://www.state.gov/g/oes/ocns/opa/rvc/index.htm> (27 May 2011).

Fig. 3. US Requests for permission to work in Russian EEZ from 1992 to 2001. Over the past 17 years 13 of 31 requests have been denied or not responded to representing an acceptance rate of 58%[10]

V. The Environmental Permitting Process

Once authorizations are received, the operations that will be carried out by the research program must then be evaluated within the context of a number of US federal environmental regulations. These include:

The National Environmental Policy Act:[11]

> The National Environmental Policy Act (NEPA) was one of the first laws ever written that establishes the broad national framework for protecting our environment. NEPA's basic policy is to assure that all branches of government give proper consideration to the environment prior to undertaking any major federal action that significantly affects the environment.
>
> NEPA requirements are invoked when airports, buildings, military complexes, highways, parkland purchases, and other federal activities are proposed. Environmental Assessments (EAs) and Environmental Impact Statements (EISs) which are assessments of the likelihood of impacts from alternative courses of action, are required

[10] I Polyakov, 'Collaborative Research in Russian Arctic Seas', Presentation at the seminar and workshop *International Scientific Collaboration and Legal Regimes in a Changing Arctic Ocean,* Fairbanks, Alaska (21 September 2009).

[11] National Environmental Policy Act of 1969, 42 USC 4321 et seq.

from all Federal agencies and are the most visible NEPA requirements.[12]

The Marine Mammal Protection Act:[13]

The Marine Mammal Protection Act (MMPA) was enacted on October 21, 1972. All marine mammals are protected under the MMPA. The MMPA prohibits, with certain exceptions, the „take" of marine mammals in U.S. waters and by U.S. citizens on the high seas, and the importation *of marine mammals and marine mammal products into the U.S.*[14]

The Endangered Species Act:[15]

The Endangered Species Act (ESA) provides a program for the conservation of threatened and endangered plants and animals and the habitats in which they are found. The lead federal agencies for implementing ESA are the U.S. Fish and Wildlife Service (FWS) and the U.S. National Oceanic and Atmospheric Administration (NOAA) Fisheries Service. The FWS maintains a worldwide list of endangered species. Species include birds, insects, fish, reptiles, mammals, crustaceans, flowers, grasses, and trees.

The law requires federal agencies, in consultation with the U.S. Fish and Wildlife Service and/or the NOAA Fisheries Service, to ensure that actions they authorize, fund, or carry out are not likely to jeopardize the continued existence of any listed species or result in the destruction or adverse modification of designated critical habitat of such species. The law also prohibits any action that causes a "taking" of any listed species of endangered fish or wildlife. Likewise, import, export, interstate, and foreign commerce of listed species are all generally prohibited.[16]

---

12 United States Environmental Protection Agency, 'Summary of the National Environmental Policy Act', see <http://www.epa.gov/regulations/laws/nepa.html> (27 May 2011).

13 Marine Mammal Protection Act of 1972, 16 USC 1361 et seq.

14 NOAA Fisheries, Offices of Protected Resources, 'Marine Mammal Protection Act (MMPA) of 1972', see <http://www.nmfs.noaa.gov/pr/laws/mmpa/> (27 May 2011).

15 The Endangered Species Act of 1973, 16 USC 1531 et seq.

16 United States Environmental Protection Agency, 'Summary of the Endangered Species Act', see <http://www.epa.gov/regulations/laws/esa.html> (27 May 2011).

and the Magnuson-Stevens Fishery Conservation and Management Act,[17] the purposes of which are:

(1) to take immediate action to conserve and manage the fishery resources found off the coasts of the United States, and the anadromous species and Continental Shelf fishery resources of the United States, by exercising

(A) sovereign rights for the purposes of exploring, exploiting, conserving, and managing all fish within the exclusive economic zone established by Presidential Proclamation 5030, dated March 10, 1983, and

(B) exclusive fishery management authority beyond the exclusive economic zone over such anadromous species and Continental Shelf fishery resources [, and fishery resources in the special areas];

(2) to support and encourage the implementation and enforcement of international fishery agreements for the conservation and management of highly migratory species, and to encourage the negotiation and implementation of additional such agreements as necessary;

(3) to promote domestic commercial and recreational fishing under sound conservation and management principles, including the promotion of catch and release programs in recreational fishing;

(4) to provide for the preparation and implementation, in accordance with national standards, of fishery management plans which will achieve and maintain, on a continuing basis, the optimum yield from each fishery;

(5) to establish Regional Fishery Management Councils to exercise sound judgment in the stewardship of fishery resources through the preparation, monitoring, and revision of such plans under circumstances (A) which will enable the States, the fishing industry, consumer and environmental organizations, and other interested persons to participate in, and advise on, the establishment and administration of such plans, and (B) which take into account the social and economic needs of the States;

(6) to encourage the development by the United States fishing industry of fisheries which are currently underutilized or not utilized by United States fishermen, including bottom fish off Alaska, and to

17 Magnuson-Stevens Fishery Conservation and Management Act, 16 USC 1801 et seq.

that end, to ensure that optimum yield determinations promote such development in a non-wasteful manner; and

(7) to promote the protection of essential fish habitat in the review of projects conducted under Federal permits, licenses, or other authorities that affect or have the potential to affect such habitat.[18]

The chief scientist along with the sponsoring agency begins a dialogue with the appropriate authorities (NOAA, EPA, Fish and Wildlife Service) aimed at evaluating the potential impact of the proposed research program within the context of the environmental regulations. These evaluations take place in consultation with representatives of the appropriate agencies (i.e. National Oceanic and Atmospheric Agency's National Marine Fisheries Service and Deptartment of Interior's Fish and Wildlife Service) and can result in a number of compliance requirements ranging from a 'categorical exclusion' if the activity is determined to have no significant environmental impact, to the need for a full environmental impact statement that is made available for comment by the public, other federal agencies and outside parties. In the case of marine mammals and other protected species, Marine Mammal/Protected Species Observers are typically required on board the research vessel and an 'Incidental Harassment Authorization' ('IHA') and/or ESA letter of concurrence, may be required. Both the NEPA EIS and MMPA IHA processes are complex, extensive and expensive, and are usually carried out by contractors with expertise in these procedures. In some cases the constraints of the environmental permitting process have delayed or prevented critical scientific studies.

In concert with federal permitting concerns, US Arctic researchers must also work closely with the indigenous communities. We attempt to involve members of the local communities in our activities (we have taken a member of the Barrow indigenous community on each of our ECS mapping legs), we make presentations to local community groups about our activities, but most importantly we must ensure that our research activities do not interfere with indigenous hunting (particularly the biannual bowhead whale hunt off the north shore of Alaska). This is done through groups like the Alaska Eskimo Whaling Commission and other marine mammal co-management organizations. Arctic research missions also often require significant local logistical support. The US is fortunate to have an extensive research infrastructure already established in Barrow, Alaska. Through coordination with the NSF and

---

18 *Ibid.* § 1801. Findings, purposes and policy [(b). purposes].

third-party contractors, logistical support for mobilization and demobilization of research expeditions can be arranged.

With increasing awareness of and concern about Arctic research, the number of regulations and the complexity of the permitting process continue to grow, sometimes having significant impact on the timing or conduct of research activities. The consultative and permitting process is still somewhat *ad hoc* and there are often large gaps in understanding of impacts and ramifications between the researchers and regulators. These will need to be addressed and perhaps future research efforts may help to develop more efficient means of mitigating some concerns (e.g. autonomous aerial vehicles for marine mammal observations).

VI. Conclusions

Unquestionably, the greatest challenge facing US Arctic researchers is *access* – both physical access through the availability of appropriate platforms and legal access through the policy of coastal States. As the community and funding agencies start to recognize the critical importance of Arctic research, demands for ship-time are sharply increasing. While there is irrefutable evidence of a steady decrease in the extent of Arctic ice,[19] for the foreseeable future, icebreakers will be required to work in the Arctic yet the availability of scientific research vessels capable of working in ice-covered waters is very limited. The bottom line is that the US (and global) scientific community need greater access to icebreakers with appropriate scientific capability; projects like the EU's *Aurora Borealas* are steps in the right direction to meet this critical demand.

Even if physical access to the Arctic is feasible through the availability of appropriate platforms, the scientific community faces further challenges in the form of environmental permitting and coastal State authorizations. US scientists must comply with an ever increasingly complex number of federal environmental regulations including the NEPA, MMPA, ESA, and the Magnuson-Stevens Fishery Conservation and Management Act. While the intend of these regulations is laudable, those responsible for implementing the regulations often do not fully

19 Intergovernmental Panel on Climate Change, *Climate Change 2007: Synthesis Report: Contribution of Working Groups I, II, and III to the Fourth Assessment Report of the Intergovernmental Panel on Climate Change* (2007).

understand the tools, techniques and environment for which the regulations are being established. Such lack of understanding can lead to the unnecessary delays and often cancellation of research cruises because regulations are difficult, if not impossible to meet within the timeframes and lead-times necessary to conduct research in the Arctic.

Finally, we are faced with the issue of legal access or authorization from the coastal States. It is the policy of the US and most other Arctic coastal States not to unreasonably deny or delay consent for marine scientific research though the United States reserves the right to participate in research activities conducted in its territorial seas and/or EEZ, and on its continental shelf. Past practice has demonstrated that this is not the view of all Arctic States as many requests for scientific research in Russian waters have often been denied or not responded to. This is cause for great concern about access for researchers to large portions of the Arctic that may be deemed extended continental shelf under Art. 76 UNCLOS. Arctic scientific research will become increasingly important to our understanding of the global impacts of climate change. We do the world and future generations a great disservice if there is not free access for research to this most critical of all regions.

Environmental Permitting Constraints on Arctic Marine Scientific Research

(Abstract)

*by Bernard Coakley**

Scientific access to a nation's exclusive economic zone ('EEZ') presumes permission to collect data in those waters. The process for obtaining permission is laid out in international law, particularly the law of the sea, and implemented in some form by each of the five Arctic coastal States in their own national permitting processes. Permissions presume compliance with all relevant national laws and may require additional permits to engage in a particular activity within their EEZs.

In Canada and the United States, native groups have a deep cultural investment in the management and preservation of their home territory. Most communities hold special hunting rights to collect valuable food resources for their communities. These rights give indigenous communities on the periphery of the Arctic Ocean special legitimacy within Canada and the United States and have given rise to additional stringent guidelines for compliance and access.

Marine geologic study of the oceans relies on acoustic imaging to map the seafloor and sediments beneath it. The frequencies used for various techniques overlap with the frequencies used by various marine mammal species to communicate, echolocate and hunt in the oceans, raising concerns about the effects of commonly used active acoustic techniques on the health and behavior of marine mammals. In the United States ('US') the Marine Mammal Protection Act ('MMPA'),[1] administered by the National Marine Fisheries Service ('NMFS'), evaluates the possible effects of scientific and industrial data acquisition and, through the issu-

* Professor, University of Alaska Fairbanks, Geophysical Institute.

1 Marine Mammal Protection Act of 1972, 16 USC 1361 et seq.

ance of an Incidental Harassment Authorization ('IHA'), permits short term disruption of animal behavior by these activities and validates mitigation plans to minimize the disruption.

The strongest restrictions are applied to the acquisition of multi-channel seismic ('MCS') reflection data. This technique is used by the oil industry to explore for hydrocarbons, but it is also used by academic scientists to study the history of the ocean basins and by governments to establish the limits of their extended continental shelves. To collect these data, it is necessary to regularly release (3 – 4 times per minute) a large volume of compressed air into the water column. This release generates low-frequency sound waves that can be used to image sedimentary layers deep below the earth's surface. These sounds can be detected 100 or more kilometers away from the 'shooting' ship.

The MMPA applies to any ship wishing to operate in the US EEZ and on US flagged ships operating anywhere in the world. The mitigation plan is built on models of the level and frequency content of the particular air gun array being used. This calculation is used to define a safety radius and a permitted number of 'takes' or sightings of marine mammals within that radius. Mitigation, as described in the application for IHA, defines shutdown criteria and ramp up procedures for beginning or restarting data acquisition. While the focus is on MCS data acquisition, other sources of radiated acoustic energy from the ship (bottom sounders, acoustic doppler current profilers etc.) are discussed in terms of possible effect on marine mammals, diving seabirds, fish and other creatures. There have been discussions about including propeller and ice breaking noise in IHA applications. At present each cruise is evaluated in isolation, but there has been discussion of evaluating the aggregate effects of individual cruises, both in terms of location and total radiated sound.

While there is a steadily increasing body of essentially anecdotal information about the effect of radiated sounds on marine mammals and other species, the efficacy of mitigation to prevent harm has not been scientifically established. The same standards and procedures are applied to well known marine mammal habitats and the central Arctic Ocean, which offers very few food resources to roaming cetaceans. Without consideration of the scientific value of the research or urgency of action as a counter argument, there has been a steady push towards more stringent standards and more elaborate monitoring programs. These requirements, set out in the IHA, can often be met by industry, but could, if imposed on a scientific cruise, effectively prohibit data acquisition.

While the requirements of IHA have the potential to shut down data acquisition by imposing stringent requirements for mitigation, this has not happened under MMPA. Cruises outside the US EEZ in the Arctic Ocean have been postponed and rescheduled, sometimes on short notice, but they have eventually been able to collect MCS data, sometimes in sensitive areas.

In coastal Canada and the US Arctic, native groups are granted an expanded, though informal, role in evaluating possible scientific cruises. This role has typically been confined to informal consultation and constructive arrangements to minimize interference with native subsistence hunting activities. However, on occasion environmental Non-Governmental Organizations ('eNGOs') have used the standing of native communities to shut down legitimate scientific programs by injecting themselves into the granting of permits through lawsuits.[2]

The time required for litigation works against a program on a tightly scheduled research vessel. In this context, the scientific value of the proposed data acquisition or the requirements of ship schedules are, at best, an ineffective counterweight to the legal arguments marshaled by eNGO lawyers, who can 'win' by merely postponing action for sufficient time.

Given the absence of overriding national need for particular research cruises and the lack of proof of the efficacy mitigation, the onus for proof of no harm falls on the research community. When we cannot prove mitigation is effective, we cannot prove it is sufficient and are driven to progressively higher standards and more restrictive requirements. How to give curiosity-driven scientific research greater significance in this discussion is the problem to be solved.

---

[2] For an example of a cruise cancelled due to eNGO lawsuits, see: LS Hollister, 'Termination of the Batholiths Marine Seismic Experiment: the Scientific Method Loses to Hearsay'; see <http://www.princeton.edu/geosciences/people/hollister/pdf/batho_agu.pdf> (24 May 2011).

Marine Scientific Research in Waters under the Jurisdiction of the Russian Federation

by Alexander S. Studenetsky[*]

The maritime doctrine of the Russian Federation up to 2020, approved by the President of the Russian Federation on 27 July 2001, classifies marine scientific research as a priority of national maritime policy.[1] The Russian Federation in the exercise of its sovereignty and jurisdiction has the exclusive power to regulate, license and conduct marine scientific research in its territorial sea, Exclusive Economic Zone and on its continental shelf. The realization and protection of the Russian Federation's national interests in the field of maritime activities are largely ensured by domestic marine science achievements.

Marine scientific research includes fundamental and applied research as well as appropriate experimental work aimed at providing knowledge of all aspects of the natural processes occurring on and under the sea bottom, in the water column and in the atmosphere. It must be emphasized that marine scientific research is not directly connected with the studying, prospecting, exploration and exploitation of non-living resources or with the research, exploration and harvesting of living resources.

Marine scientific research in the Russian Federation is regulated by the UN Convention on the Law of the Sea (1982),[2] ratified by the Russian

---

* Ministry of Education and Science of the Russian Federation.

1 Russian Federation, 'Maritime Doctrine of Russian Federation 2020', Approved by President Vladimir Putin (27 July 2001), Pr-1387, see <http://www.oceanlaw.org/downloads/arctic/Russian_Maritime_Policy_2020.pdf> (28 June 2011).

2 United Nations Convention on the Law of the Sea (concluded 10 December 1982, entered into force 16 November 1994) 1833 UNTS 397.

Federation in 1997,[3] as well as by the following federal laws: 'On internal marine waters, the territorial sea and contiguous zone of the Russian Federation',[4] 'On the exclusive economic zone of the Russian Federation'[5] and 'On the continental shelf of the Russian Federation'.[6]

In accordance with Russian legislation, the regulation of marine scientific research in waters under the jurisdiction of the Russian Federation is a matter for the federal executive agency responsible for science and technology. This is currently the Ministry of Education and Science of the Russian Federation.

In order to put the above laws into practice with respect to marine scientific research, the Government of the Russian Federation on 30 July 2004 by its Decision No. 391 approved the 'Regulations for the implementation of marine scientific research in the internal marine waters, within the territorial sea, in the exclusive economic zone and on the continental shelf of the Russian Federation'.[7] The Regulations are now applied to all marine waters under the jurisdiction of the Russian Federation, including the Arctic region. This document governs all features of marine scientific research, policy harmonization and the procedure for issuing authorizations to conduct marine scientific research.

According to existing provisions of international maritime law and the law of the Russian Federation, foreign applicants have the possibility annually to submit requests to conduct marine scientific research in waters under the jurisdiction of the Russian Federation. It should be borne in mind that marine scientific research should be carried out exclusively for peaceful purposes, in order to promote scientific knowledge of the marine environment. Such research may not pose a threat to the defense and security of the Russian Federation and shall take place only with the consent of the Russian Federation.

---

[3] Federal Law of the Russian Federation No. 30- Ф3 (26 February 1997) promulgated on 3 March 1997 in the Russian Federation Code of Laws for 1997 (part 9, Art.1013).

[4] Federal Law of the Russian Federation No. 155- Ф3 (31 July 1998) promulgated in August 1998 in the Russian Federation Code of Laws for 1998 (part 31, Art. 3833).

[5] Federal Law of the Russian Federation No. 191- Ф3 (17 December 1998) promulgated on 24 December 1998 in the "Rossiyskaya Gazeta" No. 244.

[6] Federal Law of the Russian Federation No. 187- Ф3 (30 November 1995) promulgated on 7 December 1995 in the "Rossiyskaya Gazeta" No. 237.

[7] Decision of the Russian Federation Government No. 391 (30 July 2004) promulgated on August 2004 in the Russian Federation Code of Laws for 2004 (part 32, Art. 3338).

The requests of foreign applicants are considered on an individual basis and in accordance with the established procedures. According to these procedures, foreign applicants should submit to the Ministry of Education and Science of the Russian Federation through diplomatic channels, not later than six months before the planned start date of the research, the request form in the Russian language and their own language, prepared in accordance with the above mentioned Regulations. It should be noted that similar timing for the consideration of requests to conduct marine scientific research is stipulated in the UN Convention on the Law of the Sea (Arts 247, 248, 252).

The authorization issued by the Ministry of Education and Science of the Russian Federation is the basis for carrying out the requested marine scientific research. This authorization may be given to applicants only after their requests have been approved by the following interested Russian federal executive agencies: the Ministry of Defense of the Russian Federation, the Federal Security Service of the Russian Federation, the Federal Service for Technical and Export Control, the Federal Service for Hydrometeorology and Environmental Monitoring, the Federal Custom Service, the Federal Agency for Fisheries, the Federal Agency for Natural Resources Management. The requests may be also be subject to the approval of other Russian federal executive agencies if this is necessary due to the nature of the request and/or because of the competence of such other Russian federal agency.

Foreign applicants may be refused an authorization if the exclusively peaceful nature of the research is open to question, or if the research:

- is relevant to studying, prospecting, exploration and exploitation of non-living resources or to studying, exploration and harvesting of living resources;
- is incompatible with the requirements on protection of the marine environment, living and non-living resources;
- includes drilling on the continental shelf, use of explosives, pneumatics or the introduction of harmful substances into the marine environment;
- includes the construction, operation or use of artificial islands, installations and structures;
- creates unjustified obstacles to the activities undertaken by the Russian Federation in the exercise of its sovereign rights and jurisdiction in the Exclusive Economic Zone and on the continental shelf.

Besides, foreign applicants may be refused an authorization if marine scientific research does not match the information on nature or purpose of studies referred to in the request and if the applicants did not fulfill their obligations to the Russian Federation in connection with earlier marine scientific research.

Foreign applicants who have the authorization to conduct marine scientific research in waters under the jurisdiction of the Russian Federation are obliged to meet a number of conditions:

- to enforce the implementation of the above-mentioned Regulations, the legislation of the Russian Federation and international treaties and agreements signed by the Russian Federation;
- not to create unjustified obstacles to activities undertaken by the Russian Federation in the exercise of its sovereign rights and jurisdiction;
- to strictly observe the authorized research program;
- to ensure the right of the Russian Federation to take part in the research;
- to refer the reports on implemented research to the Ministry of Education and Science of the Russian Federations;
- to pass the collected data and samples to the Russian Federation data funds;
- to inform immediately the Ministry of Education and Science of the Russian Federation as well as the Russian coastal surveys about any material change in the research program;
- to ensure, if necessary, the participation of the Russian Federation observer in the research activities.

According to information presented by the Ministry of Education and Science of the Russian Federation, the average annual number of marine scientific expeditions in waters under the jurisdiction of the Russian Federation is around 100-120. Marine scientific research in the Arctic region accounts for about half of all marine expeditions, i.e. approximately 50-60 annually.

Climate Change and Traditional Knowledge

(Speech)

by Klemetti Näkkäläjärvi[*]

Respected Chairman and honored participants,

On behalf of the Finnish Saami Parliament, I am pleased to attend this seminar and bring the perspective of indigenous peoples into the discussion.

Representatives of the dominant culture have commonly regarded the Arctic region as being sparsely populated. It should be understood that the Arctic region is, in fact, a densely populated area, where people strive to secure sufficient and ecologically sustainable livelihoods for both animals and themselves. The view of the Arctic region as an infertile and barren land is misleading. In reality, some of the most dense seasonal biomass concentrations can be found there, although their exploitation is not as simple as in other areas, such as at the equator. Arctic indigenous peoples have populated their regions in a way that is purposeful and environmentally sustainable. Now, climate change is about to alter this balance.

In the Arctic region, climate change will cause all indigenous peoples similar problems. It will have an impact on the resource base of the entire Arctic region, preventing the traditional, sustainable use of the environment by indigenous peoples. Changes in the flora, fauna and climate, combined with the loss of entire habitats, will force indigenous peoples to seek new ways of adapting. Meanwhile, these changes are obliging States to seek ways of preserving the Arctic region and its indigenous cultures for future generations.

Reindeer grazing will become more difficult with the increase in snow cover and crusty snow. People will be less able to earn their livelihoods

* President, Saami Parliament, Helsinki, Finland.

and live in their traditional territories. Such changes will not only affect material culture but also the cultural foundation, and the entire sphere of social life of indigenous peoples. Forced relocation from traditional habitats will erode the foundation of indigenous peoples' culture and compel them to adapt to a new environment. Climate change will affect the cultural and social life of indigenous peoples.

Climate change is not the only factor that threatens the future of Arctic indigenous peoples. Oil and gas drilling, especially in Russia, threatens their resource base. The future of traditional livelihoods is not secured either. The profitability of traditional ways of making a living is low and other competing land use models threaten the resource base. The sealskin markets have declined rapidly because of decisions by the European Union and the United States on banning the seal skin trade. This is sad news because harvesting seals is an essential part of Inuit culture. Traditional harvesting has to be supported and promoted so that cultural knowledge and livelihood patterns can survive. I hope that the European Union can help the Inuit to preserve their culture.

The Saami are the only indigenous people in the European Union. Saami people have their own representative organs in Norway, Finland and Sweden and a joint cooperative organ called the Saami Parliamentary Council. Climate change issues are at the core of our cooperation and we have a common climate policy strategy. The Saami people demand that the rise in the global average temperature be limited to 1.5°C.

The traditional knowledge of indigenous peoples helps them to adapt to climate change and the knowledge about the weather, snow and climate are more important than ever. Language lies at the core of indigenous cultures; it enables people to earn a living and maintain the community's memory. For example, Saamis are experts in reading nature and have a very special and distinct terminology for environmental conditions and phenomena. The Saami language has a vast store of terminology and appellatives for snow, which creates certainty when navigating and moving through the landscape. We have, for instance, around 1000 words to describe our natural environment and 1500 words to describe reindeer in the North Saami language.

The link between language, livelihoods and the environment can disappear for several reasons, such as the increasing use of indigenous peoples' traditional territories for the industries of the dominant culture, fewer opportunities to engage in traditional livelihoods and the spread of the dominant culture's social structures.

Arctic indigenous livelihoods rely on distinctive ecological structures. They are based on detailed knowledge of the territory and nature, as well as the close and mutually beneficial relationship between animals and people, sharing the same living environment.

For example, reindeer-herding Saami adapt directly to their natural environment, live in harmony with nature and migrate according to the seasons. Mobility and social flexibility form their basic structures, enabling reindeer Saami to exploit their living environment extensively and in sustainable ways, while eliminating the ecological crises caused by changes in plant and animal life. The reindeer Saami community continues to rely on kinship rights in the organization of its society. These aspects of the ecosystem guarantee the community's safety and continuity.

The Arctic attracts political, military, commercial and logistical interest, as well as interest in the energy sphere. For indigenous peoples, this is not at all welcome. The sensitive nature of the Arctic environment cannot sustain large numbers of people or the wide-scale exploitation of energy resources.

The melting of sea-ice and the opening of new sea-routes makes all-year seafaring possible. This produces environmental threats, because increased traffic, off-shore drilling and oil shipping increase the risk of environmental disasters, like the Exxon Valdez or Gulf of Mexico disasters. Bearing this in mind, it is extremely important to concentrate on seeking and maintaining cooperation in the Arctic. It would be most welcome if a legally binding agreement could be concluded to control sea traffic and offshore drilling in the Arctic in an environmentally sustainable way.

Increased military presence in the Arctic also puts the environment at risk. Nuclear weapons, nuclear waste and military exercises are threats to the Arctic environment and its flora and fauna. The demilitarization of the Arctic would be important for indigenous peoples as well as for the Arctic environment.

It is important that the Arctic States have complied with their international obligations. It is important for Arctic indigenous peoples and Arctic cooperation that the United States ratifies the Biodiversity Convention.[1] It is also necessary that the United States ratifies the UN

1 Convention on Biological Diversity (concluded 5 June 1992, entered into force 29 December 1993) 1760 UNTS 79.

Convention on the Law of the Sea.[2] For indigenous peoples, it is very important that human rights obligations are also fulfilled. In this context the ratification of the ILO Convention concerning Indigenous and Tribal Peoples[3] and the implementation of the UN Declaration on the Rights of Indigenous Peoples[4] are key elements.

Indigenous peoples are frequently viewed as objects of decision-making. They are expected to adapt to the changes brought about by the industrial and post-industrial world, climate change and other competing forms of land use. From indigenous peoples' point of view, this never-ending requirement to adapt is an extension of colonialism, disguised in fine words. From this perspective, indigenous peoples can develop and adapt only under the conditions set by the dominant culture. Securing the rights and self-determination of indigenous peoples would provide a solution to this. However, the reluctance of States to recognize increased autonomy as a solution to adapting to climate change is a common problem. Instead, research and the increased participation of indigenous peoples are continuously cited as the preferred methods. The most efficient solution would be to secure indigenous peoples' self-determination at both national and international level.

Climate change poses a significant threat to the future of Arctic indigenous peoples. Thus far, its effects have remained small but they will accumulate in the near future. Simultaneously, indigenous peoples are not only seeking new ways to adapt their cultures, livelihoods and languages to the changes brought upon them by climate change, but are also fighting for the existence and future of their cultures. Indigenous peoples do not have sufficient rights to make decisions on their own issues, and in the main lack adequate funding for preserving their cultures and languages. They are engaged in a struggle against massive social pressures.

For indigenous peoples, climate change forms part of a large tangle of problems caused by industrialized nations and the assimilation policies

---

[2] United Nations Convention on the Law of the Sea (concluded 10 December 1982, entered into force 16 November 1994) 1833 UNTS 397.

[3] ILO 'Convention No 169 concerning Indigenous and Tribal Peoples in Independent Countries' (adopted 27 June 1989, entered into force 5 September 1991) 1650 UNTS 383.

[4] UNGA Res 61/178 'United Nations Declaration on the Rights of Indigenous Peoples' (13 September 2007) GAOR 61st Session Supp 49 vol 3, 15.

of nation-States. As has often been said, climate change will confront the Arctic peoples with the biggest challenge they have ever faced.

Securing the future of indigenous cultures in the Arctic region will require significant reductions in emissions, wide-scale research into the effects of climate change on indigenous peoples, and methods of mitigating the negative impacts of climate change. First and foremost, we need to find ways to secure the continued existence of their cultures. Thank you for your interest!

Extended Continental Shelf Issues in the Arctic Ocean: A Modern 'Land Grab' Or an Example of Cooperation between the Arctic Coastal States?

(Extended abstract)

by Christian Marcussen[*]

I. Introduction

The Arctic has made the headlines in the international press with regards to extended continental shelf issues and the United States Geological Survey ('USGS') assessment of substantial undiscovered hydrocarbon resources in the Arctic.

There is a widespread misunderstanding among the general public regarding the extension of the continental shelf and the perception of an ongoing 'Arctic land grab'. This misunderstanding is in part due to press coverage of an expedition of two Russian mini submarines at the North Pole on 2 August 2007 and the planting of the Russian flag. However, as Sergei Lavrov, Foreign Minister of the Russian Federation, explained in August 2007, 'the ownership of the shelf in the Arctic Ocean is defined on the basis of the United Nations Convention on the Law of the Sea (UNCLOS)'[1] – a statement the other Arctic coastal nations have agreed to.

---

[*] Senior Advisor at the Geological Survey of Denmark and Greenland ('GEUS'). The views and opinions expressed in this paper are the author's own, and do not reflect the official positions of any agency or government.

[1] Cited by RIA Novosti, 'Russia Guided by International Law in its Polar Shelf Probe', Press release (3 August 2007), see <http://en.rian.ru/world/20070803/70295919.html> (14 October 2011).

II. Assessment of Arctic Hydrocarbon Resources

The main findings of the USGS assessment of undiscovered Arctic hydrocarbon resources released in the summer of 2008 are:[2]

- About 13% of the world's undiscovered oil, 30% of undiscovered gas and 20% of undiscovered natural gas liquids may be found in the Arctic. This is equivalent to three years of world oil consumption and 20 years of world gas consumption.
- The Arctic is gas prone: more gas than oil is likely to be found.
- 84% of the estimated resources are expected offshore. Most of the resources are within the exclusive economic zone ('EEZ' – maritime area over which a coastal State exercises sovereign economic rights) of the Arctic coastal States (see Fig. 1).
- Important resources areas include the West Siberian Basin (gas) and Arctic Alaska (oil).
- Other areas include Arctic Canada, the Barents Sea, and the North-East Greenland shelf.
- The issues related to resource exploitation in the Arctic include very high costs (development and transport) and a very vulnerable environment.

III. The Ilulissat Declaration

As a consequence of the Russians planting a flag at the North Pole in 2007, Denmark and Greenland called for a meeting in Greenland in 2008 of the five Arctic coastal States (Canada, Denmark/Greenland, Norway, the Russian Federation and the United States), which led to the signature of the Ilulissat Declaration.[3] The Declaration stipulates commitment to international law (UNCLOS),[4] to an orderly settlement of any possible overlapping claims, and to close cooperation with respect to the collection of scientific data concerning the continental shelf.

---

[2] DL Gautier et al., 'Oil and Gas Resource Potential North of the Arctic Circle', in AM Spencer et al. (eds), *Arctic Petroleum Geology: Geological Society Memoir No. 35* (2011) 151–161.

[3] Ilulissat Declaration (28 May 2008) (2009) 48 ILM 382.

[4] United Nations Convention on the Law of the Sea (concluded 10 December 1982, entered into force 16 November 1994) 1833 UNTS 397.

IV. Art. 76 United Nations Convention on the Law of the Sea

UNCLOS came into force in 1994. At present, a total of 162 States – excluding the US – have ratified UNCLOS. Art. 76 UNCLOS specifies the criteria for an extension of the continental shelf of coastal States beyond 200 nm (EEZ). It provides the coastal State with certain sovereign rights to living and non-living resources on and below the sea bed. Coastal States that have ratified UNCLOS have ten years after ratification to prepare a submission for an extended continental shelf to the Commission on the Limits of the Continental Shelf ('CLCS'). The CLCS does not resolve disputes or overlapping submissions. If overlaps between the submissions of circumpolar neighbors become apparent, they will have to be resolved in accordance with international law.

A coastal State's entitlement to extended sovereign rights depends on the width of its continental margin. To qualify, the State must meet the criteria specified in Art. 76, which are based on a consideration of seafloor morphology, bathymetry and of underlying sediment thickness. To meet these criteria, it is necessary to assemble and analyze data for reliably determining the locations of three undersea features: (1) the foot of slope, defined as the point of maximum change of gradient at the base of the continental slope; (2) the 2500 m isobath; and (3) the location of the so-called Gardiner Points where the sediment thickness equals one percent of the distance back to the foot of slope.

These three features are used as points of departure for defining the two formula lines and two constraint lines of Art. 76. The formula lines are used to determine the outer limit of the continental margin and are derived from an analysis of seafloor morphology and sediment thickness. The constraint lines are used to preclude exaggerated or unwarranted claims relating to the width of the continental margin and are defined by their distances from the 2500 m isobath and from the coastal State's territorial sea baseline. When combined, the formula and constraint lines prescribe the outer limit of the coastal State's extended continental shelf (see Fig. 2).

Art. 76 requires a coastal State to document the foregoing process in a submission by: (a) describing the data sets that have been assembled for delimitation purposes; (b) presenting and discussing the results of their analysis; and (c) listing the geographical coordinates of the State's proposed outer limit(s). The submission is then examined by the CLCS, which reviews its contents and its conclusions in order to develop a set of recommendations that may or may not confirm the submitting State's entitlement to an extended continental shelf.

By October of 2011, the CLCS had received a total of 57 submissions and issued 14 recommendations. In addition, the CLCS received more than 40 'Preliminary Information Notes' on potential future submissions.[5] Under the current *modus operandi* of the CLCS, the consideration of the 57 submissions received to date could extend to 2030.

V. Status of Work on Extended Continental Shelf Issues in the Arctic Ocean

The current status of submissions in the Arctic Ocean is as follows (see Fig. 3):

- The Russian Federation forwarded a submission in 2001. In 2002, the CLCS recommended a revised submission. Additional field work is ongoing with further expeditions in the coming years. A revised submission is expected for 2013.
- Norway submitted its claim in the North Atlantic and in the Arctic Ocean in 2006, and in 2009 it was recommended by the CLCS with some changes.
- In September 2010, Norway and Russia signed the 'Treaty concerning Maritime Delimitation and Cooperation in the Barents Sea and the Arctic Ocean'.[6] The Treaty entered into force on 7 July 2011.[7]
- Canada's deadline for submission is at the end of 2013 and Denmark's at the end of 2014.

---

[5] Commission on the Limits of the Continental Shelf, see <http://www.un.org/Depts/los/clcs_new/clcs_home.htm> (9 August 2011).

[6] Treaty between Norway and the Russian Federation concerning Maritime Delimitation and Cooperation in the Barents Sea and the Arctic Ocean (done 15 September 2010, entered into force 7 July 2011), reprinted in T Henriksen and G Ulfstein, 'Maritime Delimitation in the Arctic: The Barents Sea Treaty', *Ocean Development and International Law* 42 (2011) 1 et seq. (11-17).

[7] Norwegian Ministry of Foreign Affairs, 'Norway and Russia Ratify Treaty on Maritime Delimititation', see <http://www.regjeringen.no/en/dep/ud/press/news/2011/maritie_delimitation.html?id=646614> (9 August 2011).

- Despite the fact that the US has not ratified UNCLOS yet, the US Extended Continental Shelf Project is actively acquiring data to establish the full extent of the continental shelf of the United States.

Two prominent ridges in the Arctic Ocean – the Lomonosov Ridge and the Alpha-Mendeleev Ridge – will play an important role in the extended continental shelf submissions in the Arctic Ocean as already seen in the 2001 submission by the Russian Federation. In their recent paper Brekke and Symonds[8] note that '[i]n cases where seafloor highs are enclosed by the foot of the slope envelope, such highs are automatically regarded as integral parts of that continental margin on a morphological basis'. Which constraint line is applicable depends on whether the seafloor high can be regarded as submarine elevation or submarine ridge. This assessment will be based on the geological characteristics of the seafloor high in addition to their already proven morphological continuity. According to Brekke and Symonds 'the CLCS makes an assessment as to what extent a seafloor high is geologically associated or continuous with the landmass of the coastal state, and to what extent it is geologically different to the surrounding deep ocean floor. This assessment is made solely on the basis of geological and geophysical evidence.'

VI. Conclusion

Acquisition of scientific data to support extended continental shelf claims in the Arctic Ocean constitutes considerable logistical challenges[9] (see Fig. 4). The Arctic coastal States are therefore actively engaged in logistical and scientific cooperation during both the data acquisition and the interpretation phase. This results in the following benefits: (a) sharing of the limited logistical resources; (b) cost savings; and (c) a better understanding of the geology of the Arctic Ocean.

---

8 H Brekke and P Symonds, 'Submarine Ridges and Elevations of Article 76 in Light of Published Summaries of Recommendations of the Commission on the Limits of the Continental Shelf', *Ocean Development and International Law* 42:4 (2011) 289-306.

9 C Marcussen and R Macnab, 'Extending Coastal State Boundaries into the Central Arctic Ocean: Outer Continental Shelves beyond 200 Nautical Miles and the Quest for Hydrocarbons', in AM Spencer et al. (eds), *Arctic Petroleum Geology: Geological Society Memoir No. 35* (2011) 715–730.

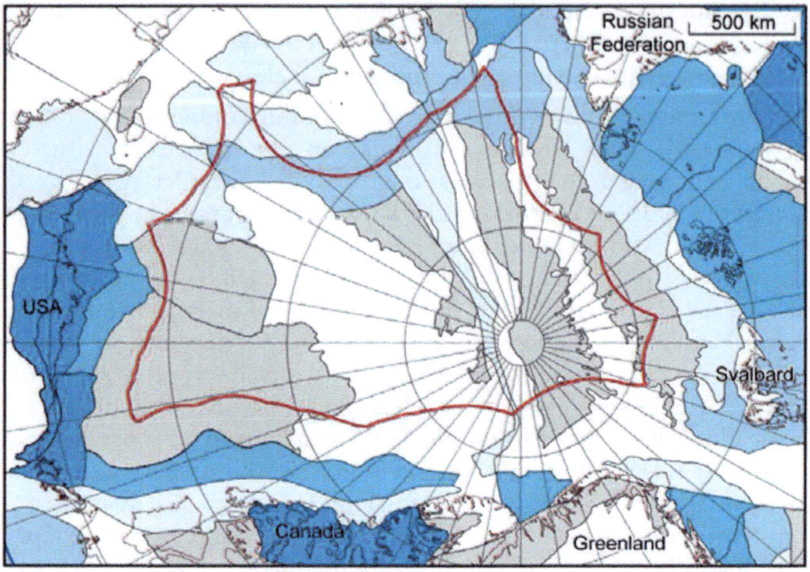

Fig. 1. Map[10] displaying the probability of major gas and oil finds north of the Arctic Circle. Areas shaded in white and grey are considered to have petroleum probabilities of less than 10%. Areas shaded in blue have probabilities that range from 10-30% (light blue) to 100% (dark blue)

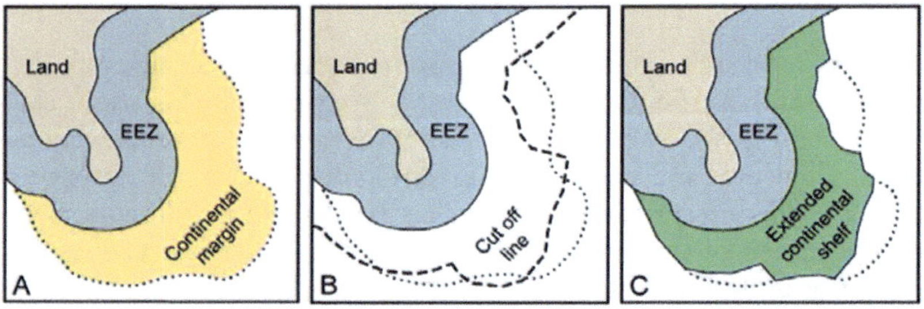

Fig. 2. Steps in the development of the outer or extended continental shelf. In (A), the *formula lines* of Art. 76 UNCLOS have been used to determine the extent of the *continental margin*. In (B), the *constraint lines* of Art. 76 have been used to develop a *cutoff line*. In (C), the cutoff line is applied to the continental margin to develop the outer or extended continental shelf, which must be defined by straight lines that are no longer than 60 nm

---

10 Adapted from Gautier, see note 2.

Fig. 3. Maritime jurisdiction and boundaries in the Arctic region. Areas with heavy colors – coastal States' exclusive economic zones; areas with light colors – claimed continental shelf beyond 200 nm (Norway and the Russian Federation); hatched areas – potential claimed continental shelf beyond 200 nm (Canada, Greenland and USA); light blue areas – unclaimed or unclaimable continental shelf; red lines – agreed boundaries and stippled grey lines – median lines[11]

---

[11] Map courtesy of the International Boundaries Research Unit ('IBRU'), 'Maritime Jurisdiction and Boundaries in the Arctic Region' (2010), see <http://www.dur.ac.uk/ibru> (9 August 2011).

Fig. 4. The Russian nuclear icebreaker *50 let Pobedy* and the Swedish icebreaker *Oden* trailing behind in severe ice conditions in the Lomonosov Ridge area north of Greenland during the LOMROG I expedition in 2007. (Photo: Hans Ramløv)

Setting the Stage: The Continental Shelf and Marine Science in the Arctic Ocean

by Ted L. McDorman[*]

I. Introduction

The intention of this brief contribution is to provide an overview of the relevant international law and practice respecting the continental shelf within and beyond 200 nm regarding the central Arctic Ocean basin. Included as section two is a truncated history of the international legal regime of the continental shelf as it sheds light on the occasional misperception that what the States are doing in the Arctic Ocean regarding the continental shelf is novel, adventurous, or spurred on by the effects of global climate change in the Arctic. This is followed by a brief summarization of the criteria and processes related to a coastal State's outer limit of the continental shelf beyond 200 nm. An interesting feature of the outer limit criteria is their clear relationship to hydrocarbon resources and a coastal State maximizing its shelf area which may contain hydrocarbon resources. Section four briefly describes the international legal regime that applies to marine scientific research ('MSR') regarding the continental shelf beyond 200 nm. Section five contains an overview of the formal actions of States respecting the continental shelf in and adjacent to the central Arctic Ocean basin. An important feature of section five is that there is significant uncertainty respecting the extent to which there will be overlapping claims by Arctic States to shelf areas

---

[*] Faculty of Law, University of Victoria, Victoria, British Columbia, Canada. The paper draws, with modification, from previously published work as noted in the text and a background document prepared for the section of the Arctic Council's Arctic Marine Shipping Assessment on international governance.

beyond 200 nm and the extent to which there will be areas of the seafloor of the central Arctic Ocean basin that are beyond areas of national jurisdiction.

The Arctic coastal States (herein defined for the purposes of this contribution as the five States that border the central Arctic Ocean basin, Canada, Denmark/Greenland, Norway, Russia and the United States), in word and deed, apply the international law of the sea to the Arctic Ocean in much the same manner as all coastal States deal with their adjacent ocean areas by reference to the law of the sea. This is the case even though the Arctic Ocean, especially the central Arctic Ocean basin, is a unique, sensitive and relatively pristine marine environment and this includes the 'special' nature of some of the coastal populations and the impact that developments in the environment (global climate change) are and may have on global weather, sea-levels, ocean currents, etc. While during the negotiation of the United Nations Convention on Law of the Sea ('UNCLOS')[1] little attention was paid to the Arctic Ocean, it was not totally absent from consideration. It has been observed that: 'The fact that the Arctic rarely received specific mention [at the negotiations] – by virtue of an unspoken "gentleman's agreement" among Arctic and non-Arctic nations – took little away from the general applicability of the Convention to the Arctic.'[2] Uwe Jenisch wrote in a paper published in 1985 that:

> [I]t has to be maintained that the Arctic clearly is regulated by the international law of the sea. Throughout the LOS Conference, there were no objections to the full application of the new law to Arctic waters. The Conference avoided any tendencies to establish a "special" or "regional" regime for polar seas, with the sole exception of art. 234 dealing with "ice-covered waters."[3]

Interestingly, in at least one venue, the idea that the Arctic Ocean 'was and remains *sui generis* in the law of the sea, and that an unresolved jurisdictional issue such as this requires specific *ad hoc* treatment at the global or regional level' was put forward as an option respecting the

1 United Nations Convention on the Law of the Sea (concluded 10 December 1982, entered into force 16 November 1994) 1833 UNTS 397.

2 Report of an unidentified Canadian international lawyer, in DM Johnston (ed.), *Arctic Ocean Issues in the 1980's* (1982) 12.

3 U Jenisch, 'The Arctic Ocean and the New Law of the Sea', in B Vukas (ed.), *Essays on the New Law of the Sea* (1985) 479 et seq. (484).

high seas beyond national jurisdiction.[4] It was an option that had little possibility of acceptance in the face of the clarity in the 1980s that the law of the sea applied and rights had been exercised in the Arctic Ocean.[5]

II. History[6]

The international legal entitlement of a coastal State to exclusive jurisdiction over the resources of its adjacent physical continental shelf is usually traced back to the 1945 Truman Proclamation on the Continental Shelf[7] or the 1942 Treaty between the United Kingdom and Venezuela respecting the Gulf of Paria, where the phrase submarine areas and 'sea-bed and sub-soil' outside territorial waters, rather than continental shelf, was used.[8] The principal legal features of the continental

---

4 Report of an unidentified speaker, in Johnston, see note 2, 17; see also K Beauchamp, 'International Legal Issues in Arctic Waters', in Canadian Arctic Resources Committee (ed.), *Ocean Policy and Management in the Arctic* (1984) 53 et seq. (54). Beauchamp goes on to conclude that: 'World opinion appears to be that this marine region should be treated as an ocean area that is unique but also subject to the law of the sea' (at 55).

5 See generally BJ Theutenberg, 'The Arctic Law of the Sea', *Nordisk Tidsskrift for International Relations* 52 (1983) 3 et seq. and D Pharand, 'The Implications of Changes in the Law of the Sea for the "North American" Arctic Ocean', in JK Gamble Jr. (ed.), *Law of the Sea: Neglected Issues* (1979) 183 et seq. (186-187).

6 Parts of this section have been drawn, with modification, from: TL McDorman, 'The Continental Shelf beyond 200 nm: Law and Politics in the Arctic Ocean', *Journal of Transnational Law & Policy* 18 (2009) 155 et seq. (162-167) and TL McDorman, 'The Outer Continental Shelf in the Arctic Ocean: Legal Framework and Recent Developments', in D Vidas (ed.), *Law, Technology and Science for Oceans in Globalisation* (2010) 499 et seq. (503-507).

7 United States, Proclamation 2667, 'Policy of the United States with Respect to the Natural Resources of the Subsoil and Sea Bed of the Continental Shelf' (28 September 1945) 59 US Statutes at Large 884, codified as Executive Order 9633, 'Reserving and Placing Certain Resources of the Continental Shelf Under the Control and Jurisdiction of the Secretary of the Interior' (28 September 1945), 10 Federal Register 12305.

8 Treaty between the United Kingdom and Venezuela relating to the Submarine Areas of the Gulf of Paria (done 26 February 1942, entered into force 22

shelf regime were set out in the 1958 Geneva Convention on the Continental Shelf ('1958 Continental Shelf Convention'),[9] including:

- that the nature of a coastal State's rights over the continental shelf is 'sovereign rights for the purpose of exploring it and exploiting its natural resources,'[10] which meant 'the mineral and other non-living resources' of the shelf 'together with' sedentary species;[11]
- that these 'sovereign rights' are exclusive to that State;[12] and
- that a coastal State rights over its adjacent continental shelf do not depend upon occupation or an express proclamation.[13]

The International Court of Justice in the 1969 *North Sea Continental Shelf Cases* introduced the concept of 'natural prolongation' in the context that a coastal State has rights over the area of the continental shelf 'that constitutes a natural prolongation of its land territory'[14] and commented 'that the rights of the coastal State in respect of the area of continental shelf [...] exist *ipso facto* and *ab initio*, by virtue of its sovereignty over the land. In short, there is [...] an inherent right.'[15]

Thus, the international legal entitlement of a coastal State to its adjacent continental shelf predates the 1982 UNCLOS, the extension of territorial seas to 12 nm and the 200 nm exclusive economic zone ('EEZ'), and any inkling of global climate change.

What was ambiguous in the 1958 Geneva Convention was the location of the outer limit of a coastal State's continental shelf. Art. 1 1958 Continental Shelf Convention provided criteria for the outer limit of the shelf – the seabed and subsoil within the envelope of waters of a depth of 200 m 'or, beyond that limit, to where the depth of the superjacent

---

September 1942) 205 LNTS 121; see generally DP O'Connell, *The International Law of the Sea*, Vol. I (edited by IA Shearer) (1982) 470.

9 Convention on the Continental Shelf (done 29 April 1958, entered into force 10 June 1964) 499 UNTS 311.

10 *Ibid.*, Art. 2(1).

11 *Ibid.*, Art. 2(4).

12 *Ibid.*, Art. 2(2).

13 *Ibid.*, Art. 2(3).

14 *North Sea Continental Shelf Cases (Federal Republic of Germany/Denmark; Federal Republic of Germany/Netherlands)* [1969] ICJ Rep 3, para. 19 and see paras 43-44 and 95-96.

15 *Ibid.*, para. 19.

waters admits of the exploitation of the natural resources [...]'. While it was believed at the time that a definite limit was intended by the wording,[16] as a result of technological, political and economic pressure primarily related to offshore hydrocarbon resource development, the exploitability criterion was seen as ambiguous and as effectively meaning that the 1958 Continental Shelf Convention did not provide a determinable outer limit of the continental shelf.[17] It was not only the outer limit of legal continental shelf that could not be agreed upon in 1958, negotiators also could not agree on the width (outer limit) of the territorial sea.[18]

Entering the Third United Nations Conference on the Law of the Sea ('UNCLOS III') in the 1970s, States such as Canada were strongly of the view that the 1958 Continental Shelf Convention and the 1969 *North Sea Continental Shelf Cases* were clear that a coastal State had exclusive national jurisdiction over its adjacent shelf area to the point where the shelf met the deep ocean floor.[19]

It is in the above context that one can appreciate the complex negotiations that took place to achieve the wording in the 1982 UNCLOS. The easy part was the substantive rights of entitlement set out in the 1958 Continental Shelf Convention, which were repeated in the UNCLOS. With the advent of the 200 nm EEZ, it was readily agreed that a coastal State had a continental shelf co-terminus with the EEZ irrespective of the existence of a physical continental shelf. One negotiating difficulty was establishing agreed wording on criteria for the resolution of overlapping national continental shelf claims. The final wording on this topic is not seen as being very helpful.[20] Another difficult issue was the detailed criteria and process for the establishment of an outer limit of a

---

[16] See generally BH Oxman, 'The Preparation of Article 1 of the Convention on the Continental Shelf', *Journal of Maritime Law and Commerce* 3 (1972) 245 et seq., 445 et seq. and 683 et seq. and ED Brown, *The Legal Regime of Hydrospace* (1971) 1-40.

[17] See RR Churchill and AV Lowe, *The Law of the Sea*, 3rd edn (1999) 147.

[18] *Ibid.*, 79.

[19] Canada, Department of External Affairs, *The Third United Nations Conference on the Law of the Sea* (1973) 13-15.

[20] See *Government of the State of Eritrea and the Government of the Republic of Yemen (Second Stage: Maritime Delimitation) (Award)* Permanent Court of Arbitration (17 December 1999) (2001) 40 ILM 983 (1003, para. 116).

coastal State's legal continental shelf beyond 200 nm set out in Art. 76 UNCLOS.

Unlike in 1958, during UNCLOS III negotiations there was a necessity to provide for a definitive outer limit of the continental shelf regime where the shelf extended beyond 200 nm. The political necessity was created by the 'Common Heritage of Mankind' and the International Seabed Authority ('ISA'), since the ISA and the Common Heritage was to apply to the mineral resources of the seafloor beyond national jurisdiction, in other words beyond the outer limits of coastal State's continental shelves.[21] The compromise that was agreed upon between those States which were of the view that exclusive national authority existed over the shelf beyond 200 nm and those States seeking to limit coastal State continental shelf authority at 200 nm involved: adoption of a complex formula for determining the outer limit of a State's continental shelf beyond 200 nm; creation of the Commission on the Limits of the Continental Shelf[22] to assist States in applying the complex formula respecting the outer limit; and revenue sharing with the international community in respecting of mineral resources exploited from the continental shelf area under national jurisdiction beyond 200 nm.[23]

While the Arctic Ocean was not at the forefront of State considerations during the negotiation of the continental shelf beyond 200 nm, it is worth noting that the Arctic Ocean was not entirely absent. During the negotiations regarding the criteria to be adopted respecting the outer limit beyond 200 nm, the United States, for example, had concerns about the possible misuse by Russia of Arctic ridges as a means of claiming large areas of the Arctic Ocean.[24] Moreover, in 1980 the United States made it clear that it viewed the Chukchi Plateau in the central Arctic Ocean basin as being a 'submarine elevation' and thus the feature was not understood to be a submarine ridge one of the key factors in the outer limit criteria.[25] At a workshop held in 1981, there was much informed discussion of the Arctic Ocean and law of the sea issues

---

[21] Compare: Arts 1(1), 133 and 136 UNCLOS, see note 1.

[22] The Commission is established pursuant to UNCLOS, see note 1, Annex II.

[23] Art. 82 UNCLOS, see note 1.

[24] See EL Miles, *Global Ocean Politics* (1998) 387-388.

[25] US Ambassador Elliot Richardson, 'Statement' (3 April 1980), in Third United Nations Conference on the Law of the Sea, *Official Records*, Vol. XIII (1981) 43.

including the observation 'that most of the sea-bed of the Arctic would be included under national jurisdiction by virtue of the formula in Art. 76 [UNCLOS].'[26] Thus, the continental shelf of the Arctic Ocean was being contemplated by the adjacent coastal States long before global climate change considerations and the 2007 Russian planting of a flag on the North Pole.

III. Outer Limit: Criteria and Process[27]

One scholar has described the key paragraphs of Art. 76 UNCLOS that sets out the criteria for determining the outer limit of the continental shelf as combining the 'influences of geography, geology, geomorphology, and jurisprudence'[28]. The complexity of the criteria reflect the importance that coastal States placed on obtaining or retaining exclusive control over any potential hydrocarbon resources in their adjacent seabed areas.

The proposal that formed the basis of the negotiations of the outer limit criteria was the so-called Irish formula, which provided that the outer limit of the legal continental shelf should be determined on the basis of sediment thickness seaward of the foot of the slope or by a 60 nm limit from the foot of the continental slope. The foot of the slope was taken as the starting point since it was a recognizable physical characteristic in large parts of the ocean floor and thus was seen as providing some ease in locating the outer edge of the margin. Since the physical continental margin consists of the continental shelf, continental slope and continental rise, the foot of the slope ensures that, at a minimum, a coastal State has legal authority over the key physical components of its adjacent offshore seafloor which are most likely to contain hydrocarbon resources, plus at least part of the continental rise. The idea for the foot-of-the-slope-plus-zone, borrowed from American geologist Hollis D. Hedberg,[29] was to provide a method of delineation that would not involve

[26] Johnston, see note 2, 17.

[27] Parts of this section have been drawn, with modification, from: McDorman [2009], see note 6, 169-171 and McDorman [2010], see note 6, 507-510.

[28] DM Johnston, *The Theory and History of Ocean Boundary-Making* (1988) 91.

[29] Professor Hedberg wrote about the foot of the slope plus zone in numerous works; see H Hedberg, *National-International Jurisdictional Boundary on*

acquiring sediment thickness information which as seen at the time as being difficult to obtain. The rationale for the sediment thickness criteria, however, was to ensure that a coastal State secured jurisdiction over all the hydrocarbon resources that might possibly exist in the offshore areas adjacent to it.[30] Essentially, if the sediment was thick enough there might exist hydrocarbon resources and, therefore, it should come under coastal State authority. Interestingly, the sediment thickness criterion was criticized by Hedberg as being

> based more on factors of economic advantage to certain coastal countries than on impartial considerations of where a boundary should most naturally, most logically and most rightfully be.[31]

The Soviet Union put forward a proposal to prevent coastal States from claiming jurisdiction on the basis of the Irish formula to areas beyond a 300 nm limit.[32] In the end, a compromise was reached that limited the extent of the Irish proposal to either 350 nm or 100 nm from the 2,500 meter isobath, whichever was further seaward. A last issue concerning the criteria to be used for the establishment of the outer limit related to ridges. The concern was that underwater ridges might be used by some coastal States to extend their jurisdiction to the middle of the ocean. The compromise that was reached distinguishes between oceanic ridges, submarine ridges and submarine elevations. The criteria are not easily applicable in any given situation because of the technical and definitional difficulties of determining the thickness of sedimentary rocks, the foot of the continental slope, the 2,500 meter isobath, and distinguishing among

---

the Ocean Floor (1975); H Hedberg, 'The National-International Jurisdiction Boundary on the Ocean Floor', *Ocean Management* 1 (1973) 83 et seq.; H Hedberg, 'Limits of National Jurisdiction over Natural Resources of the Ocean Bottom', in LM Alexander (ed.), *The Law of the Sea: National Policy Recommendations* (1970), 159 et seq. and H Hedberg, 'Relation of Political Boundaries on the Ocean Floor to the Continental Margin', *Virginia Journal of International Law* 17 (1976) 57 et seq.

30 FA Eustis III, 'Method and Basis of Seaward Delimitation of Continental Shelf Jurisdiction', *Virginia Journal of International Law* 17 (1976) 107 et seq. (125); H Hedberg, 'Discussion', in E Miles and JK Gamble Jr. (eds), *Law of the Sea: Conference Outcomes and Problems of Implementation* (1977) 215.

31 Hedberg [1977], *ibid.*, 215.

32 The Soviet proposal is discussed in BH Oxman, 'The Third United Nations Conference on the Law of the Sea: The Seventh Session (1978)', *American Journal of International Law* 73 (1979) 1 et seq. (19-21).

submarine ridges, oceanic ridges, and submarine elevations that are natural components of the continental margin.

The UNCLOS creates a procedure to be followed by a coastal State in the determination of its outer limit of the continental margin where it extends beyond 200 nm. The procedural steps are derived from Art. 76(8) which provides that a coastal State is to submit information supporting its proposed outer limit of the 'legal' continental shelf to the Commission on the Limits of the Continental Shelf. The Commission is to consider the submitted material and make recommendations to the submitting State regarding the information received and the relevant Art. 76 criteria. It is important to note that the role of the Commission is only to review the information provided and make recommendations to the submitting State. The Commission does not have the legal authority to determine or impose its views respecting the location of the outer limit of the continental margin on a coastal State. In other words, the Commission is not a court. It is the coastal State, and not the Commission, that establishes the outer limit of its continental margin beyond 200 nm.[33] It is also important to note that the work and product of the Commission is without prejudice to bilateral boundary disputes between States.[34]

---

[33] The US government, for example, has stated: 'Ultimate responsibility for the delimitation [of the outer limit of the continental margin] lies with the coastal State itself.' President Clinton, 'Message from the President of the United States transmitting United Nations Convention on the Law of the Sea, with Annexes, done at Montego Bay, December 10, 1982 (the Convention) and the Agreement Relating to the Implementation of Part XI of the United Nations Convention on the Law of the Sea of 10 December 1982, adopted at New York, July 28, 1994 (the Agreement) and signed by the United States, subject to ratification, on July 29, 1994,' Senate, Treaty Document 103-39, 103rd Congress, 2nd Session IV (1994) 40, reprinted in (1995) 34 ILM 1393; see also United Nations, Office of Legal Affairs, Division of Ocean Affairs and the Law of the Sea, *The Law of the Sea: Definition of the Continental Shelf* (1993) 29 and Committee on Legal Issues of the Outer Continental Shelf, in International Law Association, *Report of the Seventy-First Conference (Berlin 2004)* (2004) 773 et seq. (785-786).

[34] Art. 76(10) UNCLOS, see note 1; see generally Committee on the Legal Issues of the Outer Continental Shelf, *ibid.*, 809-813; AG Oude Elferink, 'Submissions of Coastal States to the CLCS in Cases of Unresolved Land or Maritime Disputes', in MH Nordquist, JN Moore and TH Heidar (eds), *Legal and Scientific Aspects of Continental Shelf Limits* (2004) 263 et seq.; and C Johnson and AG Oude Elferink, 'Submissions to the Commission on the Limits of the Continental Shelf in Cases of Unresolved Land and Maritime Disputes: The Significance of Article 76(10) of the Convention on the Law of the Sea', in D

IV. MSR and the Continental Shelf

A fundamental conundrum regarding marine scientific research respecting the legal continental shelf prior to existence of 200 nm zones and as regards the shelf area beyond 200 nm is that in both cases the waters above the legal shelf subject to the exclusive authority of the coastal State are subject to the regime of the high seas and the freedom of marine scientific research.[35] While a coastal State has exclusive jurisdiction over the resources of its adjacent shelf, it does not have jurisdiction over the water column above the shelf. This is further complicated by the reality of the sensitivity of a coastal State respecting MSR undertaken by other States (including their science community) that may be perceived as interfering with or gathering information concerning the principal resource of interest in the continental shelf – hydrocarbons. This latter issue is not unique to the shelf as it bedevils the MSR regime within the EEZ as well.

Two provisions in the 1958 Continental Shelf Convention explicitly dealt with MSR.[36] Together they have been described as constituting 'the first legal regime for marine scientific research'.[37] The ambiguities and apparent conflicts within the 1958 provisions need not be detailed here but notice can be made that in one provision the consent of a coastal State was to be obtained for all research concerning the continental shelf and that such consent was not normally to be withheld,[38]

Freestone, R Barnes and DM Ong (eds), *The Law of the Sea: Progress and Prospects* (2006) 161 et seq.

[35] See Art. 3 1958 Continental Shelf Convention, see note 9; Art. 78(1) UNCLOS, see note 1, and respecting the high seas Art. 87(1)(f) UNCLOS. While MSR is not mentioned as a high seas freedom in the 1958 Geneva Convention on the High Seas (done 29 April 1958, entered into force 3 January 1963) 450 UNTS 11, Art. 2, 'it is commonly accepted that scientific research is one of the "other" freedoms of the high seas recognized by the 1958 GCHS according to general principles of international law, although no specific mention of the freedom is made in Article 2', M Gorina-Ysern, *An International Regime for Marine Scientific Research* (2003) 246.

[36] Art. 5(1) and (8) 1958 Continental Shelf Convention, see note 9; see generally Gorina-Ysern, *ibid.*, 252-273, and 223-227 and AHA Soons, *Marine Scientific Research and the Law of the Sea* (1982) 56-77.

[37] Gorina-Ysern, *ibid.*, 26.

[38] Art. 5(8) 1958 Continental Shelf Convention, see note 9.

and in the other provision exploration on the shelf by a coastal State was not to interfere with MSR.[39]

For the purposes here, the relevant provision of the UNCLOS is Art. 246 which deals with MSR in a coastal State's EEZ and 'on' a State's continental shelf and sets out a requirement for coastal State consent.[40] Within 200 nm, 'in normal circumstances' that coastal State is to grant consent.[41] However, consent may be withheld if, amongst other things, the proposed MSR 'is of direct significance for the exploration and exploitation of natural resources',[42] or 'involves drilling into the continental shelf'.[43] Art. 246(6) deals specifically with the shelf area beyond 200 nm and provides that a coastal State may not withhold consent to MSR except for activities that may take place in specific areas clearly identified by the coastal State as ones where exploration and exploitation of resources take place. As described by one author, Art. 246(6) appears to indicate that MSR is 'little restricted' on the continental shelf beyond 200 nm.[44] During the UNCLOS negotiations this provision generated considerable consternation amongst certain coastal States.[45] It has been noted that the provision was ultimately found to be acceptable 'as long as the coastal State power to designate areas of particular interest to them was not subject to compulsory settlement of disputes [...]'[46] Indeed, Art. 297(2)(a)(i) exempts from compulsory dispute settlement 'the exercise by the coastal State of a right or discretion in accordance with article 246'. More generally, it has been noted that the exemption from

---

39 *Ibid.*, Art. 5(1); see generally Soons, see note 36, 56-57.

40 Art. 246(2) UNCLOS, see note 1; see also Arts 245 and 252 UNCLOS concerning implied consent.

41 *Ibid.*, Art. 246(3).

42 *Ibid.*, Art. 246(5)(a).

43 *Ibid.*, Art. 246(5)(b).

44 A Kirchner, 'The Outer Continental Shelf: Background and Current Developments', in TM Ndiaye and R Wolfrum (eds), *Law of the Sea, Environmental Law and Settlement of Disputes* (2007) 593 et seq. (605).

45 Kirchner, *ibid.*, commented: 'that coastal States with a broad continental shelf were rather unwilling [...] to accept the restriction of their jurisdiction to authorize the conduct of MSR projects on the outer continental shelf. [...] [T]hey are, in practice, very reluctant to incorporate the provisions of article 246(6) [...] in their national legislation' (at 605); see also Gorina-Ysern, see note 35, footnote 118, 313.

46 Gorina-Ysern, see note 35, footnote 118, 313.

compulsory dispute settlement of the exercise by a coastal State of its discretion to withhold consent to MSR 'favors the coastal States over those States [...] conducting marine research'.[47] It has been noted that coastal States may simply not make the distinction between the MSR regime that is to apply inside 200 nm from that which is to apply outside 200 nm.[48]

It is important to recognize that, as written, the MSR regime that applies to the continental shelf beyond 200 nm is favorable to foreign States. Moreover, while it can be surmised that operationally a coastal State may have a favorable position respecting controlling foreign State MSR on its shelf area beyond 200 nm, each coastal State can implement the MSR regime in such a manner that it is favorable to foreign State MSR by exercising its discretion to grant consent freely. Moreover, there is nothing in the UNCLOS that inhibits cooperation amongst States as to how to implement and apply the MSR consent regime in a particular area or for a particular activity.

V. Maritime Boundaries and Disputes

As this contribution is one of 'stage setting', the following overview of maritime boundary agreements and disputes is not detailed.

1. Bilateral Agreements

There are three maritime boundary agreements that deal with shelf areas within the central Arctic Ocean basin. The 1990 United States – Russian Federation Agreement,[49] which also deals with the Bering Sea,

47 Gorina-Ysern, see note 35, footnote 118, 313.

48 Kirchner, see note 44, 605.

49 Agreement between the United States of America and the Soviet Socialist Republics on the Maritime Boundary (done 1 June 1990, provisionally entered into force 15 June 1990) (1990) 29 ILM 941. The Agreement is not yet in force because of opposition within the Russian Federation; see generally AG Oude Elferink, 'Arctic Maritime Delimitations: The Preponderance of Similarities with Other Regions', in AG Oude Elferink and DR Rothwell (eds), *The Law of the Sea and Polar Maritime Delimitation and Jurisdiction* (2001) 179 et seq.

establishes the boundary for the territorial sea and the 200 nm zones of the two States in the Arctic Ocean and indicates, in Art. 2, that the line extends into the Chukchi Sea should the continental margin of both States extend beyond 200 nm. The Agreement is, therefore, a complete boundary for the two States in the Arctic.

In 2006, Denmark (Greenland) and Norway (Svalbard) completed a continental shelf and fisheries zone maritime boundary agreement.[50] The line appears to follow an equidistance line utilizing the coasts of both Greenland and Svalbard. This Agreement supports the view of Norway that Svalbard is capable of generating offshore zones and is to be taken into account in maritime boundary delimitation.

The 1957 Norwegian – Russian Federation Agreement,[51] delineates the maritime boundary between Norway and the Russian Federation for a distance of 24.35 nm within the Varanger Fjord. The exciting new agreement is that between Norway and the Russian Federation entered into in 2010.[52] The Agreement is the result of nearly 40 years of negotiations and creates a single line for the EEZ and shelf area beyond 200 nm through the Barents Sea and into the central Arctic Ocean basin. It is clear that there was compromise on both sides in the reaching of the accord.[53]

It is worth noting that there are a number of maritime boundary agreements in the southern areas adjacent to the central Arctic Ocean basin.

(182-183) and EG Verille, 'United States – Soviet Union', in JI Charney and LM Alexander (eds), *International Maritime Boundaries*, Vol. I (1996) 447-460.

50 Agreement between Norway and Denmark together with the Home Rule Government of Greenland on the Other Hand, concerning the Delimitation of the Continental Shelf and the Fisheries Zones in the Area between Greenland and Svalbard (done 20 February 2006, entered into force 2 June 2006), reprinted in AG Oude Elferink, 'Maritime Delimitation Between Denmark/Greenland and Norway', *Ocean Development and International Law* 38 (2007) 375 et seq. (378-380).

51 Agreement between Norway and the Soviet Union concerning the Sea Frontier in the Varanger Fjord (done 27 February 1957, entered into force 17 March 1958) 312 UNTS 289.

52 Treaty between the Norway and the Russian Federation concerning Maritime Delimitation and Cooperation in the Barents Sea and the Arctic Ocean (done 15 September 2010), reprinted in T Henriksen and G Ulfstein, 'Maritime Delimitation in the Arctic: The Barents Sea Treaty', *Ocean Development and International Law* 42 (2011) 1 et seq. (11-17).

53 See generally Henriksen and Ulfstein, *ibid.*, 1-10.

The Iceland – Norway (Jan Mayen) Agreements are interesting since the two States first agreed on a maritime boundary for fishing zones,[54] then resorted to a conciliation commission to assist respecting a continental shelf boundary,[55] which led to the adoption of the water column line as the bilateral shelf boundary and created a joint development zone for hydrocarbon development.[56] The 1995 Denmark (Greenland) – Norway (Jan Mayen) Continental Shelf and Fisheries Zone Boundary Agreement[57] is the implementation of the 1993 International Court of Justice case between the two States[58] that was initiated by Denmark.[59] In 1997, Denmark (Greenland) and Iceland agreed on a boundary for both the 200 nm fishery zone and continental shelf.[60] This led to agreements on a trijunction point between the three States in the area.[61]

---

[54] SJ Rolston and TL McDorman, 'Maritime Boundary Making in the Arctic Region,' in DM Johnston and PM Saunders (eds), *Ocean Boundary Making: Regional Issues and Developments* (1988) 16 et seq. (33-34).

[55] Report and Recommendations to the Governments of Iceland and Norway of the Conciliation Commission on the Continental Shelf Area between Iceland and Jan Mayen (May 1981) (1981) 20 ILM 797.

[56] Agreement on the Continental Shelf between Iceland and Jan Mayen (Iceland-Norway) (done 22 October 1981, entered into force 2 June 1982) (1982) 21 ILM 1222.

[57] Agreement between Denmark and Norway concerning the Delimitation of the Continental Shelf Area between Jan Mayen and Greenland and Concerning the Boundary between the Fishery Zones in the Area (done 18 December 1995, entered into force 18 December 1985) *Law of the Sea Bulletin* 31 (1996) 59 et seq.

[58] *Case Concerning Maritime Delimitation in the Area between Greenland and Jan Mayen (Denmark v. Norway)* [1993] ICJ Reports 38.

[59] See generally DH Anderson, 'Denmark (Greenland) – Norway (Jan Mayen)', in JI Charney and LM Alexander (eds), *International Maritime Boundaries*, Vol. III (1998) 2507 et seq.

[60] Agreement between Denmark along with the Local Government of Greenland and Iceland on the Delimitation of the Continental Shelf and Fishery Zone in the Area between Greenland and Iceland (done 11 November 1997, entered into force 27 May 1998) *Law of the Sea Bulletin* 39 (1999) 35 et seq.; see generally DH Anderson, 'Denmark (Greenland) – Iceland', in JI Charney and RW Smith (eds), *International Maritime Boundaries*, Vol. IV (2002) 2941 et seq.

[61] Additional Protocol to the Agreement of 28 May 1980 between Iceland and Norway concerning Fishery and Continental Shelf Questions and the Agreement Derived therefrom of 22 October 1981 on the Continental Shelf between Jan Mayen and Iceland (done 11 November 1987, entered into force 27

Canada and Denmark have a continental shelf boundary from Davis Strait to the Lincoln Sea.[62] The boundary terminates in the Robeson Channel before entering the Arctic Ocean and, thus, is incomplete. There is also a small gap in this maritime boundary as a result of the sovereignty dispute over Hans Island.[63] It is reported that the two States may be close to an agreement respecting this somewhat insignificant islet.[64]

In September 2006, Denmark (Faroe Islands), Iceland and Norway completed 'Agreed Minutes' respecting a maritime boundary for those areas of overlapping claims of the continental shelf that may exist beyond the 200 nm zones of each State.[65] The Agreed Minutes establishes the shelf boundary between the three States, subject to each of the States documenting to the Commission that the continental margin adjacent to each of the three States extends to the boundary limit. In the event that one or more of the States does not document that its area of continental shelf beyond 200 nm corresponds with the area determined in the Agreed Minutes, the maritime boundaries in the Minutes are to be adjusted. Unlike the maritime boundary agreements noted above, this tripartite maritime boundary agreement is provisional pending the completion of the Commission process. It does indicate that in this area

May 1998) *Law of the Sea Bulletin* 39 (1999) 38 and Additional Protocol to the Agreement of 18 December 1995 between Norway and Denmark concerning the Delimitation of the Continental Shelf in the Area between Jan Mayen and Greenland and the Boundary between Fishery Zones in the Area (done 11 November 1997, entered into force 27 May 1998) *Law of the Sea Bulletin* 39 (1999) 37.

[62] Agreement between Canada and Denmark relating to the Delimitation of the Continental Shelf between Greenland and Canada (done 17 December 1973, entered into force 13 March 1974) 950 UNTS 147.

[63] See R Huebert, 'The Return of the Vikings: The Canadian – Danish Dispute over Hans Island – New Challenges for the Control of the Canadian North', in F Berkes et al. (eds), *Breaking Ice: Renewable Resources and Ocean Management in the Canadian North* (2005) 319 et seq.

[64] J Ibbitson, 'Canada and Denmark Make Headway in Dispute Over Hans Island', *Toronto Globe and Mail* (27 January 2011) A-5.

[65] Agreed Minutes on the Delimitation of the Continental Shelf beyond 200 Nautical Miles between the Faroe Islands, Iceland and Norway in the Southern Part of the Banana Hole of the Northeast Atlantic (done 20 September 2006), see <http://www.regjeringen.no/nb/dep/ud/dok/lover_regler/retningslinjer/2006/Agreed-Minutes.html?id=446839> (2 August 2011).

of the sub-Arctic the three States have essentially resolved their outer continental shelf boundaries.

2. Disputes

There are only two situations in the central Arctic Ocean basin where there are clear situations of overlapping claim disputes. One is between Canada and the United States in the Beaufort Sea and the other is between Canada and Denmark (Greenland) in the Lincoln Sea. Both involve overlapping 200 nm zone claims. As regards the continental shelf beyond 200 nm in the central Arctic Ocean basin, there are as yet no overlapping claim disputes as none of the States have officially indicated the extent of their shelves beyond 200 nm.

In the Beaufort Sea, defined by the coasts of Alaska and Yukon-Northwest Territories, Canada and the United States has approximately 6250 square nm of overlapping claimed territorial sea and 200 nm zones.[66] Canada has delineated its 200 nm zone in the area using the 141st west meridian,[67] relying on Art. III 1825 Russia-Great Britain Treaty,[68] which provides for a boundary between the two States along the 141st meridian 'in its prolongation as far as the Frozen ocean' ('*dans son prolongement jusqu'à la Mer Glaciale*', the authentic language of the Treaty is French). The US position is that maritime boundary is an equidistance line.[69] Canada and the United States appear to have adopted informal policies of preventing drilling or other hydrocarbon-

---

[66] DH Gray, 'Canada's Unresolved Maritime Boundaries', *Geomatica* 48(2) (1994) 131 et seq. (135).

[67] Canada, Fishing Zones of Canada (Zone 6) Order, *Consolidated Regulations of Canada* (1978), chapter 1549, 13747-13750.

[68] Convention between Great Britain and Russia Concerning the Limits of their Respective Possessions on the North-West Coast of America and the Navigation of the Pacific Ocean (done 16 February 1825), reprinted in C Parry (ed.), *Consolidated Treaty Series*, Vol. 75 (1969) 95 et seq. This Agreement is binding on the United States as a result of the US acquisition of Alaska from Russia in 1867; see Art. 1 Convention Ceding Alaska between Russia and the United States, (done 30 March 1867, entered into force 20 June 1867), reprinted in C Parry (ed.), *Consolidated Treaty Series*, Vol. 134 (1969) 331 et seq.

[69] United States, Department of State, 'Exclusive Economic Zone and Maritime Boundaries,' Public Notice 2237 (23 August 1995) 60 Federal Register 43825 et seq.

related activity from taking place in the disputed area within the Beaufort Sea.[70] It is quite likely that there is an area of continental shelf beyond 200 nm adjacent to the Beaufort Sea that may be subject to overlapping Canada and US claims. It has been noted that beyond 200 nm strict equidistance rather than an extension of the 141st meridian becomes more favorable to Canada and less favorable to the United States.[71] The two States are cooperating in the collection of data and mapping respecting the seafloor area (possible continental shelf) adjacent to the Beaufort Sea.[72] It is reported that Canada and the United States are in the early stages of discussions about the Beaufort Sea.[73]

As already noted, the 1973 Canada – Denmark Continental Shelf Agreement[74] does not extend into the Lincoln Sea. It is reported that both States accept that equidistance should be used to delineate their overlapping territorial sea and 200 nm zone claims in the Lincoln Sea.[75] The area in dispute within 200 nm is described as being a modest 65 square nm split between two areas,[76] which is said to arise because of the differing views of the two States over the base points to be used in determining the equidistance line.[77] It is likely that the shelf area in the Lincoln Sea extends beyond 200 nm from each State. It is reported that Canada and Denmark (Greenland) have been engaged in negotiations on some of the above matters.[78]

---

70 Gray, see note 66, 135 refers to the two States having 'established a moratorium on exploration' in the disputed area.

71 R Boswell, 'Canada, U.S. flip-flop positions in Beaufort Sea boundary dispute', *Canwest News Service (Montreal Gazette)* (8 March 2010).

72 See United States, Department of State, '*U.S.-Canada Joint Expedition to Survey the Extended Continental Shelf in the Arctic*' (26 July 2010), see <http://www.state.gov/r/pa/prs/ps/2010/07/145175.htm> (2 August 2011).

73 R Boswell, 'Work Underway to Resolve Beaufort Boundary Dispute', *Postmedia News* (26 July 2010).

74 Canada – Denmark Continental Shelf Agreement, see note 62.

75 D Pharand, 'Delimitation Problems of Canada (Second Part)', in D Pharand and U Leanza (eds), *The Continental Shelf and the Exclusive Economic Zone: Delimitation and Legal Regime* (1993) 171 et seq. (179).

76 Pharand, *ibid*., 179 and see Gray, see note 66, 138.

77 Gray, see note 66, 138; see also Oude Elferink [2001], see note 49, 194-195.

78 R Boswell, 'Canada, Denmark Start Talks to Resolve Border Dispute', *Canwest News Service* (27 March 2010).

Only Norway has submitted information to the Commission,[79] received recommendations from the Commission[80] and announced that it will proceed with outer continental shelf delineation on the basis of the recommendations of the Commission.[81] The shelf area involved is a small slice beyond 200 nm north of Svalbard.

Russia submitted information to the Commission in 2001, including a now infamous map, indicating their proposed outer limits.[82] The Commission has requested that additional information be provided respecting the continental shelf outer limit proposed by Russia in the Arctic Ocean.[83] Russia has indicated that it will be providing such information.

It is anticipated that Canada will submit information to the Commission on its proposed continental shelf outer limit in the Arctic Ocean in 2013 and Denmark (Greenland) is expected to follow suit in 2014. Both Canada and Denmark (Greenland) have 'unofficial' charts and maps on websites and elsewhere, which provide pictorial guidance of possible shelf areas beyond 200 nm that may be enclosed by proposed outer limits. The United States is not a party to the UNCLOS and thus is under

---

[79] Commission on the Limits of the Continental Shelf, 'Continental Shelf Submission of Norway in respect of Areas in the Arctic Ocean, the Barents Sea and the Norwegian Sea: Executive Summary' (December 2006), see <http://www.un.org/Depts/los/clcs_new/submissions_files/nor06/nor_exec_sum.pdf> (2 August 2011).

[80] Commission on the Limits of the Continental Shelf, 'Summary of the Recommendations of the Commission on the Limits of the Continental Shelf in regard to the Submission Made by Norway in respect of Areas in the Arctic Ocean, the Barents Sea and the Norwegian Sea on 27 November 2006' (27 March 2009), see <http://www.un.org/Depts/los/clcs_new/submissions_files/nor06/nor_rec_summ.pdf> (2 August 2011).

[81] Norway, Ministry of Foreign Affairs, 'Extent of Norway's Continental Shelf in the High North Clarified', Press release No 025/09 (15 April 2009), see <www.regjeringen.no> (2 August 2011).

[82] Russian Federation, Continental Shelf Submission, Executive Summary, attached to Commission on the Limits of the Continental Shelf, 'Receipt of the Submission Made by the Russian Federation to the Commission on the Limits of the Continental Shelf' (20 December 2001) CLCS.01.2001.LOS (Continental Shelf Notification), see <http://www.un.org/Depts/los/clcs_new/submissions_files/submission_rus.htm> (2 August 2011).

[83] See UNGA 'Report of the Secretary-General – Oceans and Law of the Sea: Addendum' (8 October 2002) UN Doc. A/57/57/Add.1, paras 311-313.

no obligation to submit information to the Commission. The United States, however, has indicated its intentions to act in a manner consistent with the wording of Art. 76[84] and, like Canada and Denmark, has made public charts and maps on websites and elsewhere indicating areas of the continental shelf beyond 200 nm in the Arctic Ocean.

The possible dispute between Norway (Svalbard) and Denmark (Greenland) regarding claimed continental shelf areas beyond the 200 nm limits in the Arctic basin is referred to in their 2006 bilateral Agreement[85] and the 2006 Norwegian submission to the Commission. In Norway's submission to the Commission it was indicated that Denmark (Greenland) does not object to the Commission considering the Norwegian proposed outer limit in this area and that a maritime boundary will be negotiated between the two States.[86]

It may be the case that the Russian Federation will have a continental margin area beyond 200 nm in the Arctic Basin that overlaps with shelf areas of both Denmark (Greenland) and Canada. In response to the 2001 Russian submission to the Commission, both Denmark and Canada explicitly noted that the Russian submission and any recommendations by the Commission were 'without prejudice to the delimitation of the continental shelf' between the States.[87] In the case of Denmark (Greenland), the seafloor area in question is most likely to be related to

---

[84] See 'United States Policy Governing the Continental Shelf of the United States of America' (17 November 1987), attachment to a Memorandum from Assistant Secretary of State John D. Negroponte to Deputy Legal Adviser Elizabeth Verille, reproduced in JA Roach and RW Smith, *United States Responses to Excessive Maritime Claims*, 2nd edn (1996) 201-202.

[85] Denmark (Greenland) – Norway (Svalbard) Agreement, see note 50, Preamble, para. 3.

[86] Continental Shelf Submission of Norway, Executive Summary, see note 79, 11 et seq.

[87] Denmark, 'Note to the Secretary-General of the United Nations' (4 February 2002) attached to UN Secretary General, 'Denmark: Notification Regarding the Submission Made by the Russian Federation to the Commission on the Limits of the Continental Shelf' (26 February 2002) CLCS.01.2001.LOS/DNK and Canada, 'Note to the Secretary-General of the United Nations' (18 January 2002) attached to UN Secretary General, 'Canada: Notification Regarding the Submission Made by the Russian Federation to the Commission on the Limits of the Continental Shelf' (26 February 2002) CLCS.01.2001.LOS/CAN, see <http://www.un.org/Depts/los/clcs_new/submissions_files/submission_rus.htm> (2 August 2011).

the Lomonosov Ridge. The seafloor area in question between Canada and Russia may also involve the Lomonosov Ridge as well as the Alpha-Mendelev Ridge. Until Canada and Denmark (Greenland) indicate the extent of their continental shelf in the central Arctic Ocean basin, it is unclear whether or not there exist overlapping claims. As already noted, this should become clearer when, as expected, the two States submit their information to the Commission on their proposed continental shelf outer limits.

VI. Conclusion

The UNCLOS *recognized*, rather than created, the international legal rights of coastal States to continental shelf areas beyond 200 nm based on natural prolongation of their land territory. This is an inherent right *recognized* in, not created by, the UNCLOS. An important legal consequence of this is that the legal authority exercisable by an adjacent coastal State over its legal continental shelf beyond 200 nm is no different than the rights it has respecting the continental shelf area within 200 nm. The UNCLOS does create two additional responsibilities for coastal States with authority over shelf areas beyond 200 nm. One is to submit information on the outer limit of the shelf beyond 200 nm to the Commission. The other is to comply with the revenue sharing obligation in Art. 82 UNCLOS.

The coastal States adjacent to the central Arctic Ocean basin in statements and action are exercising their legal rights over the continental shelf in the same manner as other States in other oceans. It is clear that during the negotiation of the UNCLOS that the seafloor of the Arctic Ocean was understood to be no different than the seafloor of other oceans.

While it is premature to assert with certainty the future legal regime/s that will apply to the seafloor of the central Arctic Ocean basin, there has been speculation about the extent of national jurisdiction on the continental shelf and the areas of floor that may be beyond national jurisdiction and subject to the jurisdiction of the ISA respecting the mineral resources of the deep ocean floor. Estimates made in 2001 indicated that most of the seafloor of the Arctic Ocean Basin would be subject to national jurisdiction, leaving two smallish areas of seafloor outside of national authority.[88] Subsequent revisions made to the 2001 study indicated that there may be four areas of seafloor outside of national authority in the central Arctic Ocean basin.[89] New information arriving regularly keeps adjusting the speculative picture as regards the Arctic Ocean.

[88] R Macnab, P Neto and R van de Poll, 'Cooperative Preparations for Determining the Outer Limit of the Juridical Continental Shelf in the Arctic Ocean: A Model for Regional Collaboration in Other Parts of the World?', *Boundary and Security Bulletin* 9 (2001) 86 et seq.

[89] R Macnab, 'The Outer Limit of the Continental Shelf in the Arctic Ocean', in Nordquist, Moore and Heidar, see note 34, 302, 304-305 and Figures 2 and 5.

Legal Issues of MSR in the Arctic: A Chinese Perspective

by Zhiguo Gao[*]

The Arctic region is the northernmost area on earth, consisting of a vast area of permanently frozen ice floating in the middle of the Arctic Ocean and surrounded by continental land masses and islands. The Arctic is not only a key area in the world climatic system and an essential actor in the physical, chemical and biological balance of the planet, but also an important region for global change research. As the Arctic is extremely vulnerable to projected climate change and since the changes in the atmosphere, ocean, and land of the Arctic will impact on other regions in the world, marine scientific research ('MSR') in the Arctic is essential for understanding and mitigating environmental change in both the Arctic and any other given region of the world.

China's interest in Arctic MSR is of only recent origin. Situated in the northern hemisphere, China is susceptible to Arctic environmental impacts and needs to understand and be prepared for such changes. Over the last decade or so, China has developed a national interest in Arctic MSR. As MSR conducted by China in the Arctic has significantly increased, the legal issues of MSR in the Arctic have increasingly aroused China's attention.

This short paper is presented in four parts. Following a brief introduction on China's MSR activities in the Arctic in the first section, the second section provides an overview of the legal regimes for MSR in the Arctic at both international and national levels. Some research findings

---

* Prof. Dr. Zhiguo Gao, BA., LL.M., LL.M., JSD, Executive Director of China Institute for Marine Affairs ('CIMA') and Judge of the International Tribunal for the Law of the Sea. The author wishes to acknowledge the research assistance provided by Mr. Dan Zhang, Research Fellow at CIMA, in the course of preparing this conference paper.

and policy considerations on the MSR in the Arctic are presented in the third section. The final section wraps up the discussion with some concluding remarks.

I. China's MSR in the Arctic

China is the most populous country in the world, with more than 1.3 billion people at the end of 2009. The land area of China is about 9.6 million square km, and the mainland is flanked to the east and south by the Bohai, Yellow, East China and South China seas, with a total maritime area of 4.73 million square km. In accordance with international law and the United Nations Convention on the Law of the Sea ('UNCLOS'),[1] roughly 3 million square km of those sea areas fall under China's jurisdiction.

As early as the 1920s, China had already established a certain relationship with the Arctic. In 1925, China became a State Party to the Spitsbergen Treaty (also referred to as the 'Svalbard Treaty'), which was originally concluded between Norway, the United States, Denmark, France, Italy, Japan, the Netherlands, Great Britain and Ireland, the British Overseas Dominions and Sweden on 9 February 1920.[2] By acceding to the Spitsbergen Treaty, China acquired such rights as equal rights for nationals and ships with regard to fishing and hunting in the territories specified in the treaty as well as in their territorial waters[3] and with regard to maritime, industrial, mining and commercial operations.[4] However, for many years after acceding to the Spitsbergen Treaty, China did not conduct any scientific research activities, let alone any other operations in the Arctic region.

Although Chinese nationals had began to attempt scientific research in the Arctic in the 1950s, it was not until 1995 that a group of Chinese scientists and journalists travelled to the North Pole on foot and conducted research on the Arctic Ocean's ice cover, climate and environment. This Arctic expedition became a national focus in China. There

---

[1] United Nations Convention on the Law of the Sea (concluded 10 December 1982, entered into force 16 November 1994) 1833 UNTS 397.

[2] Treaty concerning the Archipelago of Spitsbergen (signed 9 February 1920, entered into force 14 August 1925) 2 LNTS 7.

[3] *Ibid.*, Art. 2.

[4] *Ibid.*, Art. 3.

are two major reasons for China to pay more attention to Arctic MSR. First, climatic and environmental processes in the Arctic exert a direct influence on China's climate and environment, which are closely linked to the sustainable development of the country's national economy. Therefore, it is imperative for China to study the climatic and environment changes in the Arctic region. Second, as an ideal place for scientific research in many fields, conduct of MSR in the Arctic will help China develop a credible international scientific capacity.

In 1993, China purchased from Ukraine the MV Xuelong ('Snow Dragon'), built originally as a polar region cargo ship by the Kherson Shipyard, Ukraine. This became the only research icebreaker in China after elementary modifications in 1994, for both the Arctic and Antarctic regions. In the following years, the MV Xuelong has served as an important platform for China's comprehensive research of the Arctic Ocean region.

The first Chinese National Arctic Research Expedition ('CHINARE') was not carried out until 1999, when a group of 124 scientists and staff took part in the 71-day research project aboard the MV Xuelong. In July 2003, the second CHINARE was organized, which lasted for 74 days. In order to better understand the pattern and effect of the rapid changes in the Arctic Ocean and their influence on the climate and environmental changes in China, the country launched its third CHINARE from 11 July 2008 to 24 September 2008, during which multidisciplinary and comprehensive studies on marine science and ice observation were conducted. The surveyed area during this expedition covered the Bering Sea, the Chukchi Sea, the Chukchi Plateau and the Canadian Basin. 112 spots were selected for in situ survey, of which 35 were in the Bering Sea and 77 in the Arctic Ocean.[5] In 2010, China started its fourth CHINARE, and a group of 122 scientists, staff, and journalists took part in the 85-day research project. The fourth CHINARE had a major focus on the impact on the Arctic marine ecosystem of the rapid changes in sea ice, observation of the large-scale sea ice melting and the associated atmospheric and oceanic processes, and the multidisciplinary study of ecological systems in the Bering Sea, Bering Strait, Chukchi

5 Chinese Arctic and Antarctic Administration ('CAA'), Annual Report on Polar Program of China 2008 (2008) 5.

Sea, Beaufort Sea, Canada Basin, the Mendeleev Ridge and other waters.[6]

Since the MV Xuelong alone can not meet China's increasing demand for conducting expeditions simultaneously in the Antarctic and Arctic, China has approved the building of a new high-tech polar expedition research icebreaker. The new polar research icebreaker is expected to be operational in 2013, and will do much to improve China's Arctic MSR capabilities.[7]

In addition to the above CHINAREs carried out by MV Xuelong, China established its first Arctic station in July 2004, named Huanghe ('Yellow River') Station, in Ny-Ålesund, Svalbard. Huanghe Station, designed to accommodate 18 personnel, can be used to conduct research on topics including atmospheric science, oceanography, glaciology and ecology, as well as sea ice research. Since 2004, China has carried out six yearly expeditions. In 2009, there were 12 scientific projects accomplished at the Huanghe Station. The scientific activities were mainly focused on environmental monitoring, climate study, glacier study, bio-ecology study and upper atmospheric physics study.[8]

China has also been active in international cooperation in MSR in the Arctic. In 1996, China has joined the International Arctic Science Committee ('IASC'), a non-governmental organization established in 1990 and composed of international science groups participating in Arctic scientific research. In 2005, China was invited to join the Ny-Ålesund Science Managers Committee, which was established in 1994 to enhance cooperation and coordination between research activities in Ny-Ålesund. In 2006, China began to apply for observer status at the Arctic Council, which was formally established by the Ottawa Declaration of 1996[9] as a high level intergovernmental forum to provide a

---

[6] China Daily, 'China's Fourth Arctic Expedition will Set Off', Press release (26 June 2010), see <http://www.china-daily.org/China-News/China-s-fourth-Arctic-expedition-will-set-off/> (13 October 2011).

[7] China Daily, 'China's New Polar Icebreaker to Launch in 2013', Press release (22 June 2011), see <http://www.chinadaily.com.cn/china/2011-06/22/content_12747950.htm> (13 October 2011).

[8] Chinese Arctic and Antarctic Administration ('CAA'), *National Annual Report on Polar Program of China 2009* (2009) 7.

[9] Arctic Council, 'Declaration on the Establishment of the Arctic Council' (19 September 1996), see <http://arctic-council.org/article/about> (13 October 2011).

means for promoting cooperation, coordination and interaction among the Arctic States, and has attended Arctic Council meetings as an ad hoc observer since 2007. In the same year, a group of Chinese polar experts began to work for the fourth International Polar Year program, a major initiative of the World Meteorological Organization and the International Council for Science on a wide range of scientific research programs in both the Arctic and the Antarctic regions.

With respect to bilateral collaboration, Norway, Canada and the USA are the three countries that have engaged with China in bilateral dialogue on Arctic issues. For instance, a Sino-Norwegian meeting was held in June 2009, a Sino-Canadian Workshop on the Arctic was organized in February 2010, and a Sino-US dialogue on Polar and Law of the Sea Issues was conducted in March 2010. The issues discussed at these meetings included climate change, polar research, Arctic policies, energy resources, sea routes and the law of the sea.

II. Legal Regimes for MSR in the Arctic

At present, there are perhaps two sets of legal regimes applicable to MSR in the Arctic: one is the United Nations Convention on the Law of the Sea, which is 'A Constitution for the Oceans'.[10] The other is the Spitsbergen Treaty of 9 February 1920, which applies in certain areas of the Arctic. Apart from the above-mentioned two multilateral treaties, relevant regulations of the Arctic States are also applicable to Arctic MSR activities.

1. The United Nations Convention on the Law of the Sea

Under the legal regime of UNCLOS, the world's ocean space has been divided into different zones of jurisdiction, inter alia, the territorial sea, the exclusive economic zone ('EEZ'), the continental shelf (and extended continental shelf), the high seas and the international deep seabed (the Area). Although other parts of UNCLOS also contain provisions relating to MSR, Part XIII UNCLOS is entirely devoted to addressing the subject of MSR, and attempts to balance the rights of

---

[10] TB Koh, 'A Constitution for the Oceans', in United Nations, *The Law of the Sea: United Nations Convention on the Law of the Sea* (1983) xxxiii.

coastal States and researching States by operating a zonal approach to rights in connection with MSR. The closer to the shore, the greater are the rights of the coastal State.[11]

The term 'MSR' is not defined in UNCLOS, despite the fact that a number of proposals were made calling for a clear definition in the Convention during the negotiation process.[12] However, the rules contained in Part XIII UNCLOS obviously apply to all ocean spaces, including the territorial sea, EEZ, the continental shelf and any extended continental shelf which may exist or be established in the Arctic.

All States, irrespective of their geographical location, and competent international organizations have the right to conduct MSR subject to the rights and duties of other States as provided for in the Convention.[13] States and competent international organizations shall promote and facilitate the development and conduct of MSR in accordance with UNCLOS.[14] A number of general principles regarding the conduct of MSR are set out by the Convention. MSR shall: (a) be conducted exclusively for peaceful purposes; (b) use appropriate scientific methods and means compatible with this Convention; (c) not unjustifiably interfere with other legitimate uses of the sea; (d) comply with all relevant regulations adopted in conformity with the Convention including those for the protection and preservation of the marine environment.[15]

In the territorial sea, MSR may only be conducted with the express consent of and under the conditions set forth by the coastal State.[16]

In the exclusive economic zone, coastal States have the right to regulate, authorize and conduct MSR. Research in the EEZ may only be con-

---

[11] KN Scott, 'Marine Scientific Research in the Southern Ocean: Balancing Rights and Obligations in a Security Related Context', Paper presented at the Conference *Responding to Contemporary Challenges and Threats to Antarctica: Legal and Policy Challenges*, University of Canterbury, Christchurch, New Zealand (5-7 July 2009) 7.

[12] United Nations, Division for Ocean Affairs and the Law of the Sea ('DOALOS'), Office of the Legal Affairs, *Marine Scientific Research: A Revised Guide to the Implementation of the Relevant Provisions of the United Nations Convention on the Law of the Sea* (2010) 4, para. 7.

[13] Art. 238 UNCLOS, see note 1.

[14] *Ibid.*, Art. 239.

[15] *Ibid.*, Art. 240.

[16] *Ibid.*, Art. 245.

ducted with the consent of the coastal State.[17] However, under normal circumstances, coastal States are supposed to grant their consent to MSR to be carried out exclusively for peaceful purposes in their exclusive economic zone in accordance with the Convention and in order to increase scientific knowledge of the marine environment for the benefit of all mankind.[18] A coastal State may withhold its consent to the conduct of an MSR project by another State or competent international organization in its exclusive economic zone where the proposed research relates to natural resources development or is likely to impact negatively on the environment.[19] On the continental shelf, coastal States have the same right to regulate, authorize and conduct MSR as in their EEZ.[20]

As a final note on MSR in relation to the continental shelf, it should be pointed out that, in accordance with the spirit of UNCLOS and perhaps as a matter of common sense, the control introduced by coastal States over MSR on their extended continental shelf should generally be less stringent than that they imposed on research activities carried out on their normal continental shelf.

Art. 143 UNCLOS emphasizes that MSR in the Area shall be carried out exclusively for peaceful purposes and for the benefit of mankind as a whole. Both the Authority and States Parties may carry out MSR, and shall promote the conduct of MSR and international cooperation in the Area.

While Art. 257 provides that all States and competent international organizations have the right to conduct MSR in the water column beyond the limits of the exclusive economic zone, freedom of scientific research is expressly referred to in Art. 87 UNCLOS as a freedom of the high seas. Under Art. 87, the freedom of scientific research is subject to Part VI on the continental shelf and Part XIII on MSR. This acknowledges the fact that the continental shelf, over which a coastal State has sovereign rights, may extend beyond 200 nm from the baselines. Freedom of scientific research is to be exercised with due regard to the interests of other States in their exercise of the freedoms of the high seas, and also with due regard to the rights related to activities in the Area.

---

17 *Ibid.*, Art. 246(1) and (2).

18 *Ibid.*, Art. 246(3).

19 *Ibid.*, Art. 246(5).

20 *Ibid.*, Art. 246(6).

The freedom envisioned in Art. 87 is not limited to MSR but also extends to such activities as hydrographic surveys.[21]

2. The Spitsbergen Treaty

Under the Spitsbergen Treaty the Contracting Parties undertake to recognize the full and absolute sovereignty of Norway; such recognition is 'subject to the stipulations of the present Treaty';[22] which means that Norway in return for the recognition of its sovereignty has to accord certain rights to the other parties to the Treaty.[23] The Treaty provides for the full and absolute sovereignty of Norway over the Archipelago of Svalbard, comprising all the islands situated between 10° and 35° longitude East of Greenwich and between 74° and 81° latitude North, especially West Spitsbergen, North-East Land, Barents Island, Edge Island, Wiche Islands, Hope Island or Hopen-Eiland, and Prince Charles Forland, together with all islands and rocks appertaining thereto. However, as mentioned above, such recognition is 'subject to the stipulations of the present Treaty.'[24]

The core elements of the Spitsbergen Treaty are the recognition of Norway's full and absolute sovereignty as well as the rights accorded to other States Parties. Such rights include equal rights of nationals and ships with regard to fishing and hunting in the territories specified in the treaty as well as in their territorial waters,[25] and with regard to equal liberty of access and entry to the waters, fjords and ports of the territories specified in the treaty, as well as equal rights for carrying out maritime, industrial, mining and commercial operations.[26]

It is not sufficiently clear whether this right of 'equal liberty of access and entry to' includes access for MSR or not. One view is that Norway may unilaterally impose any regulations on scientific research unless and until a convention stipulating conditions for research is concluded.

---

[21] United Nations, Division for Ocean Affairs and the Law of the Sea ('DOALOS'), Office of the Legal Affairs, see note 12, 16, para. 56.

[22] Art. 1 Treaty concerning the Archipelago of Spitsbergen, see note 2.

[23] *Ibid.*, Art. 2.

[24] *Ibid.*, Art. 1.

[25] *Ibid.*, Art. 2.

[26] *Ibid.*, Art. 3.

The other view is that Norway is prohibited by the Spitsbergen Treaty from unilaterally imposing conditions on scientific research. The reality appears to lie in between these two contacting views.[27] To date, eleven institutions from ten countries have established research stations at Ny-Ålesund, Svalbard, three of which are permanently manned.

The two different legal instruments have different areas of geographical application. While the MSR rules of UNCLOS obviously apply to all ocean spaces, the relevant provisions of the Spitsbergen Treaty are restricted to certain land and sea territories. Moreover, the research rights provided in the Spitsbergen Treaty are much wider than the MSR stipulated in UNCLOS.

3. National Regulations

All Arctic States are parties to UNCLOS, except for the United States. Each of the five Arctic Ocean littoral States now follows some variation of the rules set out in Part XIII UNCLOS. They all require scientists to obtain coastal States' permission before conducting MSR in the territorial sea, exclusive economic zone or on the continental shelf.[28]

Canada implements an MSR clearance procedure requiring permission for foreign vessel entry into Canadian ports and for the conduct of MSR activities in Canadian zones of maritime jurisdiction. Canada also exercises prescriptive and enforcement jurisdiction to prevent pollution and environmental damage by foreign vessels, including icebreakers conducting MSR activities in its Arctic waters.[29]

The instructions issued by the Government of Denmark require applications for permission to conduct MSR to be made through diplomatic

---

[27] Y Takei, 'Polar Complications in the Law of the Sea: A Case Study of the Regime for Research and Survey Activities in the Arctic Ocean', see <http://www.gmat.unsw.edu.au/ablos/ABLOS10Folder/S3P2-P.pdf> (13 October 2011) 7.

[28] B Baker and H Eicken, 'Marine Research Access in the Arctic Ocean: Background for Potential Guidelines in a Changing Arctic', unpublished White Paper (10 March 2010), see <http://www.iarc.uaf.edu/workshops/2009/4/> (8 June 2011) (click on 'download whitepaper') 2; the paper is also attached as an Appendix to the contribution of B Baker, in this volume.

[29] A Anand, *Marine Scientific Research Governance in the Arctic Ocean* (2008) 91.

channels at least 30 days before the start of the research cruise. Sometimes permission for research in the territorial sea is conditional on participation of a Danish observer on board.[30]

The Norwegian MSR regulations apply to foreign non-military ships and international organizations, and regulate the conduct of MSR not related to natural resources. Consent is required six months in advance from Norway's Department of Fisheries for conducting MSR in its internal waters, the territorial sea, EEZ and on the continental shelf.[31]

Russia exercises jurisdiction over the conduct of MSR activities by foreign researchers in maritime areas under its jurisdiction. Clearance for foreign MSR vessels must be submitted through official channels at least six months prior to the intended activities.[32]

The USA does not claim jurisdiction over MSR in its EEZ. However, MSR activities relating to its EEZ or continental shelf resources are regulated. MSR in the territorial sea is subject to prior consent.[33]

Although the above surveyed national regulations do not apply solely to the Arctic Ocean, MSR in the sea areas of the Arctic Ocean under the jurisdiction of these coastal States are certainly subject to these national regulations.

4. The Antarctic Treaty System

Apart from the regimes contained in UNCLOS, the Spitsbergen Treaty and the various national regulations, a fourth regime that is relevant to the issue of MSR in the Arctic is the so-called Antarctic Treaty System ('ATS'). The Antarctic Treaty System refers to the Antarctic Treaty,[34] its associated separate international instruments in force and the measures in effect under those instruments.[35] The 1959 Antarctic Treaty established a legal regime for the area, including all ice shelves and the sur-

---

30 *Ibid.*, 91.

31 *Ibid.*, 92.

32 *Ibid.*

33 *Ibid.*, 92-93.

34 Antarctic Treaty (signed 1 December 1959, entered into force 23 June 1961) 402 UNTS 71.

35 Art. 1(e) Protocol on Environmental Protection to the Antarctic Treaty (done 4 October 1991, entered into force 14 January 1998) (1991) 30 ILM 1455.

rounding waters, south of 60° south latitude. The promotion of scientific research for peaceful purposes is a fundamental aim of the Antarctic Treaty. Under the Antarctic Treaty System, freedom of scientific investigations in Antarctica and cooperation toward that end shall continue.[36] The Contracting Parties are under an obligation to exchange information regarding plans for scientific programs, exchange scientific personnel between expeditions and stations, and to publish scientific observations and results.[37]

The 1991 Protocol on Environmental Protection to the Antarctic Treaty[38] provides for more detailed provisions with regard to scientific research within the Antarctic Treaty Area. Planning and permits are the two most important systems of control introduced in the Protocol. Activities, including MSR, in the Antarctic Treaty area shall be planned and conducted so as to limit adverse impacts on the Antarctic environment and dependent and associated ecosystems.[39] In some circumstances, special permits are required to conduct MSR. For example, the 1991 Environmental Protocol stipulates that no species of animal or plant not native to the Antarctic Treaty area shall be introduced onto land or ice shelves, or into water in the Antarctic Treaty area except in accordance with a permit.[40] Moreover, when MSR is likely to have more than a minor or transitory impact, an initial environmental evaluation or a comprehensive environmental evaluation shall be prepared.[41]

III. Some Findings and Recommendations

1. Comparison of the UNCLOS and ATS Regimes

In contrast with the broad provisions of UNCLOS on MSR, the MSR regime set forth in the Antarctic Treaty System is more sophisticated and highly regulated. The ATS regime contains not only the most effec-

---

36 Art. 2 Antarctic Treaty, see note 34.

37 *Ibid.*, Art. 3.

38 Protocol on Environmental Protection to the Antarctic Treaty, see note 35.

39 *Ibid.*, Art. 3(2)(a).

40 *Ibid.*, Art. 4(1) Annex II.

41 *Ibid.*, Arts 2 and 3 Annex I.

tive regulation of MSR to date, but also represents the current best MSR practice. The potential codification and application of the norms and principles of the ATS in the Arctic region should be further explored.

2. Correlation between UNCLOS and the Spitsbergen Treaty

The UNCLOS and the Spitsbergen Treaty regimes are not compatible. Differences exist between the national regulations on MSR. Views differ on the relationship between UNCLOS and the Spitsbergen Treaty. One commentator points out that 'Needless to say, there is no such thing in the UNCLOS as a provision which has "expressly permitted" a derogation from UNCLOS' own provisions in order to apply a system such as that of the 1920 Spitsbergen Treaty in the EEZ or on the shelf.'[42] But the complexity of the issue is perhaps not as simple as derogation from UNCLOS provisions. The Spitsbergen Treaty is a multilateral legal instrument in force. More importantly, it provides for the full and absolute sovereignty of Norway in the Archipelago of Svalbard. So UNCLOS perhaps cannot override automatically the Spitsbergen Treaty. The relationship between these two treaties and the validity of the latter with respect to MSR need to be studied and examined more carefully.

3. Potential for More Restricted MSR Access

Once all the five Arctic States have established their EEZ, only 1.2 million square km (465,000 square miles) of unclaimed ocean space will be left in the Arctic. Provisional analysis also suggests that the extended continental shelves of the Arctic States could encompass most of the central Arctic Ocean basin, leaving only two smallish areas of the sea floor outside of national jurisdiction, subject to the common heritage of mankind regime. With the larger part of the Arctic marine area falling under national jurisdiction, only a small portion of the Arctic Ocean would remain open to all States for MSR under the high seas freedom of UNCLOS.

---

[42] CA Fleischer, 'The New International Law of the Sea and Svalbard', Paper presented at the Norwegian Academy of Science, 150th Anniversary Symposium (25 January 2007) 11.

Moreover, since MSR is likely to produce some or a transitory impact on the marine environment, it is customary for the Arctic coastal States to withhold consent to MSR in their EEZ or continental shelf. In fact, some coastal States have already enacted more stringent regulations for ships operating in the Arctic Ocean. Unless research vessels are excluded from the scope of such laws and regulations, they would also be applicable to these vessels.[43]

4. Areas of Common Interest

It is widely acknowledged that the Arctic States enjoy their sovereignty, sovereign rights and jurisdiction in their territorial sea, EEZ, and continental shelf in the Arctic Ocean. However, other States are entitled to conduct certain activities in the Arctic, in accordance with the relevant provisions of UNCLOS.

The issues of the Arctic arouse the attention of many non-Arctic States as well as the international community. For instance, China, the Republic of Korea, Italy, Japan and the European Union ('EU') applied for permanent observer status of the Arctic Council in 2009, but no decision has yet been taken. The EU, China and other countries may share a legitimate and common interest in and concern about issues such as peace and security in the Arctic, environmental protection and sustainable development, obtaining permanent observer status of the Arctic Council, and access to freedom of MSR in the Arctic.

IV. China's Arctic Policy Considerations

China's research activities in the Arctic are only of recent origin. Although the country has not yet elaborated any kind of national Arctic Strategy, China needs to develop an overarching Arctic policy to guide its interests and activities in the Arctic. Such an emerging policy may include the following major pillars: MSR, environmental protection, commercial and strategic interests, and international cooperation.

China's polar research programs need international cooperation, with both the Arctic and non-Arctic States, for the purposes of strengthening dialogue, enhancing scientific understanding, promoting informa-

[43] Takei, see note 27, 6.

tion exchange, and expanding and deepening collaboration. It is felt that bilateral dialogue could be undertaken with the EU as well as with Germany. This could be done at either the official level between the competent agencies, or the academic level between research institutions.

V. Concluding Remarks

Changes in the Arctic have a great impact on the living environment and conditions for all mankind. Thus MSR in the Arctic region is essential for understanding and dealing with the issue of climatic change and its associated environmental and development problems. In addition, the Arctic is a unique and ideal place for conducting scientific research in many fields. In this respect, access to MSR in the Arctic Ocean is important to the Arctic States as well as non-Arctic States.

UNCLOS sets out for the first time an extensive set of provisions for MSR. The principle of 'common heritage of mankind' in general and freedom of research in particular, as enshrined in the Convention, should be upheld and maintained in the Arctic Ocean. Despite the broad acceptance of the principles of MSR in UNCLOS, there is a need for uniform and consistent application of these provisions by the littoral States. The relevant provisions of both UNCLOS and the Antarctic Treaty System may serve as a sound basis for the further development of a legal framework for MSR in the Arctic Ocean.

It can be anticipated that international interest in scientific research in the Arctic will continue to grow. The future objective and task in this regard is to determine how to strike a proper balance between the rights of the Arctic coastal States on the one hand and the needs of non-Arctic States and the international community on the other hand.

'Common Interests' as an Evolving Body of International Law: Applications to Arctic Ocean Stewardship

by Paul Arthur Berkman[*]

I. Beyond National Interests

The 20th century transformed our civilization into a global society with transboundary perspectives and responsibilities on a planetary scale. While the impacts of human progress began extending around the world well beforehand, as with agricultural transformations of landscapes across continents, it was the two 'world' wars that inexorably introduced humankind to the interconnectedness among all nations and peoples on earth (Fig. 1).

---

* Bren School of Environmental Science & Management, University of California, Santa Barbara, California, United States.

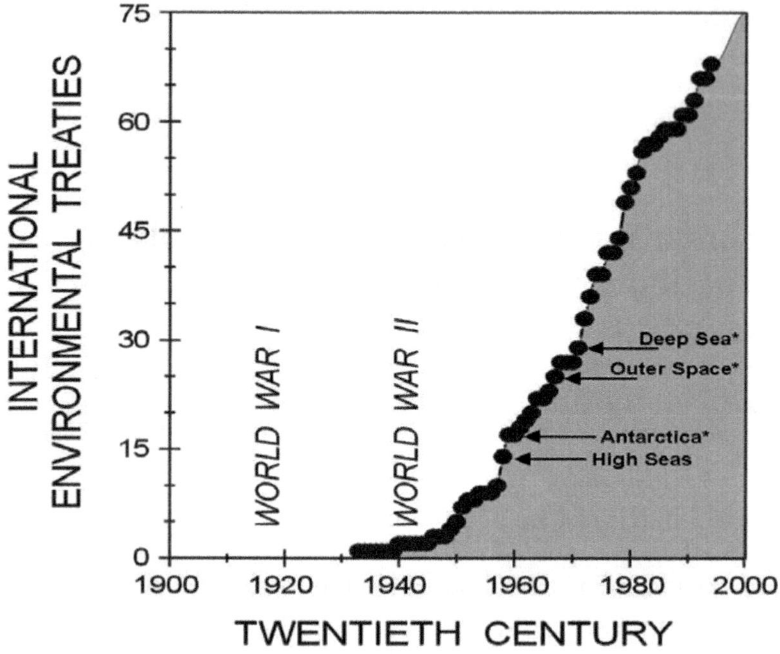

Fig. 1. Emergence of our global society during the 20th century.[1] In stark contrast to the 'world' wars during the first half of the 20th century – nearly 95% of the multilateral ecosystem and environmental regimes in force (dots) have been signed after 1950. Institutions to establish 'international spaces' beyond sovereign jurisdictions (arrows) further highlight the development of common interests among all humankind, including regions that shall be used peaceful purposes without nuclear weapons (asterisks). Remote sensing and digital technologies along with accelerating transport of commodities across the earth also are contributing to our global interconnectedness

Following World War II – with motivation to forever prevent such horrors from returning to our world – the United Nations emerged in 1946 'to maintain international peace and security' and 'to be a center for harmonizing the actions of nations in the attainment of these common ends.'[2] This concept of 'common' has been embodied in legal[3] and so-

[1] Figure adapted from PA Berkman, *Environmental Security in the Arctic Ocean: Promoting Cooperation and Preventing Conflict* (2010).

[2] Art. 1 Charter of the United Nations (adopted 26 June 1945, entered into force 24 October 1945) 145 BSP 805.

[3] TFT Pluknett, *A Concise History of the Common Law*, 5th edn (2001).

cietal[4] systems for centuries, however, it has been only during the past few decades when common strategies for all humankind have begun emerging to resolve impacts and issues that have global relevance (Table 1).

4 G Hardin, ‘Tragedy of the Commons’, *Science* 162 (1986) 1243 et seq.

TABLE 1: Representative International Institutions that are Relevant to All Humankind with Explicit References to Common Interests

Agreement Name	Done At	Entry into Force	Common Interests
Charter of the United Nations[5]	San Francisco 26.6.1945	24.10.1945	*'common interest'* *'common ends'*
Antarctic Treaty[6]	Washington, DC 1.12.1959	23.6.1961	*'matters of comon interest'*
Treaty on Principles Governing the Activities of States in the Exploration and Use of Outer Space, Including the Moon and other Celestial Bodies[7]	London, Moscow Washington, DC 27.1.1967	10.10.1967	*'common interest of all mankind'*
Treaty on the Prohibition of the Emplacement of Nuclear Weapons and Other Weapons of Mass Destruction on the Seabed and the Ocean Floor and in the Subsoil[8]	London, Moscow Washington, DC 11.2.1971	18.5.1972	*'common interest of mankind'*
United Nations Convention on the Law of the Sea[9]	Montego Bay 10.12.1982	16.11.1994	*'common heritage of mankind'*
United Nations Framework Convention on Climate Change[10]	New York 9.5.1992	21.3.1994	*'common concern of humankind'*
Convention on Biological Diversity[11]	Rio de Janeiro 5.6.1992	29.12.1993	*'common concern of humankind'*

---

5 Charter of the United Nations, see note 2.

6 Antarctic Treaty (signed 1 December 1959, entered into force 23 June 1961) 402 UNTS 71.

7 Treaty on Principles Governing the Activities of States in the Exploration and Use of Outer Space, Including the Moon and other Celestial Bodies (signed 27 January 1967, entered into force 10 October 1967) 610 UNTS 205.

8 Treaty on the Prohibition of the Emplacement of Nuclear Weapons and Other Weapons of Mass Destruction on the Sea-Bed and the Ocean Floor and in the Subsoil Thereof (concluded 11 February 1971, entered into force 18 May 1972) 955 UNTS 115.

9 United Nations Convention on the Law of the Sea (concluded 10 December 1982, entered into force 16 November 1994) 1833 UNTS 397.

10 United Nations Framework Convention on Climate Change (with Annexes) (adopted 9 May 1992, entered into force 21 March 1994) 1771 UNTS 107.

11 Convention on Biological Diversity (concluded 5 June 1992, entered into force 29 December 1993) 1760 UNTS 79.

The keystone of common progress for all humankind is our natural world with environments and ecosystems crossing as well as extending beyond national boundaries. Such global progress is reflected by the geometric growth of multilateral environmental treaties and conventions since World War II (Fig. 1), providing solutions both to promote cooperation and to prevent conflict – two sides of the coin of peace – in our world.

Special among the international regimes are the institutions that have emerged for areas beyond sovereign jurisdictions, across nearly 70% of the earth in the high seas and deep sea as well as Antarctica. These 'international spaces'[12] reflect a jurisdictional transition in our civilization with inclusion of all humankind in stark contrast to the sovereign principles that have isolated societies throughout history within the 30% of the earth that has been defined by national boundaries. Looking beyond the earth, international spaces also include outer space with its celestial bodies.

Forever after we will have a jurisdictional dichotomy with nations and international spaces reflecting national interests and common interests, respectively. The universal common interest for all humankind is peace.

II. Applying Common Interests

Identification, elaboration and implementation of common interests involve processes with ongoing dialogues among relevant stakeholders. At the level of international spaces, the common-interest process is exemplified by the continuing dialogue initiated by Art. IX 1959 Antarctic Treaty[13] for the:

> purpose of exchanging information, consulting together on matters of common interest pertaining to Antarctica, and formulating and considering, and recommending to their Governments, measures in furtherance of the principles and objectives of the Treaty [...].

---

[12] J Kish, *The Law of International Spaces* (1971); PA Berkman, 'President Eisenhower, the Antarctic Treaty, and the Origin of International Spaces', in PA Berkman et al. (eds), *Science Diplomacy: Antarctica, Science and the Governance of International Spaces* (2011) 17 et seq.

[13] Antarctic Treaty, see note 6.

In the Arctic, which has completely different circumstances than the Antarctic,[14] discussions about common interests have been ongoing through the high-level forum of the Arctic Council, even though it is 'without legal personality'.[15] With its establishment through the 1996 Ottawa Declaration,[16] the Arctic Council began to:

> provide a means for promoting cooperation, coordination and interaction among the Arctic States, with the involvement of the Arctic indigenous communities and other Arctic inhabitants on common Arctic issues, in particular issues of sustainable development and environmental protection in the Arctic (Art. 1(a), footnote omitted).

Motivated by common interests, the United Nations World Commission on Environment and Development expressed vision for 'Our Common Future'[17] with sustainable development as the integrating framework:

> Sustainable development is not a fixed state of harmony, but rather a process of change in which the exploitation of resources, the direction of investments, the orientation of technological development, and institutional change are made consistent with future as well as present needs (Overview, para. 30).

Fundamentally, 'common Arctic issues' reflect an understanding of shared interests among the Arctic States and indigenous peoples as well as with other stakeholders who will influence sustainable development in the region with particular emphasis in the Arctic Ocean (Fig. 2). The diversity of nations who have expressed interests in the Arctic Ocean is reflected in Table 2, including representative States from all of the continents with indigenous human populations.

---

[14] See Table 6, in PA Berkman, *Environmental Security in the Arctic Ocean: Promoting Cooperation and Preventing Conflict* (2010).

[15] ET Bloom, 'Establishment of the Arctic Council', *American Journal of International Law* 93 (1999) 712 et seq. (714).

[16] Arctic Council, 'Declaration on the Establishment of the Arctic Council' (19 September 1996), see <http://arctic-council.org/article/about> (26 July 2011).

[17] UN World Commission on Environment and Development, 'Our Common Future' (4 August 1987) UN Doc A/42/427, Annex (Brundtland Report).

Fig. 2. Concentric levels of responsibilities to manage human activities in the Arctic Ocean. Eight States north of the Arctic Circle (including those with and without coastlines in the Arctic Ocean) and six indigenous peoples organizations signed the 1996 Ottawa Declaration that established the Arctic Council. Additional involvement of non-Arctic States in Arctic organizations and engagement of non-State actors, especially industry, reflects the interplay of global civil society in the Arctic Ocean[18]

For the Arctic Ocean as well as elsewhere on earth – the sustainability challenge is to balance economic prosperity, environmental protection and social equity in a manner that offers hope and inspiration for generations to come. However, the 'common Arctic issues' are incomplete. Explicit use of the term 'peace' as a common interest in the 1996 Ottawa Declaration was specifically excluded, even though the concept of 'peaceful uses of the Arctic' had been considered as a key purpose for the Arctic Council since shortly after the Cold War.[19]

---

[18] Figure adapted from PA Berkman, *Environmental Security in the Arctic Ocean: Promoting Cooperation and Preventing Conflict* (2010).

[19] Arctic Council Panel, 'To Establish and International Arctic Council: A Framework Report' (1991), see <http://www.carc.org/pubs/v19no2/2.htm> (26 July 2011); D Pharand, 'The Case for an Arctic Region Council and a Treaty Proposal', *Revue générale de droit* 23 (1992) 163 et seq.

TABLE 2: INTERNATIONAL PARTICIPATION IN ARCTIC ORGANIZATIONS

	ARCTIC ORGANIZATION[2]													
	AC	AMEC	BEAC	FARO	IASC	NACG	NC	NEAF[3]	NF	OSPA[3]	PB	SAR	SCPA	SPIT
Afghanistan														X
Albania														X
Argentina														X
Australia														X
Austria														X
Belgium [4]						X				X				X
Bulgaria [4]								X						X
Canada	X			X	X	X			X		X	X	X	X
Chile														X
China	X			X	X				X					X
Cuba								X						
Czech Republic [4]														X
Denmark [4,5]	X		X	X	X	X	X	X		X	X	X	X	X
Dominican Republic														X
Egypt														X
Estonia [4]						X								X
Finland [4]	X		X	X	X	X	X		X	X		X	X	X
France [4]	X			X	X	X				X				X
Germany [4]	X			X	X	X				X				X

TABLE 2: INTERNATIONAL PARTICIPATION IN ARCTIC ORGANIZATIONS

	ARCTIC ORGANIZATION[2]													
	AC	AMEC	BEAC	FARO	IASC	NACG	NC	NEAF[3]	NF	OSPA[3]	PB	SAR	SCPA	SPIT
Greece [4]														X
Hungary [4]														X
Iceland	X		X	X	X	X	X	X	X	X		X	X	X
India														X
Ireland [4]						X				X				X
Italy [4]	X			X	X									X
Japan				X	X				X					X
Korea (South)	X			X	X				X					
Latvia [4]						X								
Lithuania						X								
Luxembourg [4]										X				
Monaco														X
Netherlands [4]	X			X	X	X				X				X
New Zealand														X
Norway	X	X	X	X	X	X	X	X		X	X	X	X	X
Poland [4]	X			X		X								X
Portugal [4]						X		X						X
Romania														X
Russian Federation	X	X	X	X	X	X		X	X			X	X	X
Saudi Arabia														X

TABLE 2: INTERNATIONAL PARTICIPATION IN ARCTIC ORGANIZATIONS														
	ARCTIC ORGANIZATION[2]													
	AC	AMEC	BEAC	FARO	IASC	NACG	NC	NEAF[3]	NF	OSPA[3]	PB	SAR	SCPA	SPIT
Serbia														X
South Africa														X
Spain [4]	X					X		X		X				X
Sweden [4]	X		X	X	X	X	X			X	X	X	X	X
Switzerland										X				X
United Kingdom [4]	X	X		X	X	X				X				X
United States	X	X		X	X	X			X		X	X	X	X
Venezuela														X
Number of Nations	17	4	6	17	16	20	5	8	8	15	5	8	8	42

1 Highlighted nations are the eight Arctic states.

2 **AC** (1996 Arctic Council); **AMEC** (1996 Arctic Military Environmental Cooperation Programme); **BEAC** (1993 Barents Euro-Arctic Council); **FARO** (1998 Forum of Arctic Research Operators); **IASC** (1990 International Arctic Science Committee); **NACG** (2007 North Atlantic Coast Guard Forum); **NC** (1952 Nordic Council); **NEA** (1980 *Convention on Future Multilateral Cooperation in North-East Atlantic Fisheries*); **NF** (1991 Northern Forum); **OSPA** (1992 *Convention for the Protection of the Marine Environment of the North-East Atlantic*); **PB** (1973 *Agreement on the Conservation of Polar Bears*); **SAR** (2011 *Agreement on Cooperation on Aeronautical and Maritime Search and Rescue in the Arctic*); **SCPA** (1994 Standing Committee of the Conference of Parliamentarians of the Arctic Region); **SPIT** (1920 *Treaty Concerning the Archipelago of Spitsbergen, and Protocol*)

3 Includes European Economic Community

4 Member of the European Union

5 Includes Greenland and the Faroe Islands as autonomous areas.

Exclusion of peace as a common interest in the Arctic remains unchanged as reflected in the 2008 Ilulissat Declaration[20] from five of the six Arctic coastal States (Russian Federation, Norway, Denmark, Canada and United States without Iceland), in which they promoted their 'stewardship role' (para. 5) while emphasizing their 'sovereignty, sovereign rights and jurisdiction in large areas of the Arctic Ocean' (para. 3). Arguably, stewardship is in the interest of all humankind and the absence of peaceful purposes as a common interest among the Arctic coastal States raises questions about achieving sustainable development in the Arctic Ocean.

It is not to say that the Arctic States are interested in conflict or that peace is excluded in all declarations. The 2009 Tromsø Declaration of the Arctic Council[21] confirms 'that in international relations the rule of law is a prerequisite for peaceful regional development' (Preamble). Similarly, the 2011 Nuuk Declaration of the Arctic Council[22] recognizes 'the importance of maintaining peace, stability and constructive cooperation in the Arctic' (Preamble). However, peace and stability in the Arctic Ocean have yet to be established explicitly as a 'common Arctic issue' among all Arctic States and peoples, opening a new era of trust to promote cooperation as well as to prevent conflict in the High North for the benefit of all.

III. Arctic Ocean Stewardship

Stewardship in the Arctic Ocean involves balancing interests, rights and responsibilities across as well as beyond sovereign jurisdictions. An entrée to this discussion actually was facilitated by the 2008 Ilulissat Declaration, in which the Arctic coastal States indicated that they 'remain committed' to the legal framework of the law of the sea and 'see no

---

20 Ilulissat Declaration (28 May 2008) (2009) 48 ILM 382.

21 Arctic Council, Tromsø Declaration on the Occasion of the Sixth Ministerial Meeting of The Arctic Council, Tromsø, Norway (29 April 2009), see <http://arctic-council.org/filearchive/Tromsoe%20Declaration-1..pdf> (26 July 2011).

22 Arctic Council, Nuuk Declaration on the Occasion of the Seventh Ministerial Meeting of the Arctic Council, Nuuk, Greenland (12 May 2011), see <http://www.arctic-council.org/filearchive/Nuuk%20Declaration%20FINAL.pdf> (26 July 2011).

need to develop a new comprehensive international legal regime to govern the Arctic Ocean' (paras 4 and 5). Law of the sea is taken to be the 1982 United Nations Convention on the Law of the Sea ('UNCLOS'), even though the United States has yet to ratify this global agreement that is in force for all other Arctic States and more than 155 nations (Fig. 3). The lasting value of UNCLOS is its inclusive and equitable framework regarding the rights and responsibilities of all nations in all ocean areas on earth, including the Arctic Ocean.

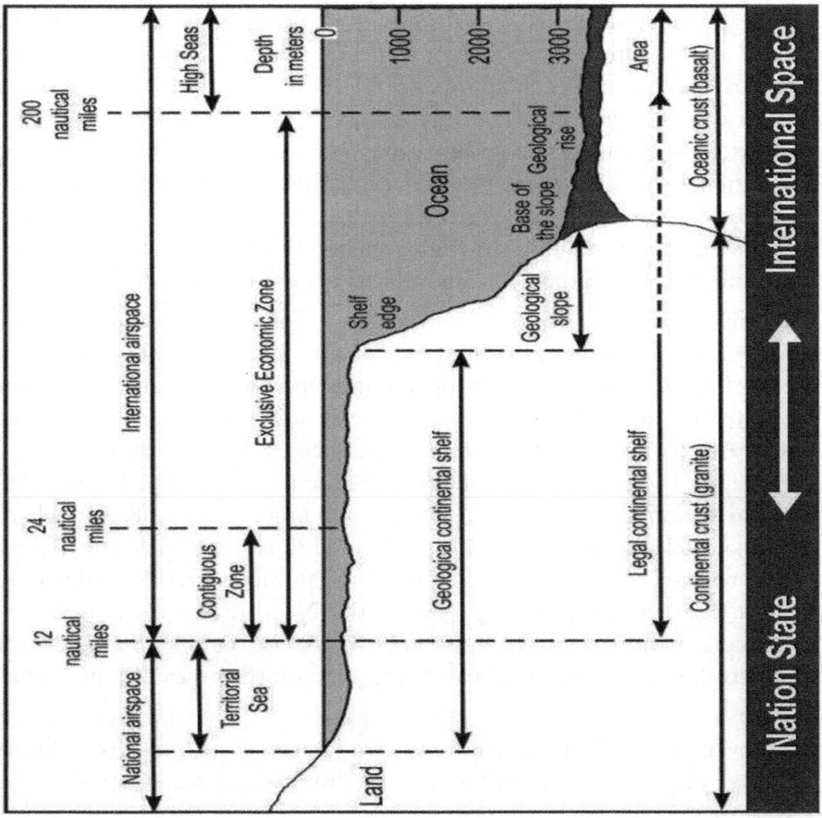

Fig. 3. Sea zones defined by the 1982 United Nations Convention on the Law of the Sea, progressing across a jurisdictional continuum from the coastal boundaries of nation States into the international spaces of the high seas and deep sea. UNCLOS promotes 'the peaceful uses of the seas and oceans, the equitable and efficient utilization of their resources, the conservation of their living resources, and the study, protection and preservation of the marine environment [...]' (Preamble).[23]

[23] Adapted, with addition of the 'Area', from United States Department of State, *Third Conference on the United Nations Convention on the Law of the Sea* (1985).

National policies further reveal individual commitments of the Arctic coastal States to the law of the sea. For example, the United States[24] and Russian Federation[25] each adopted policies in 2009 regarding their sovereignty, sovereign rights and jurisdiction over the territorial sea, contiguous zone, exclusive economic zone and continental shelf emanating from their respective coastal boundaries. However, these national policies are without reference to the international sea zones beyond sovereign jurisdictions, namely the high seas beyond the exclusive economic zone and the area of the deep sea beyond the continental shelf.

The high seas generally overlies the Area (Fig. 2). In some locations, however, the high seas also may overlie the continental shelf, as in the Arctic Ocean. Indeed, following recommendations in 2009 from the Commission on the Limits of the Continental Shelf ('CLCS') instituted under Art. 76 UNCLOS Norway[26] has been acknowledged to have a continental shelf that extends beyond 200 nm in the Arctic Ocean. Additional submissions to the CLCS are pending from the Russian Federation (2001), Denmark (2009) and Iceland (2009) in the Arctic Ocean. Such proposals and activities like the 2007 public-private expedition to plant a Russian flag on the seafloor at the North Pole clearly are demonstrating national interests among the Arctic coastal States as the world watches the Arctic Ocean being sliced into pieces of a geopolitical pie.

Moreover, national interests among Arctic as well as non-Arctic States are escalating in the Arctic Ocean as the diminishing sea ice awakens vi-

---

[24] United States National Security Presidential Directive and Homeland Security Presidential Directive NSPD 66/HSPD 25 'Arctic Region Policy' (9 January 2009) (2009) 48 ILM 274; see also <http://www.fas.org/irp/offdocs/nspd/nspd-66.htm> (27 July 2011).

[25] 'Osnovi gosudarstvennoi politiki Rossiyskoi Federatsii v Arktike na period do 2020 goda i dalneishuju perspektivu' ('The Fundamentals of the State Policy of the Russian Federation in the Arctic in the Period Up to 2020 and Beyond') Utverzhdeni Presidentom Rossiyskoi Federatsii (adopted by the President of Russia) (18 September 2008) promulgated on 30 March 2009 in the 'Rossiyskaya Gazetta', see also <http://www.scrf.gov.ru/documents/98.html> (27 July 2011).

[26] See Commission on the Limits of the Continental Shelf, 'Summary of the Recommendations of the Commission on the Limits of the Continental Shelf in regard to the Submission Made by Norway in respect of Areas in the Arctic Ocean, the Barents Sea and the Norwegian Sea on 27 November 2006' (27 March 2009), see <http://www.un.org/Depts/los/clcs_new/submissions_files/nor06/nor_rec_summ.pdf> (27 July 2011).

sions of new oil and gas, mineral, trade and fishery opportunities. Historically, the Arctic Ocean has been dominated by multi-year ice that accretes over many years. This has changed. Rather than projecting out to the mid-21st century when the Arctic Ocean may be open water during the summer[27] – viewing the sea ice largely in terms of its minimum lateral extent – we can now see that the sea ice composition already has crossed a threshold with more than 50% newly forming each year (Fig. 4). In effect, we already have a new Arctic Ocean.

[27] MM Holland, CM Bitz and B Tremblay, 'Future Abrupt Reductions in the Summer Arctic Sea Ice', *Geophysical Research Letters* 33 (2006) L23503.

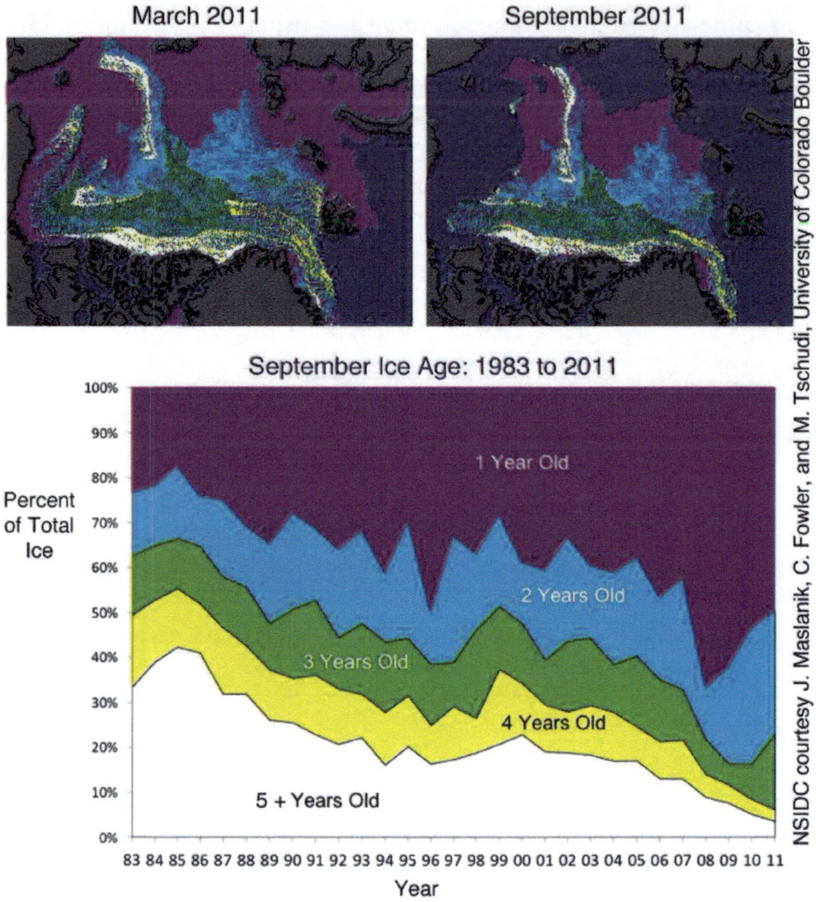

Fig. 4. Arctic sea ice age composition during the 2011 winter (left upper) and summer (right upper) with historical changes in summer sea ice age composition from 1983 to 2011 (below).[28] It is noteworthy that winter sea ice along the Russian Coast, from the Bering Strait to the Barents Sea, is now less than 1 year old. In contrast, most of the multi-year sea ice in the Arctic Ocean remains adjacent to the North American coast

The Arctic Ocean is undergoing the largest environmental state-change on earth, transforming the North Pole from a sea ice cap that has persisted for millennia to an annual sea ice system. It is like a room with the floor, ceiling and walls with inflow and outflow through the doors.

---

[28] From the National Snow and Ice Data Center, see <http://nsidc.org> (8 October 2011).

If the ceiling were removed, the dynamics of the entire room system would be altered as we are exposed to the outside weather. Like removing the ceiling to a room – the fundamental shift in the sea-surface boundary of the Arctic Ocean will create a new natural system with different dynamics than anything previously experienced by humans in the region. Risks of political, economic and cultural instabilities are inherent consequences of the environmental state-change.

For example, as recognized during the middle of the winter in March 2010, there is first year sea ice from the Bering Strait to the Barents Sea (Fig. 4), providing a passage for ice-strengthened vessels (versus icebreakers) to transit across the Arctic Ocean during the winter as well as the summer. While Arctic shipping remains mostly destinational at the moment, opening Arctic Ocean trade routes with year-round transpolar shipping could emerge quickly (i.e., within this decade) given the necessary economic justifications based on price differentials of commodities between Asian and European markets.[29]

To consider common interests in the Arctic Ocean, one useful way forward is to draw a clear distinction between the sea floor and the overlying water column. Ecologically and legally distinct, the sea floor and overlying water column reveal alternative jurisdictional configurations for Arctic and non-Arctic nations alike to respond to the inherent risks that are emerging with the environmental state-change in the Arctic Ocean.

In the Arctic Ocean, the high seas is delimited without dispute in the central Arctic beyond the exclusive economic zones, accepted by all States either as parties of UNCLOS or through customary international law. Moreover, even if the sea floor all the way to the North Pole would be defined as continental shelf, there would still be high seas in the center of the Arctic Ocean as an unambiguous and perpetual international space. The jurisdictional parallel between the static sea floor and dynamic water column is in terms of national interests and common interests (Fig. 5).

---

[29] T Nilsen, 'The Future History of the Arctic is Now', *Barents Observer* (2 September 2010), see <http://www.barentsobserver.com/the-future-history-of-the-arctic-is-now.4814307-116320.html> (29 July 2011).

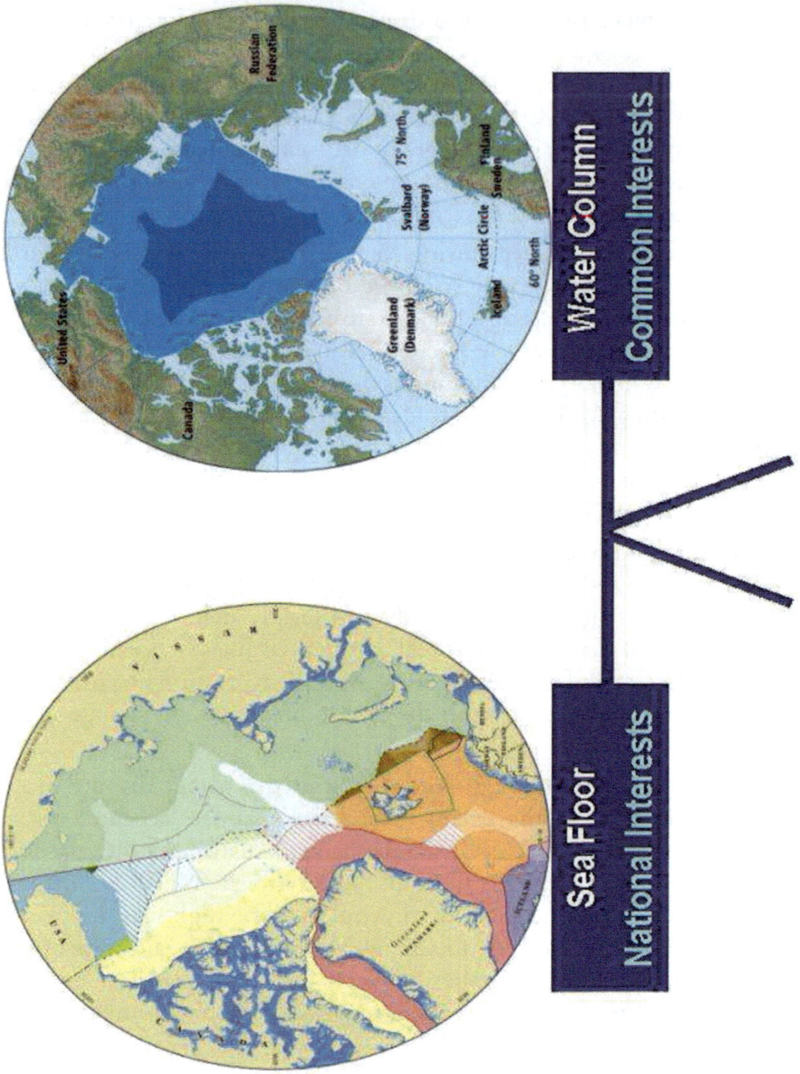

Fig. 5. Balancing jurisdictions in the Arctic Ocean, as provided by the law of the sea. Left: National interests in the seafloor with sovereign areas and outer continental shelf claims (different colors). Right: Common interests in the overlying water column with the high seas (dark blue) as an unequivocal international space in the central Arctic Ocean surrounded by exclusive economic zones (light blue).[30]

---

30 Modified from: PA Berkman and OR Young, 'Governance and Environmental Change in the Arctic Ocean', *Science* 324 (2009) 339 et seq.

Arctic Ocean stewardship requires balanced perspectives. The coastal States have central rights and responsibilities in the Arctic Ocean from their jurisdictions toward the North Pole (Fig. 2). Without contravening the 'sovereignty, sovereign rights and jurisdiction' of the coastal States – the international community has rights and responsibilities in the Arctic Ocean in the international space of the high seas (Fig. 5), outward from the North Pole toward the coastal periphery. The dichotomy of rights and responsibilities in the Arctic Ocean, as established by the law of the sea (Fig. 3), underscores the challenge of balancing national interests and common interests to achieve stewardship in this globally relevant region.

The missing ingredient is statesmanship by the leaders of nations who are the only individuals that can establish the political will to both promote cooperation and prevent conflict in the Arctic Ocean for the lasting benefit of all. Such statesmanship appears rarely, bringing capacity to put out the brush fires of the moment and the vision to offer hope for future generations. In this 'age of the Arctic' – in the midst of spiraling international urgencies globally – 'matters relating to the Arctic and the High North must be addressed at the highest political level', as noted by Hans Corell (former Legal Counsel of the United Nations) in his speech on *Common Concern for the Arctic* in September 2008 in Ilulissat, Greenland.[31]

Common interests are an evolving area of international law with global relevance to regions, resources and impacts that extend across as well as beyond sovereign jurisdictions. As a starting point for the Arctic, sustainable development and environmental protection already have been agreed as 'common Arctic issues' by the Arctic States and indigenous peoples organizations in the 1996 Ottawa Declaration that established the Arctic Council. With the environmental state-change in the Arctic Ocean and more than 40 nations involved in various Arctic institutions (Table 2) – further consideration of common interests has now become a key to establish integrated adaptation and mitigation strategies that will resolve increasingly urgent issues in the Arctic Ocean for the benefit of all.

---

[31] H Corell, 'Chairman's Conclusions', in Nordic Council of Ministers (ed.), *Common Concern for the Arctic: Conference arranged by the Nordic Council of Ministers 9-10 September 2008, Ilulissat, Greenland* (2008) 15 et seq. (31).

The International Seabed Authority and the Arctic

by Michael W. Lodge[*]

I. Introduction

I have been asked to speak on the role of the International Seabed Authority ('Authority') with respect to the Arctic Ocean. This is often posed as a sensitive political question, usually in the context of a media narrative which speaks of a scramble by the Arctic States to assert extensive national claims in order to preclude rich seafloor mineral and hydrocarbon resources coming within the jurisdiction of the Authority. Alternatively, a comparison is drawn with Antarctica to make the suggestion that there needs to be a comprehensive international treaty for the protection of the Arctic environment.

I hope to show that the true position bears little resemblance to these scenarios. Apart from its extreme climatic conditions, and vulnerability to environmental stresses, the Arctic Ocean is no different to any other of the world's oceans.[1] Present indications are that there are parts of the Arctic Ocean basin which lie beyond national jurisdiction and therefore form part of the Area, as defined in the 1982 United Nations Convention on Law of the Sea ('UNCLOS').[2] It is universally accepted that the

---

[*] Legal Counsel, International Seabed Authority. The views expressed in this paper are those of the author and do not necessarily reflect those of the Authority.

[1] JN Moore, 'The UNCLOS Negotiations on Ice-Covered Areas', in MH Nordquist, JN Moore and TH Heidar (eds), *Changes in the Arctic Environment and the Law of the Sea*(2010) 17 et seq.

[2] United Nations Convention on the Law of the Sea (concluded 10 December 1982, entered into force 16 November 1994) 1833 UNTS 397.

Arctic is subject to the same legal regime as any other part of ocean space, namely the UNCLOS. The regime set out in Part XI UNCLOS and the 1994 Agreement Relating to the Implementation of Part XI of the Convention ('1994 Agreement')[3] is therefore applicable to these areas. A number of consequences flow from this, which are identified and discussed in this paper.

The paper begins by recalling the purposes for which the International Seabed Authority was established and its powers and functions under the UNCLOS and the 1994 Agreement. The existence of areas beyond national jurisdiction in the Arctic is identified. The paper then goes on to describe the consequences which flow from the applicability of the UNCLOS regime to the Arctic Ocean as a whole in three particular areas, namely the conduct of 'activities in the Area', the implementation of Art. 82 UNCLOS with respect to the outer continental shelf, and the conduct of marine scientific research in the Area.

The paper concludes by reflecting on the importance attached to international cooperation in the context of the UNCLOS, and suggests that opportunities exist for strengthening cooperation between States and competent international organizations, including the Authority, in the Arctic Ocean. Some preliminary suggestions are offered as to the modalities for such cooperation.

II. The International Seabed Authority

The International Seabed Authority is the institution created by Part XI UNCLOS and the 1994 Agreement to administer the regime for the Area, which is defined in Art. 1(1) UNCLOS as 'the seabed and ocean floor and subsoil thereof, beyond the limits of national jurisdiction'. The regime for the Area is based on the Declaration of Principles Governing the Seabed and the Ocean Floor and the Subsoil Thereof, beyond the Limits of National Jurisdiction ('Declaration of Principles') adopted by the General Assembly of the United Nations in 1970.[4] The

---

[3] Agreement Relating to the Implementation of Part XI of the United Nations Convention on the Law of the Sea of 10 December 1982 (done 28 July 1994, entered into force 28 July 1996) 1836 UNTS 41.

[4] UNGA Res 2749 (XXV) 'Declaration of Principles Governing the Seabed and the Ocean Floor, and the Subsoil Thereof, beyond the Limits of National Jurisdiction' (17 December 1970) GAOR 25th Session Supp 28, 24.

Declaration of Principles establishes the international seabed area and its resources as the common heritage of mankind, to be administered by an international machinery and in accordance with an international regime to be established by a treaty. Part XI UNCLOS was designed to give effect to the Declaration of Principles by elaborating a comprehensive regime for the deep seabed and establishing the International Seabed Authority as the organization through which States Parties to the UNCLOS are to organize and control 'activities in the Area', particularly with a view to administering the resources of the Area.

Under Art. 133 UNCLOS, the term 'resources' in this context is limited to 'solid, liquid and gaseous mineral resources *in situ* in the Area at or beneath the seabed.' Known mineral resources to occur in the Area to date include polymetallic (manganese) nodules, seafloor massive sulphides, cobalt-rich ferromanganese crusts and gas hydrates. The term 'activities in the Area' is also defined in Art. 1(3) UNCLOS as 'all activities of exploration for, and exploitation of, the resources of the Area.'[5] The use of these terms makes it clear that, despite the sweeping language used in the Declaration of Principles, the Authority is an international organization with precisely defined and limited powers and

[5] In its recent Advisory Opinion on the Responsibilities and Obligations of States Sponsoring Persons and Entities with Respect to Activities in the Area, the Seabed Disputes Chamber of the International Tribunal for the Law of the Sea gave important guidance as to what activities are included in the notion of 'activities in the Area' as referred to in Art 139, Annex III, Art. 4(4), Art. 145 and Annex III, Art. 17(2)(f), and Annex IV, Art. 1(1) UNCLOS. In particular, the Chamber considered that the phrase 'activities in the Area' includes the recovery of minerals from the seabed and their lifting to the water surface, as well as activities directly connected to such operations, but excludes the processes through which metals are extracted from the minerals recovered from the seabed and transportation to points on land from the part of the high seas superjacent to the part of the Area in which the contractor operates. On the other hand, transportation within that part of the high seas, when directly connected with extraction and lifting, should be included in activities in the Area. This would include, for example, transportation between the ship or installation where the lifting process ends and another ship or installation where the evacuation of water and the preliminary separation and disposal of material to be discarded take place (at paras 94 et seq.); ITLOS, Seabed Disputes Chamber, *Responsibilities and Obligations of States Sponsoring Persons and Entities with Respect to Activities in the Area (Advisory Opinion),* ITLOS Case No. 17 (1 February 2011), see also <http://www.itlos.org/case_documents/2011/document_en_379.pdf> (13 June 2011).

functions. As a creature of the UNCLOS, these powers and functions are to be found exclusively in the UNCLOS and the 1994 Agreement.

Thus, in addition to managing, in the sense of licensing and permitting, exploration for and exploitation of deep seabed minerals, the UNCLOS and the 1994 Agreement also confer certain specific tasks and responsibilities on the Authority. These include the conduct of marine scientific research in the Area (Art. 143 UNCLOS), the acquisition and transfer of technology and scientific knowledge relating to activities in the Area (Arts 144 and 273 UNCLOS) and the distribution of revenues from the continental shelf beyond 200 nm from the baseline from which the territorial sea is measured (Art. 82(4) UNCLOS).

The Arctic Ocean and its waters are, like any other ocean, subject to the regime of the UNCLOS and the 1994 Agreement.[6] This implies, amongst other things, that any areas of the seabed beyond national jurisdiction in the Arctic Ocean would be considered part of the Area and subject to the legal regime set out in Part XI UNCLOS and the 1994 Agreement, as well as the detailed rules, regulations and procedures of the Authority on such matters as mineral exploration.

How much of the seabed of the Arctic Ocean lies beyond national jurisdiction remains to be seen. All indications are that the geomorphological characteristics of the seabed are such that the majority of it will form part of the continental shelves of five littoral States. However, only two of the Arctic States have so far made their submissions to the Commission on the Limits of the Continental Shelf ('CLCS') in relation to the delineation of the continental shelf beyond 200 nm. The Russian Federation made its submission in 2001, whilst Norway made

---

6 P Taksøe-Jensen, 'An International Governance Framework for the Arctic: Challenges for International Public Law', ZaöRV 69 (2009) 625 et seq. The pre-eminence of the UNCLOS as the applicable legal framework for the management of the Arctic Ocean was also recognized in the Ilulissat Declaration and the Tromsø Declaration; Ilulissat Declaration (28 May 2008) (2009) 48 ILM 382; Arctic Council, Tromsø Declaration on the Occasion of the Sixth Ministerial Meeting of The Arctic Council, Tromsø, Norway (29 April 2009), see <http://arctic-council.org/filearchive/Tromsoe%20Declaration-1.pdf> (15 June 2011). Even the United States, which is not a State Party to the UNCLOS, has agreed with this position, as evidenced by its support for the Ilulissat Declaration; see A Proelss and T Müller, 'The Legal Regime of the Arctic Ocean', ZaöRV 68 (2008) 651 et seq.

its submission in 2006.[7] The CLCS issued recommendations in respect of the Norwegian submission in 2009.[8] With regard to the Russian submission, the CLCS issued recommendations in 2002,[9] which included a recommendation that the Russian Federation prepare a revised submission in respect of the Central Arctic Ocean based on the findings contained in the recommendations of the CLCS. No revised submission has yet been made by the Russian Federation.

With regard to the other three Arctic States, Canada and Denmark (in respect of Greenland) have until 2013 and 2014, respectively, to make their submissions to the CLCS. The United States of America is not yet a party to the UNCLOS, but in due course will have the prescribed period of ten years from the date of its ratification to make a submission.

It is clear, therefore, that the process of delineating the outer limits of the continental shelf in the Arctic Ocean will be a lengthy one, even without taking into account the need for delimitations between States. Estimates of where the outer limits will lie vary between authors and the method of construction used, but all published sources seem to agree that there will be at least two separate parts of the Arctic Ocean beyond national jurisdiction which, therefore, will form part of the Area.[10] It is worth restating, to avoid doubt, that, although the outer

---

7 Commission on the Limits of the Continental Shelf, 'Receipt of the Submission Made by Norway to the Commission on the Limits of the Continental Shelf' (27 November 2006) CLCS.07.2006.LOS (Continental Shelf Notification); Commission on the Limits of the Continental Shelf, 'Receipt of the Submission Made by the Russian Federation to the Commission on the Limits of the Continental Shelf' (20 December 2001) CLCS.01.2001.LOS (Continental Shelf Notification).

8 Commission on the Limits of the Continental Shelf, 'Summary of the Recommendations of the Commission on the Limits of the Continental Shelf in regard to the Submission Made by Norway in respect of Areas in the Arctic Ocean, the Barents Sea and the Norwegian Sea on 27 November 2006' (27 March 2009), see <http://www.un.org/Depts/los/clcs_new/submissions_files/nor06/nor_rec_summ.pdf> (15 June 2011).

9 For a short summary of the CLCS recommendation concerning Russia, see UNGA, 'Oceans and the Law of the Sea: Report of the Secretary General' (8 October 2002) UN Doc A/57/57/Add.1, paras 38-41.

10 See, for example, the stereographic polar projection of the Arctic Ocean, illustrating areas within 200 nm of the baselines and areas currently submitted by coastal States in respect of continental shelves beyond 200 nm, prepared by the UNCLOS Group at the National Oceanography Centre, Southampton, United Kingdom, see<http://www.unclosuk.org/downloads/PMH_9June_3.

limit of the continental shelf, established in accordance with Art. 76 UNCLOS, effectively determines the limits of the Area and hence the scope of application of Part XI UNCLOS,[11] the Authority has no role to play in the process of determining the limits of the continental shelf. Those limits are established by the coastal State and, provided they are established on the basis of the recommendations of the CLCS, shall be final and binding (Art. 76(8) UNCLOS). The only function of the Authority in this regard is to receive the charts or lists of geographical coordinates showing the outer limit lines of the continental shelf and the lines of delimitation drawn in accordance with Art. 83 UNCLOS drawn up by coastal States pursuant to Art. 84(2) UNCLOS.

Whatever the size of the Area in the Arctic Ocean, the fact that such areas exist and that the Arctic Ocean as a whole is subject to the legal regime set out in the UNCLOS and the 1994 Agreement means that certain consequences necessarily follow.

III. Conduct of 'Activities in the Area' in the Arctic Ocean

The first consequence is that any activities of mineral exploration and exploitation in the Area in the Arctic must be conducted in accordance with the regime set out in Part XI UNCLOS and the 1994 Agreement. In other words, pursuant to Art. 140 UNCLOS, such activities must be conducted for the benefit of mankind as a whole and, under Art. 153(3) UNCLOS, in accordance with a formal written plan of work approved by the Council of the Authority after review by the Legal and Technical Commission and drawn up in the form of a contract. Given the particular geographical configuration of the Arctic Ocean, it is likely that any such activities in the Area would also need to be conducted in accordance with Art. 142 UNCLOS, which requires that due regard be paid to the rights and legitimate interests of any coastal State across whose jurisdiction resource deposits in the Area lie and provides a system for consultations and prior consent in such cases.

---

pdf> (13 June 2011) and an alternative construction showing maritime boundaries and jurisdictions in the Arctic produced by the International Boundaries Research Unit, Durham, United Kingdom, see <http://www.dur.ac.uk/resources/ibru/arctic.pdf> (13 June 2011).

[11] MH Nordquist et al. (eds), *United Nations Convention on the Law of the Sea 1982: A Commentary*, Vol. VI (2002) 69.

Whether such activities are likely to occur in the Area is a different question. The nature and distribution of mineral resources, including hydrocarbons, in the Arctic Ocean basin are not well known or understood. Preliminary indications are that it is perhaps unlikely that exploitable mineral resources exist in areas beyond national jurisdiction, but it is difficult to be definitive. A July 2008 fact sheet from the United States Geological Survey indicated that '[t]he extensive Arctic continental shelves may constitute the geographically largest unexplored prospective area for petroleum remaining on Earth,' estimating that 'approximately 84 per cent of the undiscovered oil and gas [in the area above the Arctic circle] occurs offshore.'[12] According to other, more conservative, commentators, however, 'most of the offshore areas with the highest probability for the discovery of hydrocarbons (oil and natural gas) are well within the national jurisdiction of Arctic Ocean littoral States and [...] the areas beyond 200M in the Arctic Ocean basis are not seen as having a high or even middling probability for the recovery of hydrocarbon resources.'[13]

Whichever scenario is accurate, the important point of principle is that, as far as deep seabed mining beyond national jurisdiction is concerned, the regime set out in Part XI UNCLOS and the 1994 Agreement is applicable to the Area.

IV. Application of Art. 82 UNCLOS to the Outer Continental Shelf in the Arctic Ocean

The second, and perhaps more immediate, consequence of the application of the UNCLOS regime to the Arctic Ocean is that the provisions of Art. 82 UNCLOS would apply to the exploitation of the non-living resources of the outer continental shelves of Arctic States. Art. 82 stipulates an obligation on States Parties to the UNCLOS to make payments or contributions in kind with respect to the exploitation of non-living

---

12 USGS Arctic Oil and Gas Report, 'Estimates of Undiscovered Oil and Gas North of the Arctic Circle' (July 2008), see <http://geology.com/usgs/arctic-oil-and-gas-report.shtml> (16June 2011).

13 TL McDorman, 'The Outer Continental Shelf in the Arctic Ocean: Legal Framework and Recent Developments', in D Vidas (ed.), *Law, Technology and Science for Oceans in Globalisation: IUU Fishing, Oil Pollution, Bioprospecting, Outer Continental Shelf* (2010) 499 et seq. (500).

resources on the outer continental shelf. This provision was negotiated into the UNCLOS during the Third United Nations Conference on the Law of the Sea ('UNCLOS III') as a *quid pro quo* for the coastal State's right in Art. 76 to define the outer limit of its continental shelf to encompass continental margin areas outside the 200nm limit.

Pursuant to Art. 82 UNCLOS, payments and contributions are to be made annually at the rate of one percent on the value or volume of all production, commencing on the sixth year of production, and increasing by one percent per year until the rate reaches seven percent on the twelfth year, and thereafter remaining at seven percent. The Authority has an important role to play in the implementation of Art. 82. Under Art. 82(4), the payments or contributions are made by the coastal State through the Authority, which is then tasked with distributing the payments and contributions to States Parties in accordance with equitable criteria, taking into account the 'interests and needs of developing States, and in particular the least developed and land-locked States, and peoples who have not yet achieved full independence or other self-governing status.'[14]

Art. 82 UNCLOS has not been implemented to date, although a small number of coastal States have already granted prospecting and exploration licenses or leases on their outer continental shelves. Responsibility for the implementation of Art. 82 rests jointly with States that exploit the non-living resources of their outer continental shelves and, with respect to the distribution of payments and contributions from such exploitation, with the Authority. In this respect, the UNCLOS provides powers and functions to the Authority's Assembly and Council to enable it to perform its responsibilities. The Council, for example, is tasked with recommending to the Assembly the rules, regulations and procedures on the equitable sharing of financial and other economic benefits made by virtue of Art. 82, taking into account the interest and needs of developing States and peoples who have not attained full independence.[15]

In preliminary studies commissioned by the Authority on the issues associated with the implementation of Art. 82 UNCLOS, it has been noted that Art. 82 has several textual ambiguities and gaps that raise

---

[14] International Seabed Authority, *Issues Associated with the Implementation of Article 82 of the United Nations Convention on the Law of the Sea*, ISA Technical Study No. 4 (2009) 15.

[15] Art. 162(2)(o)(i) UNCLOS.

questions requiring clarification. There is a need for interpretation of explicit stipulations and for inferences to be drawn of implicit requirements to facilitate practical implementation. In this regard, a group of eminent legal experts meeting at Chatham House in London in 2009 noted that both outer continental shelf States and the Authority will have several issues to resolve in order to discharge their respective implementation responsibilities.[16] Given the long temporal scope of offshore non-living resource exploration and exploitation and the expectations set out in Art. 82, it is advisable for these States and the Authority to approach future implementation demands in a phased manner. Potential or anticipated phases of the implementation tasks could include: a pre-production period covering the stage of prospecting, exploration and development of licenses or leases, but before commencement of commercial production; a grace period covering the first five years of royalty-free production, during which the Authority would be putting procedures in place for receiving payments and contributions in kind and for their distribution; and the royalty period when payments and contributions in kind would be expected to commence and the Authority would distribute the benefits to beneficiaries on an ongoing basis.

One suggestion that was made at the Chatham House meeting, which may be of particular interest in the context of the Arctic Ocean, was that, given the likely long-term relationship between producing States and the Authority, it would be advisable for a producing State and the Authority to enter into a special agreement for the purposes of implementing Art. 82. It was suggested that the Authority should consider developing a model agreement for this purpose.[17]

A particular concern for the implementation of Art. 82 UNCLOS was the hypothetical scenario of non-living resources straddling the limits of the outer continental shelf, or different coastal State jurisdictions.[18] In the case of a unitized development of this nature, the cooperating States concerned would need to agree on the respective shares of pro-

---

16 See C Pascal and M Lodge, 'A Fair Deal on Seabed Wealth: The Promise and Pitfalls of Article 82 on the Outer Continental Shelf', Chatham House Briefing Paper (February 2009), see <http://www.chathamhouse.org.uk/files/13323_bp0209seabed.pdf> (16 June 2011).

17 *Ibid.*, 7.

18 International Seabed Authority, *Non-Living Resources of the Continental Shelf Beyond 200 Nautical Miles: Speculations on the Implementation of Article 82 of the United Nations Convention on the Law of the Sea*, ISA Technical Study No. 5 (2010) 39.

duction, which would then serve as a basis for computing the payments and contributions in kind under Art. 82. An even more complex situation would be where the resource straddles the Area where the producing State, as well as the contractor concerned, would need to cooperate with the Authority in the implementation of Art. 82.

The future implementation of Art. 82 is an issue which has been incorporated into the Authority's work program for the period 2011 to 2013. In particular, the Assembly has requested the Secretariat to convene a workshop of experts from Member States of the Authority to consider practical options for implementation of Art. 82 for consideration by the Authority's Legal and Technical Commission. Such options may include the development of model agreements as described above, the development of implementation guidelines to assist the governments of States potentially affected by Art. 82 and the development of a preliminary framework for the equitable distribution of payments and contributions.

In view of the geographic configuration of the Arctic Ocean, with the likelihood of substantial overlaps between the continental shelf jurisdictions of the five Arctic Ocean littoral States, it is suggested that the future implementation of Art. 82 is an area where close cooperation between the Arctic States and the Authority would be beneficial.

V. Conduct of Marine Scientific Research in the Area

The third general consequence of the existence of parts of the Area in the Arctic Ocean is that consideration must be given to the implementation of the provisions of the UNCLOS relating to marine scientific research in the Area. There is a long history of international cooperation in scientific research, including marine scientific research, in the Arctic region, much of which predates the UNCLOS. Examples include the 1920 Spitsbergen Treaty,[19] the 1973 Agreement on the Conservation of Polar Bears[20] and the Comité Arctique.[21] Now it is under-

---

[19] Treaty Concerning the Archipelago of Spitsbergen (concluded 9 February 1920, entered into force 14 August 1925) 2 LNTS 7.

[20] Agreement on the Conservation of Polar Bears (concluded 15 November 1973, entered into force1 November 1976) (1974) 13 ILM 13.

stood and generally accepted that the provisions of the UNCLOS, in particular Part XIII, establish the basic legal parameters for the conduct of marine scientific research and that these provisions apply to the Arctic Ocean, as they do to all other parts of the world's oceans. In recent years, given the complex processes occurring in the Arctic region as a result of climate change, the freedom of marine scientific research, in accordance with the UNCLOS, is an issue that has been highlighted as deserving of particular attention.[22]

Under the UNCLOS, all States and competent international organizations have the right to conduct marine scientific research (Art. 238). In addition, there is an obligation on States and international organizations to promote and facilitate the development and conduct of marine scientific research (Art. 239) and to make the results of such research available through publication, in particular to developing States (Arts 242-244).

The UNCLOS contains special provisions relating to the conduct of marine scientific research in the Area. Under Art. 256, all States, irrespective of their geographical location, and competent international organizations have the right, in conformity with the provisions of Part XI, to conduct marine scientific research in the Area. It is worth noting that this provision, unlike others in the UNCLOS, is not limited to 'States Parties' but refers to the right of 'all States' and is thus intended to set out a universal right applicable to all States.[23]

Turning to Part XI UNCLOS, Art. 143(1) stipulates that marine scientific research in the Area is to be carried out exclusively for peaceful purposes and for the benefit of mankind as a whole. Art. 143(2) defines the role of the Authority with respect to marine scientific research. It states that the Authority may carry out marine scientific research concerning the Area and its resources, and may enter into contracts for that

---

[21] M Jacobsson, 'International Law and Scientific Research in the Arctic – The Role of Science in Law and the Role of Law in Science', *ZaöRV* 69 (2009) 683 et seq.

[22] Taksøe-Jensen, see note 6, 632; Ilulissat Declaration, see note 6, para. 7.

[23] Under Art. 36 Vienna Convention on the Law of Treaties (concluded 23 May 1969, entered into force 27 January 1980) 1155 UNTS 331,'[a]right arises for a third State from a provision of a treaty if the parties to the treaty intend the provision to accord that right either to the third State, or to a group of States to which it belongs, and the third State assents thereto. Its assent shall be presumed so long as the contrary is not indicated, unless the treaty otherwise provides.'

purpose. The Authority is under an obligation to promote and encourage the conduct of marine scientific research in the Area and to coordinate and disseminate the results of such research and analysis when available.

Art. 143(3) deals with the position of States Parties with respect to marine scientific research in the Area. It may be distinguished in this regard from Art. 256 UNCLOS. Art. 143 provides that States Parties may carry out marine scientific research in the Area (which is implicit from Art. 256 anyway), but goes on to say that States Parties must fulfill their general duty to promote international cooperation in marine scientific research in a number of specific ways, including by participating in international programs for such purposes and effectively disseminating the results of research and analysis when available, through the Authority or other international channels. Art. 143(3)(b) is particularly interesting in that it provides that one of the ways in which States Parties may fulfill their obligation is to

> [ensure] that programmes are developed through the Authority or other international organizations as appropriate for the benefit of developing States and technologically less developed States with a view to:
>
> (i) strengthening their research capabilities;
>
> (ii) training their personnel and the personnel of the Authority in the techniques and applications of research;
>
> (iii) fostering the employment of their qualified personnel in research in the Area.

This provision is not merely hortatory. It should be implemented in good faith by States Parties. In the case of the Arctic, it seems obvious that the primary duty of implementation rests on those States Parties who are actively engaged in carrying out marine scientific research in the Arctic. In addition, relevant international organizations and bodies involved in such research, such as the International Arctic Science Committee, the European Polar Board and the Arctic Council, may act as intermediaries for the purpose of implementation of Art. 143.

Of course, in implementing Art. 143, a distinction can be made between scientific research which is specifically aimed at the collection of scientific data concerning the delimitation of the limits of the continental shelf, and that aimed more broadly at gaining a better understanding of the nature of the marine environment or the effects of climate change. It is the latter which is of legitimate interest to all States.

VI. Conclusions: The Role of the ISA in Arctic Ocean Governance

Whilst recognizing the need to pursue strengthened cooperation in order to fully implement the relevant provisions of the UNCLOS framework, it seems doubtful that there is a need for new international instruments or regimes. There would appear to be ample opportunity to strengthen international cooperation in a manner that is complementary to the implementation of existing instruments and that does not undermine the role of existing mechanisms.[24]

In this regard, as envisaged by the relevant provisions of Part XI UNCLOS, the Authority may act not only as a vehicle for the dissemination of the results of marine scientific research and analysis, particularly on the marine environment of the Arctic, but also as an intermediary for the development of the programs referred to in Art. 143 UNCLOS that aim to strengthen the research capabilities of developing States and technologically less developed States.

One way in which this could be achieved in practice is through a memorandum of understanding between the Authority and a competent regional organization or institution, such as the Arctic Council. The Arctic Council is not an international organization *per se*, but a form of cooperation *sui generis*. Nevertheless, its main functions of ensuring the protection of the environment and coordination of 'common Arctic issues, in particular issues of sustainable development and environmental protection in the Arctic'[25] are fully consistent with the responsibility of the Authority to ensure the effective protection of the marine environment from the harmful effects of deep seabed mining. A number of organizations are already observers to the Arctic Council, including several United Nations programs and bodies.[26]

---

[24] Taksøe-Jensen, see note 6, 625-633.

[25] Arctic Council, 'Declaration on the Establishment of the Arctic Council' (19 September 1996), see <http://arctic-council.org/article/about> (16 June 2011), para. 1(a); see also R Wolfrum, 'The Arctic in the Context of International Law', *ZaöRV* 69 (2009) 533 et seq. (542).

[26] As of the end of 2010, nine international organizations are observers to the Arctic Council: International Federation of Red Cross and Red Crescent Societies, International Union for the Conservation of Nature ('IUCN'), the Nordic Council, the Nordic Environment Finance Corporation, North Atlantic Marine Mammal Commission, Standing Committee of the Parliamentarians of the Arctic Region, the United Nations Economic Commission for Europe,

A good example of how regional cooperation within and between States and competent international organizations can be organized in order to effectively combine sustainable management and use of resources with environmental protection can be seen in the North-East Atlantic. Here, the Convention for the Protection of the Marine Environment of the North-East Atlantic ('OSPAR')[27] establishes a regional mechanism for the protection of the marine environment. As part of this mechanism, the OSPAR Commission cooperates closely with other relevant organizations in the region, including the North-East Atlantic Fishery Commission ('NEAFC') and the Authority. In this regard, a memorandum of understanding between OSPAR and the Authority was signed in 2010 following approval by the governing bodies of both organizations.[28] Also in 2010, as a result of a lengthy process of consultation and cooperation, both OSPAR and NEAFC put in place innovative measures to manage a number of maritime areas beyond national jurisdiction. Discussions are ongoing as to how the Authority should respond, in respect of the Area, to the measures adopted by OSPAR for the water column beyond national jurisdiction, but the point is that a framework exists for such discussions. Moreover, the framework that has been established through the mechanism of a memorandum of understanding fully reflects the respective competences of each organization. In this way, States with an interest in the region are in a better position to give effect to the obligations of cooperation inherent in the UNCLOS.

the United Nations Development Programme ('UNDP') and the United Nations Environment Programme ('UNEP').

[27] Convention for the Protection of the Marine Environment of the North-East Atlantic (opened for signature 22 September 1992, entered into force 25 March 1998) (1993) 32 ILM 1069.

[28] Memorandum of Understanding between the OSPAR Commission and the International Seabed Authority, annexed to International Seabed Authority, 'Request for Observer Status in Accordance with Rule 82, Paragraph 1 (d) of the Rules of Procedure of the Assembly on Behalf of the OSPAR Commission' (12 March 2010) UN Doc ISBA/16/A/INF/2.

The Regime for Marine Scientific Research in the Arctic: Implications of the Absence of Outer Limits of the Continental Shelf beyond 200 Nautical Miles

by Alex G. Oude Elferink[*]

I. Introduction

Coastal States have the right to regulate marine scientific research in their maritime zones. All coastal States are entitled to a 200 nm exclusive economic zone. The continental shelf also extends at least to this distance. The determination of the 200 nm limit is a relatively straightforward process. It only requires measuring a distance of 200 nm from the baselines of the coastal State.[1] The continental shelf may also extend beyond the 200 nm limit. In this case, the establishment of the outer limit is much more complex. Art. 76 United Nations Convention on the Law of the Sea ('UNCLOS')[2] requires a coastal State to gather and pro-

---

* Deputy Director, Netherlands Institute for the Law of the Sea, School of Law, Utrecht University, The Netherlands. The author would like to thank the participants at the conference and Yoshinobu Takei for their comments on an earlier version of this paper. The content of this paper remains the sole responsibility of the author.

1 Difficulties may arise if the coastal State and third States disagree over the legality of the baselines of the coastal State. For instance, third States may consider that straight baselines are not in conformity with international law or that an island of a coastal State is not entitled to an exclusive economic zone and continental shelf.

2 United Nations Convention on the Law of the Sea (concluded 10 December 1982, entered into force 16 November 1994) 1833 UNTS 397.

cess complex data sets and submit information on its outer limits to the Commission on the Limits of the Continental Shelf ('CLCS'). Only after the CLCS has reviewed this information and issued recommendations, a coastal State is in a position to establish the outer limits of the continental shelf beyond 200 nm on the basis of the Commission's recommendations. As Art. 76(8) UNCLOS indicates, these limits shall be final and binding.

Art. 76 is also highly relevant for the Arctic. A significant part of the Arctic Ocean is beyond the 200 nm limit of the five coastal States, Canada, Denmark, Norway, the Russian Federation and the United States of America. The geographical North Pole is over 170 nm from the nearest 200 nm limit and certain areas are more than 300 nm from the 200 nm limit. The water column of the entire area beyond 200 nm is governed by the freedom of the high seas, which includes the freedom of scientific research. However, most of this area may be part of the continental shelves of the coastal States. All five coastal States are engaged in gathering the data which is required to determine outer limits of the continental shelf beyond 200 nm in accordance with Art. 76 UNCLOS. Available information suggests that most of the seabed of the Arctic Ocean is part of the continental shelves of the five coastal States.

Experience with the implementation of Art. 76 to date indicates that it may take a couple of decades for all States Parties to the Convention to establish final and binding limits. It is likely that this will also be true for the Arctic Ocean. Thus far only Norway has received recommendations from the CLCS which allow it to establish final and binding limits.[3] The Russian Federation made a submission to the Commission in December 2001.[4] The submission concerned four areas, the largest of which is located in the Arctic Ocean. The Commission recommended

---

[3] Commission on the Limits of the Continental Shelf, 'Summary of the Recommendations of the Commission on the Limits of the Continental Shelf in regard to the Submission Made by Norway in respect of Areas in the Arctic Ocean, the Barents Sea and the Norwegian Sea on 27 November 2006' (27 March 2009), see <http://www.un.org/Depts/los/clcs_new/submissions_files/nor06/nor_rec_summ.pdf> (2 August 2011).

[4] Russian Federation, Continental Shelf Submission, Executive Summary, attached to Commission on the Limits of the Continental Shelf, 'Receipt of the Submission Made by the Russian Federation to the Commission on the Limits of the Continental Shelf' (20 December 2001) CLCS.01.2001.LOS (Continental Shelf Notification), see <http://www.un.org/Depts/los/clcs_new/submissions_files/submission_rus.htm> (2 August 2011).

that the Russian Federation make a revised submission as regards the Arctic Ocean, based on the findings contained in its recommendations.[5] The Russian Federation has since gathered additional data and has indicated that it intends to make a further submission to the Commission. According to a recent report, the Russian Federation intends to resubmit information in 2014.[6]

At present there are some 40 submissions waiting to be considered by the Commission. The Rules of Procedure of the Commission[7] do not indicate how the Commission will queue new or revised submissions resulting from recommendations to the coastal State. The provision on the queuing of submissions contained in Rule 51 suggests that a new or revised submission would be queued after the last submission in the line. The other Arctic coastal States have not yet made a submission to the Commission. Canada and Denmark are in principle required to make their submission by 2013 and 2014 respectively. The United States is not a party to the Convention and is not entitled to employ the procedure for establishing the outer limits of the continental shelf involving the Commission, but will be able to do so if it accedes to UNCLOS. These submissions will be queued at the end of the line of submissions before the Commission. The projected time line for dealing with submissions by the Commission indicates that submissions by the Arctic coastal States may only be considered in the 2030s.

The absence of final and binding limits raises the question of what legal regime applies to the continental shelf beyond 200 nm in the meantime. Is a coastal State entitled to exercise rights over this part of the continental shelf before final and binding limits have been established? To answer this question, section II of this paper will first briefly consider the relevant UNCLOS provisions, after which it will discuss State practice and jurisprudence. As will become apparent, no clear answer emerges from this analysis. Section III of this paper will seek to suggest how this issue might be addressed by coastal States and third States. In this connection, two questions will be addressed. First, what rules exist to determine the extent of the continental shelf beyond 200 nm before the

---

5 UNGA, 'Oceans and Law of the Sea: Report of the Secretary-General: Addendum' (8 October 2002) UN Doc. A/57/57/Add.1, para. 41.

6 RIA Novosti, 'Russia Uses New Research Data to Enhance Arctic Territorial Claim', Press release (13 November 2010), see <http://en.rian.ru/russia/20101113/161323182.html> (12 September 2011).

7 The current version of the Rules of Procedure is contained in CLCS, 'Rules of Procedure' (17 April 2008) CLCS/40 Rev.1.

Art. 76 process has been finalized? Second, what regime is applicable to this area? Section IV looks at the implications of the findings of the preceding sections for the regime of marine scientific research in the Arctic Ocean.

II. The Regime Applicable to the Continental Shelf beyond 200 Nautical Miles in the Absence of Final and Binding Limits

1. The UN Convention on the Law of the Sea

The Convention does not explicitly address the question of whether or not a coastal State is entitled to exercise its rights over the continental shelf beyond 200 nm before final and binding limits have been established. However, a number of provisions support the former view. Art. 77(3) UNCLOS provides that the 'rights of the coastal State over the continental shelf do not depend on occupation, effective or notional, or any express proclamation.' In other words, the rights of the coastal State do not depend on the process of establishing the outer limits of the continental shelf in accordance with Art. 76 UNCLOS. Secondly, Art. 76 itself points to this same conclusion. Art. 76(1) provides that the continental shelf extends to the outer edge of the continental margin or to a distance of 200 nm where the outer edge of the continental margin does not extend to that distance. Art. 76(4) requires the coastal State to determine the outer edge of the continental margin wherever the continental margin extends beyond 200 nm. It is this process of establishing the outer edge of the continental margin which is the subject of the procedure involving the CLCS. As Art. 76(4) indicates, this process is premised on the presence of a pre-existing continental shelf entitlement. The task of the Commission is not to validate the entitlement to the continental shelf beyond 200 nm, but only to determine whether the outer limits of this entitlement have been established in accordance with Art. 76. The entitlement either exists or does not exist. Obviously, in the latter case the Commission should conclude that it cannot recommend that a coastal State establish outer limits beyond 200 nm.

Art. 76(2) also supports the view that the entitlement to the continental shelf is not dependent upon the establishment of outer limits on the basis of the recommendations of the CLCS. Para. 2 provides that the continental shelf of a coastal State shall not extend beyond the limits provided for in Art. 76(4)-(6). Para. 2, which does not refer to paras 7 to 9 of Art. 76, thus indicates that a continental shelf entitlement already exists up to the limits contained in paras 4 to 6, before a coastal State has implemented paras 7 to 9.

2. State Practice and Jurisprudence

Relatively little information is readily available on State practice with respect to the regime of the continental shelf beyond 200 nm in the absence of final and binding limits. The legislation of many States defines the continental shelf by reference to the two criteria contained in Art. 76(1) UNCLOS. Such legislation does not seem to make a distinction between the two parts of the continental shelf as far as the exercise of rights is concerned. An example from an Arctic State is provided by the Federal Law on the Continental Shelf of the Russian Federation of 30 October 1995.[8] Art. 1 of the Law refers to the 200 nm limit and provides that where the continental margin extends beyond that distance 'the outer edge of the continental shelf coincides with the outer edge of the continental margin determined in accordance with the rules of international law'. As regards marine scientific research, the Federal Law contains a provision which is explicitly applicable to the continental shelf beyond 200 nm. Art. 25(6) of the Law implements Art. 246(6) UNCLOS for the Russian Federation. There is no indication that this provision does not apply in the absence of final and binding limits to the continental shelf of the Russian Federation.

Canada has taken a similar position on the relationship between the regime of the continental shelf and the absence of final and binding limits. Information from the Department of Foreign and International Trade of Canada indicates that Canada's rights over its continental shelf do not depend upon its submission to the CLCS. According to the Department, the UNCLOS continental shelf regime is a codification of customary international law, and both now and before ratifying the Convention 'Canada exercises continental shelf jurisdiction over the

---

[8] Federal Law of the Russian Federation No. 187- FZ (30 November 1995) promulgated on 7 December 1995 in the "Rossiyskaya Gazeta" No. 237; English translation available at <http://www.un.org/Depts/los/LEGISLATIONANDTREATIES/PDFFILES/RUS_1995_Law.pdf> (12 September 2011).

full extent of its continental shelf both within and beyond 200 miles'.[9] Canada has granted licenses for the exploration of petroleum resources on the continental shelf beyond 200 nm on the Grand Banks.[10] Canada at present does not seem to exercise jurisdiction over marine scientific research on the continental shelf beyond 200 nm.[11]

The United States' position in respect of the continental shelf beyond 200 nm is expressed in a policy statement, which was adopted by an Interagency Group on the Law of the Sea and Ocean Policy in November 1987.[12] It states that the United States 'has exercised and shall continue to exercise jurisdiction over its continental shelf in accordance with and to the full extent permitted by international law as reflected in Article 76, paragraphs (1), (2) and (3)' UNCLOS.[13] The policy statement indicates that the determination of the outer limits of the continental shelf of the United States will be deferred to a later date. In order to ensure that the United States' practice is consistent with international law before these outer limits are determined, the policy requires that:

> an agency planning any leasing or licensing activity on the continental shelf beyond 200 nautical miles [...] shall provide notice to the Department of State for transmittal to the Interagency Group with a brief description of the location and type of activity. [...] The Interagency Group shall have 45 days to comment on the proposed action.[14]

The purpose of this process would seem to be to avoid that a United States' agency would exercise continental shelf jurisdiction in areas

9 Canada's extended continental shelf; Frequently asked questions, see <http://www.international.gc.ca/continental/faq.aspx?lang=eng&menu_id=24> (12 September 2011).

10 *Ibid.*

11 Oceans Act, SC (1996) ch 31 (Canada), sections 13-14, 17-18 and 44, see <http://laws.justice.gc.ca/PDF/Statute/O/O-2.4.pdf> (12 September 2011); see also A Anand, *Marine Scientific Research Governance in the Arctic Ocean* (2008) 124 et seq.

12 See 'United States Policy Governing the Continental Shelf of the United States of America' (17 November 1987), attachment to a Memorandum from Assistant Secretary of State John D. Negroponte to Deputy Legal Adviser Elizabeth Verville, reproduced in JA Roach and RW Smith, *United States Responses to Excessive Maritime Claims*, 2nd edn (1996) 201-202.

13 *Ibid.*, 201.

14 *Ibid.*, 202.

beyond the potential outer limits of the continental shelf. The United States has allowed oil and gas activities in certain areas beyond 200 nm.[15]

The United States' policy is to encourage marine scientific research.[16] The current policy of the United States requires prior consent for marine scientific research in a number of specified cases.[17] These cases do not cover the continental shelf of the United States beyond 200 nm. As is indicated by the leasing and licensing policy of the United States, the United States does consider that it can exercise jurisdiction over the continental shelf beyond 200 nm in the absence of final and binding limits based on recommendations of the CLCS.

An interesting exchange of views on the regime of the continental shelf beyond 200 nm in the absence of final and binding limits has taken place in the context of the 1992 Convention for the Protection of the Marine Environment of the North East Atlantic ('OSPAR Convention').[18] The parties to the OSPAR Convention have been working on the designation of a number of marine protected areas ('MPAs') in areas beyond national jurisdiction. After Iceland lodged a submission with the CLCS in 2009 it became apparent that one of these areas, the proposed Charlie Gibbs Fracture Zone MPA, overlapped to a considerable extent with the continental shelf of Iceland.[19] This raised the question of how

---

[15] See for instance the map 'BOEMRE Gulf of Mexico OCS Region: Blocks and Active Leases by Planning Area' (1 October 2011), prepared by the Bureau of Ocean Energy Management, Regulation and Enforcement of the Department of the Interior, see <http://www.gomr.boemre.gov/homepg/lsesale/mau_gom_pa.pdf> (4 October 2011). The map identifies a number of active leases in an area beyond 200 nm in the Western Gulf of Mexico.

[16] Statement by President Reagan accompanying the 'United States: Proclamation on an Exclusive Economic Zone' (10 March 1983), reprinted in (1983) 22 ILM 461.

[17] Ocean and Polar Affairs ('OPA') within the Department's Bureau of Oceans and International Environmental and Scientific Affairs ('OES'), 'Marine Scientific Research Authorizations', see <http://www.state.gov/g/oes/ocns/opa/rvc/index.htm> (12 September 2011).

[18] Convention for the Protection of the Marine Environment of the North-East Atlantic (signed 22 September 1992, entered into force 25 March 1998) 2354 UNTS 67.

[19] The outer limits of the Reykjanes Ridge in Iceland's 2009 submission extend more than 700 nm from Iceland's baselines. Iceland had previously taken the position that its continental shelf in this area did not extend beyond 350 nm.

the parties to the OSPAR Convention should proceed with the designation of the proposed Charlie Gibbs Fracture Zone MPA. After an extensive discussion, the OSPAR Commission designated an MPA for the part of the Charlie Gibbs Fracture Zone beyond the outer limits of the continental shelf contained in the submission of Iceland to the CLCS.[20]

In the discussions concerning the consequences of the overlap of the proposed MPA with Iceland's continental shelf, the parties to the OSPAR Convention set out their views on the relationship between the regime of the continental shelf beyond 200 nm and the establishment of its outer limits. Belgium, Spain and Germany submitted that a coastal State cannot claim continental shelf rights beyond 200 nm before the outer limits of that area have been established in accordance with the procedure set out in Art. 76 UNCLOS.[21] Belgium indicated that until that time the area concerned would fall beyond the limits of national jurisdiction. In a similar vein, Germany considered that until such time a coastal State is not in a position to claim rights beyond 200 nm 'since that would result in a situation where substantial parts of the seabed of the high seas could not be used by third states or the international community regardless of the 200 [nm] boundary'.[22] Germany submitted that protective measures for such an area could be taken by the international community and not by the coastal State.[23] On the other hand, Spain stressed that any decision in the framework of the OSPAR Con-

The proposed Charlie Gibbs Fracture Zone MPA lay beyond 350 nm from Iceland's baselines. Iceland had not informed the other parties to the OSPAR Convention of the overlap between the proposed MPA and Iceland's continental shelf before it lodged its submission.

[20] See e.g. OSPAR Commission, 'Decision 2010/2 on the Establishment of the Charlie-Gibbs South Marine Protected Area' (2010) OSPAR 10/23/1-E, Summary Record of the Meeting of the OSPAR Commission (20-24 September 2010) Annex 36.

[21] See OSPAR Commission, 'Spain – Position on Icelandic Submission for Extension of its Continental Shelf in Relation to OSPAR Projected MPA BNJ (Charlie Gibbs)', Ad Hoc Working Group Charlie 09/01 Add.1-E (11 November 2009); OSPAR Commission, 'Compilation of Responses from Contracting Parties on the Way Forward with Respect to the Charlie Gibbs Fracture Zone Presented by Secretariat', Ad Hoc Working Group Charlie 09/01 Rev.1 (11 November 2009) Annex 1 – Belgium; Annex 4 – Germany, para. 11.

[22] *Ibid.*, Annex 4 – Germany, para. 10.

[23] *Ibid.*, para. 11.

vention should take into account the future potential rights of Iceland.[24] This latter view was also expressed by France: the designation of an MPA should not prejudice Iceland's potential sovereign rights.[25] The Netherlands, while recognizing that the rights of the coastal State were inherent and not potential, stressed that 'the exercise of these rights [...] can only take place [...] after the coastal State has established final and binding limits on the basis of the recommendations of the CLCS'.[26] The Netherlands also submitted that an area under consideration by the CLCS remained an area beyond national jurisdiction:

> What other purpose would the CLCS procedure have? Is it not to protect the Area and avoid indiscriminate, unilateral actions by individual coastal states claiming sovereign rights by way of an extended continental shelf?[27]

Other parties to the OSPAR Convention took the opposite view. They observed that the right of the coastal State over the continental shelf is inherent and does not depend on the recommendations of the CLCS.[28] Portugal observed that the fact that a coastal State had not yet received recommendations from the Commission did not mean that it could only exercise sovereign rights up to the 200 nm limit.[29] However, in further elaborating its views, Portugal significantly curtailed this conclusion:

> the coastal state, and obviously any other state or international organization, shall refrain to take any economic activity beyond

---

[24] OSPAR Commission, 'Spain – Position on Icelandic Submission for Extension of its Continental Shelf in Relation to OSPAR Projected MPA BNJ (Charlie Gibbs)', see note 21.

[25] OSPAR Commission, 'Compilation of Responses from Contracting Parties on the Way Forward with Respect to the Charlie Gibbs Fracture Zone Presented by Secretariat', see note 21, Annex 3 – France.

[26] OSPAR Commission, 'Comments on Document JL(2) 10/3/1 submitted by the Netherlands', JL(2) 10/3/Info.2 submitted to the Group of Jurists/Linguists ('JL'), London (25-26 August 2010) 4.

[27] OSPAR Commission, 'Report of the WG-Charlie Meeting in November 2009', Ad Hoc Working Group Charlie 10/1 (15 February 2010), para. 3.7.

[28] See e.g. OSPAR Commission, 'Compilation of Responses from Contracting Parties on the Way Forward with Respect to the Charlie Gibbs Fracture Zone Presented by Secretariat', see note 21, Annex 8 – Norway; Annex 11 – Portugal, paras 3-4; see also *ibid.*, Annex 5 – Iceland.

[29] See *ibid.*, Annex 11 – Portugal, para. 3.

> 200 [nm] concerning the exploitation of the resources as referred to in UNCLOS article 133 until favorable recommendations are granted by the CLCS.[30]

At the same time, Portugal considered that a coastal State had the right and the duty to protect and conserve ecosystems and biodiversity in accordance with international law and for the purpose of the future exploitation of this area.[31]

The OSPAR Commission has designated a number of high seas MPAs which overlap with the continental shelf as defined in the submission of a coastal State to the CLCS. The decisions on the management of these MPAs recognize that the coastal State is competent to establish the programs, measures and agreements which are necessary for the achievement of the conservation vision and conservation objectives regarding the seabed of the MPA concerned. The decisions also provide that the designation does not create any precedent regarding the establishment by the Commission of other MPAs in waters superjacent to areas of the seabed subject to submission to the CLCS or prejudice the sovereign rights and obligations of coastal States regarding the continental shelf.[32]

The rights of the coastal State over the continental shelf beyond 200 nm have also been addressed in a number of resolutions of the General Assembly of the United Nations and decisions of the Meetings of States Parties to UNCLOS. A number of instruments in addressing the work of the CLCS refer to the language contained in Art. 77(3) UNCLOS.[33] These general references to Art. 77(3) do not explicitly address the exercise of rights over the continental shelf beyond 200 nm by the coastal State in the absence of final and binding limits. An explicit reference to the exercise of jurisdiction is however contained in the General As-

---

[30] *Ibid.*, para. 6.

[31] *Ibid.*

[32] See e.g. OSPAR Commission, 'Decision 2010/5 on the Establishment of the Josephine Seamount High Seas Marine Protected Area' (2010) OSPAR 10/23/1-E, Summary Record of the Meeting of the OSPAR Commission (20-24 September 2010) Annex 42.

[33] See e.g. UNGA, 'Oceans and the Law of the Sea' (4 December 2009) UN Doc. A/RES/64/71, para. 40; 'Decision Regarding the Workload of the Commission on the Limits of the Continental Shelf and the Ability of States, Particularly Developing States, to Fulfil the Requirements of Article 4 of Annex II to the United Nations Convention on the Law of the Sea, as well as the decision contained in SPLOS/72, para. (a)' (20 June 2008) SPLOS/183, second consideration.

sembly resolutions on fisheries adopted in 2008 to 2010. Para. 104 of the 2008 resolution provides that the General Assembly:

> Recalls that nothing in paragraphs 83 to 86 of resolution 61/105 is to prejudice the sovereign rights of coastal States over their continental shelf or the exercise of the jurisdiction of coastal States with regard to that shelf under international law as reflected in the Convention.[34]

Resolution 61/105 did not include this without prejudice provision. In the debate on the draft of the 2008 resolution, Argentina explained that:

> it considered it to be appropriate, at this moment, to urge the inclusion of paragraph 104 of the draft resolution on sustainable fisheries so as to prevent any interpretation seeking to ignore the exclusive nature of the rights of coastal States over the areas of their continental shelf situated beyond the 200-mile limit.[35]

The International Court of Justice ('ICJ') and arbitral tribunals have in a number of instances considered the relationship between continental shelf entitlement and the process of establishing the outer limits beyond 200 nm. The court of arbitration in the case concerning *Delimitation of Maritime Areas (St Pierre and Miquelon)* refrained from delimiting the continental shelf beyond 200 nm.[36] After referring to the Art. 76 pro-

---

[34] UNGA, 'Sustainable Fisheries, Including through the 1995 Agreement for the Implementation of the Provisions of the United Nations Convention on the Law of the Sea of 10 December 1982 Relating to the Conservation and Management of Straddling Fish Stocks and Highly Migratory Fish Stocks, and Related Instruments' (5 December 2008) UN Doc. A/RES/63/112, para. 104; a similar paragraph is contained in UNGA, 'Sustainable Fisheries, Including through the 1995 Agreement for the Implementation of the Provisions of the United Nations Convention on the Law of the Sea of 10 December 1982 Relating to the Conservation and Management of Straddling Fish Stocks and Highly Migratory Fish Stocks, and Related Instruments' (4 December 2009) UN Doc. A/RES/64/72, para. 115 and UNGA, 'Sustainable Fisheries, Including through the 1995 Agreement for the Implementation of the Provisions of the United Nations Convention on the Law of the Sea of 10 December 1982 Relating to the Conservation and Management of Straddling Fish Stocks and Highly Migratory Fish Stocks, and Related Instruments' (7 December 2010) UN Doc. A/RES/65/38, para. 119. The latter two resolutions explicitly refer to Art. 77 UNCLOS.

[35] UNGA Plenary Meeting (4 December 2008) GAOR Sixty-third Session 63rd Plenary Meeting, UN Doc. A/63/PV.63, 4.

[36] Case concerning Delimitation of Maritime Areas (Decision) Court of Arbitration for the Delimitation of Maritime Areas between Canada and France (10 June 1992) (1992) 31 ILM 1149 (1171-1173, paras 75-82).

cess, the award indicates that its refusal to pronounce itself on this issue 'cannot signify nor may be interpreted as prejudging, accepting or refusing the rights that may be claimed by France, or by Canada, to a continental shelf beyond 200 nautical miles'.[37] This finding seems to imply that according to the court of arbitration, the rights over this part of the continental shelf do not exist prior to the completion of the Art. 76 process.

The tribunal in *Barbados v. Trinidad and Tobago* considered itself to be in a position to address the delimitation of the continental shelf beyond 200 nm in the absence of outer limits established in accordance with Art. 76.[38]

In its judgment of 2007 in the maritime delimitation case between Nicaragua and Honduras, the ICJ observed:

> The Court may accordingly, without specifying a precise endpoint, delimit the maritime boundary and state that it extends beyond the 82nd meridian without affecting third-State rights. It should also be noted in this regard that in no case may the line be interpreted as extending more than 200 nautical miles from the baselines from which the breadth of the territorial sea is measured; any claim of continental shelf rights beyond 200 miles must be in accordance with Article 76 of UNCLOS and reviewed by the Commission on the Limits of the Continental Shelf established thereunder.[39]

This finding of the Court may be open to different interpretations. Interestingly, this matter was not argued by the parties and in view of the geography of the area does not seem to have practical relevance. It could well be argued that the Court holds that it cannot pronounce itself on the delimitation of the continental shelf beyond 200 nm before the Art. 76 process has been completed. Another reading would be that the Court considered that the parties to the case cannot use the delimi-

---

37 *Ibid.*, paras 79-80.

38 *Barbados and the Republic of Trinidad and Tobago (Award) Permanent Court of Arbitration* (11 April 2006) 45 ILM 800, para. 368. The tribunal did not establish a boundary in the area beyond 200 nm because it concluded that the circumstances of the case mandated a boundary which ended at 200 nm of the baselines of Trinidad and Tobago and within 200 nm of the baselines of Barbados, see paras 381-382 and 385.

39 *Territorial and Maritime Dispute between Nicaragua and Honduras in the Caribbean Sea (Nicaragua v Honduras)* [2007] ICJ Rep 659, para. 319.

tation effected by the Court as a basis to claim rights that do not exist under Art. 76 UNCLOS.

State practice does not reveal a similar hesitance in dealing with the delimitation of the continental shelf beyond 200 nm in the absence of outer limits. A recent example is provided by the 2010 agreement between Norway and the Russian Federation, which is concerned with the Barents Sea and the Arctic Ocean.[40]

III. Dealing with the Uncertainties Resulting from the Absence of Final and Binding Outer Limits

1. Determination of the Extent of the Continental Shelf beyond 200 Nautical Miles

The first question to consider is whether continental shelf rights exist beyond 200 nm in the absence of final and binding limits. The judgment of the ICJ in the maritime delimitation case between Nicaragua and Honduras and the award in the arbitration between France and Canada both suggest that this is not the case, although the former judgment is open to different interpretations. A similar view has been adopted by a number of parties to the OSPAR Convention in the discussion concerning the potential overlap between a proposed MPA and Iceland's continental shelf beyond 200 nm. Other parties to the Convention have taken the position that rights to the continental shelf beyond 200 nm are inherent and that the coastal State can exercise these rights in the absence of final and binding outer limits. The decisions of the OSPAR Commission on the designation of a number of MPAs which are located within continental shelf limits submitted to the CLCS also reflect this view. It is also reflected in the 2008 to 2010 General Assembly resolutions on fisheries.

The view that continental shelf rights beyond 200 nm do not exist in the absence of final and binding limits is unconvincing for a number of reasons. First of all, it mischaracterizes the process involving the coastal

---

[40] Treaty between Norway and the Russian Federation concerning Maritime Delimitation and Cooperation in the Barents Sea and the Arctic Ocean (done 15 September 2010, entered into force 7 July 2011), reprinted in T Henriksen and G Ulfstein, 'Maritime Delimitation in the Arctic: The Barents Sea Treaty', *Ocean Development and International Law* 42 (2011) 1 et seq. (11-17).

State and the CLCS. As was explained previously, this process is concerned with establishing the outer limits of the continental shelf beyond 200 nm and not with validating a claim to this area. Secondly, as was also set out above, Arts 76(2) and 77(3) UNCLOS indicate that the entitlement to the continental shelf is not dependent on the establishment of outer limits. This latter point is confirmed by the definition of the continental shelf in Art. 1 Convention on the Continental Shelf[41] and in customary law as defined by the International Court of Justice in the *North Sea continental shelf cases*.[42] Both imply the existence of an entitlement beyond 200 nm and do not require the determination of final and binding outer limits for the entitlement to exist. There is no indication that UNCLOS intended to diverge from established law in this respect. It could be argued that State practice also confirms this view. There is a significant amount of practice in respect of the continental shelf beyond 200 nm. However, certain States take the view that these continental shelf rights do not exist prior to the establishment of outer limits in accordance with Art. 76 UNCLOS. A more detailed analysis of State practice would be required to determine its significance with greater certainty.

If the rights to the continental shelf beyond 200 nm do not depend on the establishment of final and binding outer limits, the next question is how the extent of this part of the continental shelf can be defined provisionally. One possible approach was suggested by Portugal in the discussions relating to the Icelandic submission in the framework of the OSPAR Convention. Portugal observed that in most cases in which the CLCS had issued recommendations, the outer limits contained therein were very close to the outer limits contained in the submission and in some areas were even more favorable to the coastal State. Portugal submitted that there should be a presumption that in the absence of final and binding limits, the outer limits of the continental shelf of a coastal State shall be considered to be the limits contained in the submission to the Commission.[43] As was also observed by Portugal, the estab-

---

[41] Convention on the Continental Shelf (done 29 April 1958, entered into force 10 June 1964) 499 UNTS 311.

[42] *North Sea Continental Shelf Cases (Federal Republic of Germany/Denmark; Federal Republic of Germany/Netherlands)* [1969] ICJ Rep 3.

[43] OSPAR Commission, 'Compilation of Responses from Contracting Parties on the Way Forward with Respect to the Charlie Gibbs Fracture Zone Presented by Secretariat', see note 21, Annex 11 – Portugal, paras 5-6.

lishment of maritime zones 'is a matter of sovereignty to be carried out in accordance with international law.'[44]

At first sight, Portugal's suggestion might seem to offer a reasonable solution. Matters are however not as straightforward as is suggested by Portugal. In a number of cases the CLCS has not adopted the limits submitted by the coastal State. In 2002, the Russian Federation was recommended to make a revised submission in respect of the outer limits of its continental shelf in the Arctic Ocean.[45] The Russian submission had included outer limit lines extending a couple of hundred nautical miles beyond the 200 nm limit. In respect of the continental shelf of Ascension Island, the Commission in its recommendations concluded that the submission of the United Kingdom did not provide a basis for establishing outer limits beyond 200 nm.[46] The United Kingdom had submitted outer limits of up to 350 nm from Ascension.

Secondly, as Portugal implicitly indicated by its reference to international law, the establishment of maritime zones is not a wholly unilateral process. The establishment of the outer limits of the maritime zones of a coastal State is a two-stage process. This is expressed as follows in a well-known observation made by the ICJ in the *Anglo-Norwegian Fisheries case* concerning the straight baselines Norway had established along its coast:

> The delimitation of sea areas has always an international aspect; it cannot be dependent merely upon the will of the coastal State as expressed in its municipal law. Although it is true that the act of delimitation is necessarily a unilateral act, because only the coastal State is competent to undertake it, the validity of the delimitation with regard to other States depends upon international law.[47]

Put differently, other States have a right to reject the limits of maritime zones established by the coastal State. This also applies if a coastal State

---

[44] *Ibid.*, para. 4.

[45] UNGA, 'Oceans and Law of the Sea: Report of the Secretary-General: Addendum', see note 5, para. 41.

[46] Commission on the Limits of the Continental Shelf, 'Summary of the Recommendations of the Commission on the Limits of the Continental Shelf in regard to the Submission Made by the United Kingdom of Great Britain and Northern Ireland in respect of Ascension Island on 9 May 2008' (15 April 2010), see <http://www.un.org/depts/los/clcs_new/submissions_files/gbr08/gbr_asc_isl_rec_summ.pdf> (4 October 2011), para. 54.

[47] *Fisheries Case (United Kingdom v Norway)* [1951] ICJ Rep 116, 132.

invokes the limits it has submitted to the CLCS as the provisional limits of its continental shelf pending the outcome of the consideration of its submission by the Commission. There is no obligation for other States to accept such provisional outer limits.[48] The conclusion that coastal States are entitled to determine and apply provisional limits does not detract from their obligation to determine the outer limits of the continental shelf beyond 200 nm in accordance with the procedure set out in Art. 76(8). This obligation is explicitly spelled out in Art. 7 Annex II to the Convention.

2. The Exercise of Rights over the Continental Shelf by the Coastal State

Is the coastal State entitled to exercise rights over the continental shelf beyond 200 nm in the absence of final and binding limits? Art. 77 UNCLOS is the principal article defining the rights of the coastal State over the continental shelf. Art. 77 does not make a distinction between the continental shelf within and beyond 200 nm and Art. 77 explicitly refers to the *exercise* of these rights. There thus does not seem to be scope for other States to object to the exercise of these rights by the coastal State over the continental shelf beyond 200 nm as such. This however is not the end of the matter.

Other States may object if a State is exercising continental shelf rights in an area which they consider to be beyond the outer limits of the continental shelf. A corollary of this position of these other States is that under their view marine scientific research in the area concerned is a freedom of the high seas and they may decide to exercise these freedoms without seeking the consent of the State claiming continental shelf rights. Whether this is an attractive option is questionable. A conflict with the claimant State may negatively impact on the execution of the research concerned.

What legal consequences would a dispute over the exercise of continental shelf rights by a State have? In the absence of final and binding limits, it will first of all be up to the States concerned to seek a solution to such a dispute. In the case of overlapping continental shelf entitle-

---

[48] The options for dealing with a dispute between a coastal State and other States concerning the outer limits of the continental shelf beyond 200 nm are briefly considered in section III.2 below.

ments, the coastal States concerned have an obligation, pending agreement on a final delimitation, to make every effort to enter into provisional arrangements and not to jeopardize or hamper the reaching of the final agreement.[49] This obligation does not exist when one State claims an area as a part of its continental shelf and another State rejects such a claim. It is moreover to be expected that a claimant State will in general not be willing to accept limitations on the exercise of its sovereign rights. If negotiations do not result in a solution, UNCLOS offers States the possibility of submitting a dispute to compulsory dispute settlement. A court or tribunal in essence will be required to decide if continental shelf rights exist in the area concerned. There is one important limitation to address this issue. A court or tribunal would not be competent to deal with matters falling within the competence of the CLCS and would not be in a position to make recommendations to coastal States. On the other hand, a court or tribunal would be competent to deal with other questions in respect of Art. 76. A court or tribunal may also be expected to take into account the stage reached by the Art. 76 process for the specific coastal State. For instance, has that State complied with its obligation to make a submission? Have recommendations been made, and what is their content? Has the State made a new or revised submission within a reasonable time and if not, what are the reasons for the absence of a new or revised submission? If a dispute has been submitted to a court or tribunal, a party could also request the indication of provisional measures pending a final decision in accordance with Art. 290 UNCLOS.

After the outer limits of the continental shelf have been determined in accordance with Art. 76 UNCLOS, it could become clear that a State has exercised continental shelf rights in an area beyond these outer limits. This would constitute an unlawful act, which, depending on the circumstances of the case, could give rise to a requirement for reparation. For instance, if a State has exploited the mineral resources of a claimed continental shelf area, which is subsequently established to be part of the Area, the International Seabed Authority would be entitled to claim damages.[50]

---

[49] Art. 83(3) UNCLOS.

[50] See Art. 137 UNCLOS.

IV. The Regime for Marine Scientific Research in the Absence of Final and Binding Outer Limits of the Continental Shelf in the Arctic Ocean

Location is the primary determinant for the regime of marine scientific research. Most of the Arctic Ocean is located within the exclusive economic zone and continental shelf of the five Arctic coastal States. Within the bounds set by international law, these coastal States are free to determine the regime of marine scientific research within their maritime zones. As the example of the United States and Canada shows, coastal States may opt not to make use of all the limitations on research that international law allows them to apply.

The extent of the continental shelf in the Arctic Ocean at present is still uncertain. Only Norway has received recommendations from the CLCS which allow it to establish final and binding limits in accordance with Art. 76 UNCLOS. Recommendations to the other four coastal States may not be forthcoming in the next couple of decades. In the absence of final and binding limits coastal States are entitled to determine the extent of their continental shelf in accordance with the substantive provisions of Art. 76. They are entitled to exercise their rights as coastal States in this area, including rights in respect of marine scientific research. Other States are not obliged to accept the outer limits a coastal State has defined unilaterally if they consider that these outer limits are not in accordance with the substantive provisions of Art. 76 and also need not accept the exercise of jurisdiction by that State in the area they consider to be beyond the outer limits resulting from the application of Art. 76. As noted in the preceding analysis, the means of settling a dispute over these matters are limited. Such a dispute may also have a negative impact on the conduct of marine scientific research.

The conveners of this conference posed a number of questions in relation to the legal consequences of uncertainty about the extent of the continental shelf in the Arctic Ocean: Would the enlarged continental shelf lead to a *de facto* limitation for science in the Arctic Ocean? Does a 'legal science *acquis*' exist, which might prevail in the future? And does the Arctic Ocean represent an area '*sui generis*' for marine science?[51]

---

51 Programme (preliminary draft as of 22 October 2011 [sic]), Arctic Science, International Law and Climate Change – Legal Aspects of Marine Science in the Arctic Ocean -, International Conference at the Federal Foreign Office in

The answer to the first of these questions has already been provided above. Coastal States at present already have the right to regulate marine scientific research on their entire continental shelf in accordance with international law. The second question suggests that this coastal State right may have been modified for the Arctic Ocean by the existing practice in respect of marine scientific research. A final answer to this question would require more information on this practice. However, it is considered highly unlikely that the regime contained in UNCLOS is not applicable unabridged to the Arctic Ocean. There is no obligation for coastal States to exercise these rights. Even if there were practice indicating that Arctic coastal States have not exercised certain rights to regulate marine scientific research, this would not preclude them from exercising those rights in the future. The practice of the United States illustrates this point. The United States does not exercise its right to assert jurisdiction over marine scientific research beyond the outer limits of its exclusive economic zone, but there is no reason to assume that it has waived these rights.

The third question posed by the conveners of the conference suggests that the characteristics of the Arctic Ocean provide a rationale for a more liberal regime for marine scientific research than that applicable to other parts of the globe. The significance of the Arctic Ocean for understanding the global climate comes to mind in this respect. In that context, international cooperation and access to this area for researchers are of key importance. The attainment of these goals stands little to gain from attempts by third States to unsettle the regime for marine scientific research contained in UNCLOS. Rather, the Convention should provide the basis for further developing a regime which will allow to effectively take advantage of the synergies of international cooperation. Questions relating to the definition of the outer limits of the continental shelf by the Arctic coastal States should rather be dealt with in the context of the implementation of Art. 76 UNCLOS and should not unnecessarily burden the debate concerning the development of the international regime for marine scientific research.

---

cooperation with the Ministry of Foreign Affairs of Finland, Berlin (17 – 18 March 2011) 3 (on file with the author).

Common Precepts of Marine Scientific Research Access in the Arctic

by Betsy Baker[*]

> An important and highly valuable legacy of IPY [the International Polar Year 2007-09] could be to reconsider access impediments in all regions of the Arctic, building on achievements made during the IPY, and through inter-governmental consultations to improve the access situation for scientists in the whole Arctic on a long-term basis.
>
> Arctic Council IPY Legacy Scoping Study[1]

A consistent message emerged during the International Conference on Arctic Marine Science, International Law and Climate Protection, convened 17-18 March 2011, at the Federal Foreign Office in Berlin: all five Arctic Ocean littoral States have procedures in place reflecting the coastal State consent requirements for marine scientific research ('MSR') set out in Part XIII UN Convention on the Law of the Sea ('UNCLOS').[2] The degree to which foreign researchers are notified of decisions and granted permission varies from one State to the next,[3] as do forms and procedures, but all five States claim to be in keeping with

---

* Associate Professor, Doctor, Vermont Law School, and Senior Fellow for Oceans and Energy, VLS Institute for Energy and the Environment; Affiliated Professor of Arctic Law and Policy, University of Alaska-Fairbanks Department of Geography.

1 W Dallmann and AH Hoel (eds), Maximizing the Legacy of IPY in the Arctic: A Scoping Study for the Arctic Council (2009) 19, Part 3.2.1.

2 United Nations Convention on the Law of the Sea (concluded 10 December 1982, entered into force 16 November 1994) 1833 UNTS 397.

3 See e.g. Appendix I: White Paper, at Table I.

Part XIII, the binding legal regime that regulates MSR in the Arctic Ocean. This is true for Canada, Denmark (on behalf of Greenland and the Faroe Islands), Norway and the Russian Federation, and for the United States, which is the only Arctic coastal State that has not yet acceded to UNCLOS.

Regulatory consistency around the pole can be a substantial basis for long term MSR access to the Arctic Ocean, including the extended continental shelves of each coastal State. Scientists and permitting authorities alike want to ensure greater consistency through more transparency and predictability in the consent process. The purpose of this paper is to propose a supplemental and non-binding statement acknowledging common precepts of Arctic marine scientific research to help improve access predictability. The statement would be open for endorsement by all States, Arctic or otherwise, interested in the conduct of MSR in the Arctic Ocean and to entities such the Arctic Council and Arctic science organizations. The agreed common precepts would not replace current rules under the UNCLOS and the international law of the sea. Instead, the precepts would offer a non-binding guide to promoting the purposes of the existing MSR regime. The precepts could be widely circulated to national and local authorities, scientists and others involved in permitting and conducting MSR in the Arctic Ocean, to help establish a common expectation of appropriate conduct for scientists and administrative authorities alike.

After a cursory introduction to relevant provisions and principles for MSR access in the UNCLOS, this paper examines policy and legal sources that the scientific and diplomatic communities can use to identify common precepts for MSR in the Arctic. First, the paper considers the national policy and strategy statements of Arctic States regarding the Arctic. It then discusses principles to which Arctic Ocean coastal States have agreed bilaterally in State-to-State science and technology cooperation agreements and in ministerial level agreements. The paper also explores a third source, statements of multilateral research organizations, treaty-based and otherwise, that may also prove relevant.

I. The Existing Legal Regime and Principles for MSR in the UNCLOS

The UN Convention on the Law of the Sea regulates, but does not define MSR. Under Part XIII UNCLOS, coastal States may regulate MSR access to both the water column and to the continental shelf within their 200 nm exclusive economic zones ('EEZ'). Seaward of the EEZ all States, and competent international organizations, have the high seas freedom 'to conduct marine scientific research in the water column beyond the limits of the exclusive economic zone' in keeping with the Convention and coastal State regulations (Art. 257 UNCLOS).[4] Coastal States have no authority to deny or grant MSR access permission to the high seas, i.e. the water column beyond the EEZ and, presumably, the sea ice there. But beneath that water column, on any continental shelf extending seaward of the EEZ over which a coastal State legitimately exercises access consent rights, it may withhold consent for those parts of the extended shelf that it has designated in advance for actual or imminent resource exploitation or detailed exploratory operations (Art. 246(6) UNCLOS). On the one hand, Art. 246(6) offers no practical hurdle to a State designating much or all of its continental shelf seaward of 200 nm as being part of this exploration and exploitation area and thus more restricted for MSR access. On the other hand, some have characterized the rights of coastal States to refuse MSR on the outer continental shelf as limited. Joanna Mossop highlights the potential for uses of the water column above a State's outer, or extended, continental shelf to compete with uses of that shelf, thus raising questions that are beyond the scope of this paper, but also relevant to MSR access, including how countries can best regulate to balance the competing uses there.[5]

---

[4] Three comprehensive works on legal issues relating to Part XIII UNCLOS are: AHA Soons, Marine Scientific Research and the Law of the Sea (1982); F Wegelein, Marine Scientific Research: the Operation and Status of Research Vessels and Other Platforms in International Law (2005) and M Gorina-Ysern, An International Legal Regime for Marine Scientific Research (2003); for individual aspects of Part XIII see e.g.: A Chircop, 'Advances in Ocean Knowledge and Skill: Implications for the MSR Regime', in MH Nordquist et al. (eds), Law, Science & Ocean Management (2007) 575 et seq. and JA Roach, 'Defining Scientific Research: Marine Data Collection', in MH Nordquist et al. (eds), Law, Science & Ocean Management (2007) 541 et seq.; see also JA Roach, 'Marine Data Collection: Methods and the Law', in JN Moore, TTB Koh and MH Nordquist (eds), Freedom of Seas, Passage Rights and the 1982 Law of the Sea Convention (2009) 171 et seq.

[5] J Mossop, 'Regulating Uses of Marine Biodiversity on the Outer Continental Shelf', in D Vidas (ed.), Law, Technology and Science for Oceans in Globalisation: IUU Fishing, Oil Pollution, Bioprospecting, Outer Continental Shelf (2010) 319 et seq. She suggests the balance will depend on circumstances.

MSR is a high seas freedom enjoyed by all States under Art. 87 UNCLOS. Thus, in the context of research in the water column above an extended continental shelf, that freedom of research is included in Art. 78's requirement that a State's exercise of its rights over the continental shelf 'must not infringe or result in any unjustifiable interference with navigation and other rights and freedoms of other States [...].' However, to balance this freedom Art. 246(8) UNCLOS provides that MSR activities in the EEZ and on the continental shelf 'shall not unjustifiably interfere with activities undertaken by coastal States in the exercise of their sovereign rights and jurisdiction provided for in this Convention.'

The first place to look for shared principles regarding MSR is UNCLOS itself, where Art. 240 identifies four 'General principles for the conduct of marine scientific research'. MSR shall be conducted (a) 'exclusively for peaceful purposes'; and (b) 'with appropriate scientific methods and means compatible with this Convention'; it shall (c) 'not unjustifiably interfere with other legitimate uses of the sea compatible with this Convention and shall be duly respected in the course of such uses'; and (d) 'shall be conducted in compliance with all relevant regulations adopted in conformity with this Convention including those for the protection and preservation of the marine environment'.

For the question of access to the continental shelf, the principle in Art. 240 UNCLOS that other uses of the sea shall duly respect MSR is particularly important. One way for other uses by the coastal State to 'duly respect MSR' is to accord it appropriate treatment not only if it is an activity already underway but also in the permitting process. Does this principle of respect for MSR in Art. 240(c) also mean that MSR is to be respected even when balancing scientific use against other coastal State activities? As we have just seen in Art. 246(8) UNCLOS, clearly yes, as long as MSR on the continental shelf does 'not unjustifiably interfere with activities undertaken by coastal States in the exercise of their sovereign rights and jurisdiction provided for in this Convention.' Coastal State sovereign rights in the continental shelf are not absolute; they are limited to exploration and exploitation of shelf resources, e.g. mineral and other non-living resources of the seabed and subsoil and living sedentary species tied to the shelf (Art. 77 UNCLOS).

In the context of marine bioprospecting she suggests several factors, including: 'Is the proposed interference with high-seas rights as minimal as possible in order to achieve the coastal state's objectives, or is a less restrictive option available to the coastal state?' (at 328).

As to what constitutes MSR, while UNCLOS does not define MSR, it does distinguish between 'marine scientific research and hydrographic surveys' in Art. 21 when specifying matters for which a coastal State may adopt laws and regulations on innocent passage through its territorial sea.[6] Thus some States consider that hydrographic surveys are excluded from the permission requirements of Part XIII UNCLOS, though other regulatory regimes may apply.[7] In his presentation at the Berlin Arctic Marine Science Conference, Uwe Nixdorf highlighted an earlier study by this author and Hajo Eicken that speculates whether understandings of what constitutes MSR will change as surveying and operational information needs increase along with increasing marine activity in the Arctic.[8]

---

6 See e.g. Roach [2009], see note 4; on non-definition of MSR see W Jilu, 'The Concept of Marine Scientific Research', in P Dutton (ed.), Military Activities in the EEZ: A U.S.-China Dialogue on Security and International Law in the Maritime Commons (2011) 65 et seq.

7 The US Department of State considers that activities not amounting to marine scientific research include:

> prospecting for and exploration of natural resources; hydrographic surveys (for enhancing the safety of navigation); military activities including military surveys; environmental monitoring and assessment of marine pollution pursuant to section 4 of Part XII of the Convention; the collection of marine meteorological data and other routine ocean observations, including through the voluntary ocean observation programs of the Joint IOC-WMO Technical Commission on Oceanography and Marine Meteorology (JCOMM) and the Argo program; and activities related to submerged wrecks or objects of an archeological and historical nature;

see US Department of State, Office of Ocean and Polar Affairs ('OPA') within the Department's Bureau of Oceans and International Environmental and Scientific Affairs ('OES'), 'Marine Scientific Research Authorizations', see <http://www.state.gov/g/oes/ocns/opa/rvc/index.htm> (8 June 2011).

8 B Baker and H Eicken, 'Marine Research Access in the Arctic Ocean: Background for Potential Guidelines in a Changing Arctic', unpublished White Paper (10 March 2010), see <http://www.iarc.uaf.edu/workshops/2009/4/> (8 June 2011) (click on 'download whitepaper'); the paper is also attached as an Appendix to this article.

II. National Strategy and Policy Documents regarding the Arctic

National policy and strategy statements regarding the Arctic are another potential source of precepts for marine scientific research access to the Arctic Ocean. In the 2008 Ilulissat Declaration the five Arctic littoral States affirmed their collective commitment to follow the 'law of the sea' for questions of 'marine scientific research' and other uses of the sea.[9] Individually each of these five States has published an Arctic policy or strategy, beginning with Norway in 2006.[10] Of the three remaining Arctic States, Finland presented its Arctic strategy document in July of 2010, acknowledging the United Nations Convention on the Law of the Sea's 'comprehensive regulation of use of the sea' in the Arctic;[11] Iceland and Sweden have not issued comparably comprehensive statements. Looking beyond national policy statements, the Inuit Circumpolar Council issued 'A Circumpolar Inuit Declaration on Sovereignty in the Arctic' in 2009 that shares some of the themes of the State policies.[12]

Promoting sovereign interests and engaging in international cooperation is a common theme in these seven strategy statements. The six national strategies specify international and regional cooperation and, specifically, international scientific cooperation, as key to their purposes.

---

[9] Ilulissat Declaration (28 May 2008) (2009) 48 ILM 382: 'Notably, the law of the sea provides for important rights and obligations concerning the delineation of the outer limits of the continental shelf, the protection of the marine environment, including ice-covered areas, freedom of navigation, marine scientific research, and other uses of the sea' (at para. 3).

[10] Norwegian Ministry of Foreign Affairs, 'The Norwegian's Government High North Strategy' (1 December 2006), see <http://www.regjeringen.no/upload/UD/Vedlegg/strategien.pdf> (8 June 2011) ('Norwegian Strategy').

[11] Finland, Prime Minister's Office, Finland's Strategy for the Arctic (2010), see also <http://www.geopoliticsnorth.org/images/stories/attachments/Finland.pdf> (8 June 2011) ('Finland's Strategy for the Arctic'); which states that '[t]he United Nations Convention on the Law of the Sea comprehensively regulates the use of the sea and creates the framework for settling territorial issues' (at 35).

[12] Inuit Circumpolar Council, 'A Circumpolar Inuit Declaration on Sovereignty in the Arctic' (April 2009), see <http://www.inuit.org/fileadmin/user_upload/File/declarations/ICC_Sovereignty_Declaration_2009_pages.pdf> (8 June 2011).

All seven also mention climate change, some with greater emphasis than others. Without detailing each of the national strategies, this paper highlights a few key points from each document specific to science organizations, marine scientific research, or questions of access for research, to see what concepts might inform a list of common precepts for MSR in the Arctic.

Of all national policies and strategies, Canada's 2010 'Statement on Canada's Arctic Foreign Policy' links the government's actions related to Arctic science most directly to the exercise of national sovereignty.[13] The Statement expands on the four pillars of Canada's Northern Strategy, announced in 2009,[14] of exercising sovereignty, promoting economic and social development, protecting the Arctic environment and improving and devolving governance: empowering the peoples of the North. Science plays the largest role in environmental protection, in the form of ecosystem-based management, addressing climate change through international cooperation and strengthening the International Polar Year ('IPY') legacy. In addition, 'Arctic science and research, including traditional knowledge' must inform standards for sustainable development; in 'no area is this more critical than in oil and gas development.'[15]

Denmark's 2008 'Proposed Strategy for Activities in the Arctic Area' focuses on Arctic and Nordic cooperation, stressing the importance of the Arctic Council and including cooperation and membership in international and regional organizations.[16] As to science specifically, Den-

---

[13] Government of Canada, 'Statement on Canada's Arctic Foreign Policy: Exercising Sovereignty and Promoting Canada's NORTHERN STRATEGY Abroad' (2010), see <http://www.international.gc.ca/polar-polaire/assets/pdfs/CAFP_booklet-PECA_livret-eng.pdf> (8 June 2011) ('Statement on Canada's Arctic Foreign Policy'), stating that 'Canada exercises its sovereignty daily through good governance and responsible stewardship. It does so through the broad range of actions it undertakes as a government [including] Arctic science and research' (at 5).

[14] Government of Canada, Minister of Indian Affairs and Northern Development and Federal Interlocutor for Métis and Non-Status Indians, Canada's Northern Strategy: Our North, Our Heritage, Our Future (2009), see also <http://www.northernstrategy.ca/cns/cns.pdf> (11 July 2011).

[15] Statement on Canada's Arctic Foreign Policy, see note 13, 11.

[16] Denmark, 'Arktis i En Brydningstid: Forslag Til Strategi for Aktiviteter i Det Arktiske Område', ['Arctic in a Time of Change: Proposed Strategy for Activities in the Arctic Area'] (May 2008), see <www.um.dk/NR/rdonlyres/962

mark's strategy speaks of ongoing resource-oriented and climate research with international partners and its membership in international Arctic research and science organizations.[17]

'Finland's Strategy for the Arctic' takes a broader view of research than the other national strategies, acknowledging investments required to maintain its 'top-level Arctic know-how' in many sectors.[18] Maintaining that level requires relevant university-level education, correctly targeted investments, and national and international cooperation.[19] International cooperation is a means of improving quality, eliminating overlapping activities, and gathering domestic and foreign resources for joint projects. The Finnish Strategy notes that research 'plays a key role in adaptation to Arctic climate change and in the utilisation of opportunities offered by the Arctic Region' and has a 'social dimension' relevant to the future of Northern Finland.[20] Its section 6 refers to Arctic Policy Tools devoted to Finland's multilateral and bilateral cooperation, with an emphasis on the Arctic Council.

The Circumpolar Inuit Declaration on Sovereignty in the Arctic 2009[21] speaks not of science access but more broadly of the increased ease of access to the Arctic and increased State focus on Arctic resources. In this connection it asserts that 'Inuit inclusion as active partners is central to all national and international deliberations on Arctic sovereignty and related questions, such as who owns the Arctic, [and] who has the right to traverse the Arctic' (Art. 3.6). The Declaration highlights the importance of Inuit knowledge throughout, specifying for example the unique Inuit knowledge of Arctic ecosystems (Art. 3.4) and that 'Inuit consent, expertise and perspectives are critical to progress on international issues involving the Arctic, such as global environmental security, sustainable development, militarization, commercial fishing, shipping, human health, and economic and social development' (Art. 3.5). In addressing climate change 'international efforts, protocols and treaties

AFDC2-30CE-412D-B7C7-070241C7D9D8/0/ARKTISK_STRATEGI.pdf> (8 June 2011), see page 13 regarding, e.g. the Arctic Council.

17 *Ibid.*, 32-33.

18 Finland's Strategy for the Arctic, see note 11, 20.

19 *Ibid.*

20 *Ibid.*, 22.

21 A Circumpolar Inuit Declaration on Sovereignty in the Arctic, see note 12.

cannot be successful without the full participation and cooperation of indigenous peoples' (Art. 3.10).

For the Russian Federation, regional and international cooperation are among the strategic priorities of State policy in the Russian Federation in the Arctic to 2020 and beyond. Other priorities include 'strengthening economic, scientific, technical and cultural cooperation as well as cross-border cooperation, including in the efficient use of natural resources and conservation of the environment in the Arctic.'[22] The 'expansion of basic and applied research in the Arctic' is seen as one means of 'improving public administration and social and economic development of the Arctic zone of the Russian Federation.'[23] Main objectives and principles of the separate document 'Russian Marine Policy through 2020',[24] which is not Arctic-specific, include 'realization and protection of sovereign rights over the continental shelf of the Russian Federation [...] and protection of freedom of the seas, including freedom of [...] research.'[25]

Only the Norwegian and US Strategies appear to mention explicitly the issue of access to other States' areas for research. In the context of fisheries research and management, Norway's Strategy states that 'Norwegian marine scientists should be able to work in the Russian zone under the same conditions as those that apply to Russian scientists in the Norwegian zone.'[26] This focus on Russia grows out of the fact that 're-

---

22 'Osnovi gosudarstvennoi politiki Rossiyskoi Federatsii v Arktike na period do 2020 goda i dalneishuju perspektivu' ('The Fundamentals of the State Policy of the Russian Federation in the Arctic in the Period Up to 2020 and Beyond') Utverzhdeni Presidentom Rossiyskoi Federatsii (adopted by the President of Russia) (18 September 2008) promulgated on 30 March 2009 in the 'Rossiyskaya Gazetta', see also <http://www.scrf.gov.ru/documents/98.html> (8 June 2011) section 7, ('Russian Federation Arctic Policy Up to 2020 and Beyond').

23 *Ibid.*, para. 7.g.

24 Russian Federation, 'Maritime Doctrine of Russian Federation 2020', Approved by President Vladimir Putin (27 July 2001), Pr-1387, see <http://www.oceanlaw.org/downloads/arctic/Russian_Maritime_Policy_2020.pdf> (8 June 2011).

25 *Ibid.*, 3.

26 Norwegian Strategy, see note 10, following a statement that the best possible basis for decisions on total allowable catch is if 'researchers from both countries have access to the whole of the Barents Sea' (at 53).

lations with Russia form the central bilateral dimension of Norway's High North policy.'[27]

The United States devotes an entire section of its Arctic Region Policy to 'Promoting International Scientific Cooperation' and deals with research access at great length.[28] It calls for research 'access throughout the Arctic Ocean and to terrestrial sites, as well as viable international mechanisms for sharing access to research platforms' and notes the particular importance of '[b]etter coordination with the Russian Federation, facilitating access to its domain [...].'[29] The Policy directs the heads of executive agencies and departments to '[a]ctively promote full and appropriate access by scientists to Arctic research sites through bilateral and multilateral measures and by other means.'[30] Speaking directly to the need for access to the seabed and subfloor to understand climate change the US Arctic Region Policy states that '[a]ccurate prediction of future environmental and climate change on a regional basis, [...] requires obtaining, analyzing, and disseminating accurate data from the entire Arctic region, including both paleoclimatic data and observational data.'[31]

Brosnan, Leschine and Miles have also recently evaluated the Arctic strategy statements of the five Arctic Ocean littoral States.[32] They analyze 'scientific research' as one of six themes common to all of the strategies studied (the five other themes are environmental concerns, resource development, sovereignty, governance and shipping).[33] The three reviewers condense the strategies' discussions of Arctic scientific research into 'just two issues': research for national priorities and research

---

27 *Ibid.*, 17.

28 United States National Security Presidential Directive and Homeland Security Presidential Directive NSPD 66/HSPD 25 'Arctic Region Policy' (9 January 2009) (2009) 48 ILM 274, see also <http://www.fas.org/irp/offdocs/nspd/nspd-66.htm> (8 June 2011), ('US Arctic Region Policy') Part III.E.

29 *Ibid.*, Part III.E.1.

30 *Ibid.*, Part III.E.5.b.

31 *Ibid.*, Part III.E.3.

32 IG Brosnan, TM Leschine and EL Miles, 'Cooperation or Conflict in a Changing Arctic?', Ocean Development & International Law 42 (2011) 173 et seq., who analyze the Arctic strategy statements of the five Arctic Ocean coastal States.

33 *Ibid.*, Table 1 (at 177-178); Table 3 (at 183-184) provides a 'Summary of scientific research issues in the strategies of the five coastal Arctic states'.

for 'improved understanding and forecasting of Arctic climate change and its physical and biological impacts.'[34] They observe that each of the five national strategy statements in their analysis includes Arctic change or climate change as a research component.

From the foregoing we can articulate four common precepts applicable to Arctic marine scientific research:

1) Improve scientific and other understanding of climate and global change;[35]
2) Exercise and strengthen sovereignty while working with neighbors;
3) Involve indigenous expertise, perspectives and knowledge; and
4) Cooperate internationally to address Arctic challenges and opportunities.

Considered as a group, then, the six national strategy documents for the Arctic and the ICC Declaration on Inuit Sovereignty reviewed briefly in this section can support an unsurprising but important conclusion: that an express acknowledgement of the need for international cooperation and a greater understanding of the Arctic should undergird any common precepts for marine scientific research access in the Arctic.

III. Bilateral Agreements (Science and Technology Cooperation Agreements, Ministerial-level Agreements, other Agreements)

The principles to which the Arctic Ocean coastal States have agreed formally in bilateral Science and Technology Cooperation Agreements

---

[34] *Ibid.* The national activities and priorities vary but generally include 'socioeconomics, human health, impacts of anthropogenic activities on the environment, and resource assessments' (at 180).

[35] K Sloan and D Hik, 'International Polar Year as a Catalyst for Sustaining Arctic Research', Sustainable Development Law & Policy 8 (2008) 4 et seq. (6), text following footnote 38: 'There is a strong consensus that scientific understanding of the changing Arctic system and its global connections and consequences requires improved Arctic observing capabilities that are linked to global observing activities.'

(e.g. between the United States and Russia,[36] Norway[37] and Denmark[38]) provide another source of common precepts for MSR in the Arctic, even though the agreements are not Arctic-specific. This author has identified elsewhere[39] the four identical 'principles' that govern the Denmark-United States and Norway-United States science and technology cooperation agreements:

1) Mutual benefit based on an overall balance of advantages;
2) Reciprocal opportunities to engage in cooperative activities;
3) Equitable and fair treatment for the participants; and
4) Timely exchange of information which may affect cooperative activities.[40]

The two agreements also specify that the parties are to treat requests for access 'with diligence, taking into account the significance of these ac-

---

[36] Agreement between the Government of the United States of America and the Government of the Russian Federation on Science and Technology Cooperation (signed 16 December 1993) TIAS 12527 ('Science and Technology Cooperation Agreement Russia-US').

[37] Agreement on Science and Technology Cooperation between the Government of the United States of America and the Government of the Kingdom of Norway (signed 9 December 2005), see <www. http://www.forskningsradet.no> (10 June 2011) ('Science and Technology Cooperation Agreement Norway-US').

[38] Agreement between the Government of the Kingdom of Denmark and the Government of the United States of America for Scientific and Technological Cooperation (signed 15 September 2009), see <http://www.euussciencetechnology.eu/link2us/st-agreements.html> (10 June 2011) ('Science and Technology Cooperation Agreement Denmark-US').

[39] B Baker, 'Reliability of Access for Marine Scientific Research (MSR) to the Arctic Ocean and the Possibility of an MSR Code of Conduct', Remarks to the PAME-Arctic Ocean Review Workshop, Washington, DC (13 September 2010); B Baker, 'Polar Science in the North and South: Tailoring Lessons from Antarctica to Improve Reliability of Legal Access for Marine Scientific Research (MSR) to the Arctic Ocean', in D König and P Stoll (eds), Coexistence, Cooperation and Solidarity – Liber Amicorum Rüdiger Wolfrum (forthcoming 2011).

[40] The two separate agreements contain an identical paragraph 3; see notes 37 and 38.

tivities to the advancement of scientific knowledge.'[41] Improved information sharing about MSR permitting processes, a common theme at the March 2011 Berlin Conference on Arctic Marine Science, could fit under the fourth principle above that calls for timely exchange. Such information sharing, like that which Sergey Priamikov and others proposed at the conference, would build on the IPY scoping study's proposed central database on permitting processes for access to each of the arctic coastal State's EEZs.[42] 'Timely exchange' could also be a criterion for informing applicants of decisions, rather than just letting permission requests wither from evident inattention.

The 1993 Russian Federation-United States[43] agreement does not contain a separate section on 'principles' but refers to cooperation based on 'shared responsibilities, contributions and benefits' (Art. 2(2)). Further, its parties 'shall facilitate entry into and exit from its territory of appropriate personnel and equipment of the other Party' (Art. 9). Each party shall also 'effectively implement [...] travel to its relevant geographic areas', and 'facilitate duty free entry for necessary materials and equipment provided pursuant to this Agreement for use in joint activities' (*id.*). While these are binding legal obligations as between Russia and the United States, they could also be re-stated as non-binding precepts common to marine science in the Arctic.

By contrast, the 2009 Russian-German Agreement on Scientific and Technical Cooperation[44] does not contain such detailed specifications regarding entry, exit and travel. The only related provision states that such questions shall be regulated according to national laws and international obligations of the two countries (Art. 7). The German-Russian agreement specifies focal points for research, in light of national priorities in science and technology, previously established relationships and common experiences in developing their cooperation. One of these focal points is marine and polar research (Art. 4).

---

[41] Art. 6(d) Science and Technology Cooperation Agreement Norway-US, see note 37; Art. 6(4) Science and Technology Cooperation Agreement Denmark-US, see note 38.

[42] See Dallman and Hoel, see note 1, 19.

[43] Science and Technology Cooperation Agreement Russia-US, see note 36.

[44] Abkommen zwischen der Regierung der Bundesrepublik Deutschland und der Regierung der Russischen Föderation über wissenschaftlich-technische Zusammenarbeit (16 July 2009), see <http://www.bmbf.de/pubRD/Abkommen_D_RUS_unterzeichnet.pdf> (10 June 2011); all translations of this Agreement are the author's own.

Each of the science and technology cooperation agreements discussed above provides for regular meetings of the contracting parties, usually using a commission established under the treaty. Regular meetings of such joint commissions are critical to the success of collaborative research, as they cultivate institutional and personal contacts, not only for planning and logistical challenges but also, as needed, for resolving any areas of disagreement. This last feature is all the more important because a coastal State's decision to grant or deny consent for MSR is not subject to compulsory dispute resolution under the UNCLOS, although conciliation may be used.[45]

Other possible sources for common precepts of marine research in the Arctic include ministerial level agreements.[46] A prime example is the non-binding 'Memorandum of Understanding between the National Oceanic and Atmospheric Administration of the Department of Commerce of the United States of America and the Russian Academy of Sciences of the Russian Federation on Cooperation in the Area of the World Oceans and Polar Regions' of 5 December 2003.[47] Art. II specifies five principles by which the '[p]arties shall conduct world oceans and polar regions cooperation', including 'shared responsibilities and shared results', 'access to and exchange of information in the field of world oceans and polar regions' scientific development' to the extent allowed by national laws, protection of intellectual property, transparency and, finally, the widest possible dissemination of research results. The principles of access and transparency are most important for ensuring reliable access, the latter principle calling for '[g]eneral transparency

---

45 Mossop, see note 5, 324; Art. 297(2) UNCLOS.

46 See e.g.: Memorandum of Understanding between the National Oceanic and Atmospheric Administration of the Department of Commerce of the United States of America and the Russian Academy of Sciences of the Russian Federation on Cooperation in the Area of the World Oceans and Polar Regions (done 5 December 2003), see <http://www.arctic.noaa.gov/aro/russian-american/noaa-ras-mou-english.pdf> (10 June 2011), and Memorandum of Understanding between the National Oceanic and Atmospheric Administration US Department of Commerce United States of America and the Department of the Environment Canada For Collaboration on Weather, Climate and Other Earth Systems for the Enhancement of Health, Safety and Economic Prosperity (done 18 January 2008), see <http://www.weather.gov/iao/ia/hom/IAOCanada.php> (10 June 2011), stating that '[w]hile the Participants intend to respect these responsibilities, this MoU is not legally binding in either domestic or international law' (at 2, Preamble).

47 *Ibid.*

of policies and programs in order to facilitate mutual understanding and the identification of opportunities for coordination and cooperation' (*id.*).

Memoranda of Understanding ['MoU(s)'] with other branches of coastal State governments may also provide common precepts. At a September 2010 workshop of the Polar Research Board of the US National Science Foundation, the clearest message for international Arctic research was that personal friendships and collegial contacts are essential and that scientists in the host country can often facilitate the permissions process.[48] For Russian projects a key point was the need to ensure support of the Russian navy for oceanographic work which, of course, is not covered under the UNCLOS's MSR requirements. However, in this connection it is worth noting continuation of the US Navy's collaboration with US-based scientists on the SCICEX cruises, under an MoU whose purpose 'is to facilitate the use of U.S. Navy submarines for scientific research in the Arctic'[49] the Navy's Office of Naval Research and the National Science Foundation intend that 'all data collected under this agreement be made publically [sic] available as soon as possible after collection, following the data policies of the respective sponsors.'[50]

In addition to the four precepts identified in part II, eight more common precepts for Arctic MSR can be gleaned from the foregoing discussion of binding and non-binding bilateral agreements. All twelve precepts are intended to complement and not to repeat requirements of the UN Law of the Sea Convention. States participating in MSR in the Arctic Ocean shall strive to:

1) Promote improved scientific understanding through MSR;
2) Exercise and promote national sovereignty through MSR cooperation with other States;
3) Continually improve transparency of MSR permitting processes;
4) Facilitate mobility of researchers according to national laws;

---

48 US National Science Foundation, Polar Research Board, Workshop on US Scientific Access to the Russian Arctic, Washington, DC (22 September 2010), author's personal notes from conference.

49 Memorandum of Agreement, 'Submarine Arctic Science Program – Phase 2', between the US Submarine Force, the Chief of Naval Research and the National Science Foundation (2000), see <http://www.csp.navy.mil/asl/Scicex/SCICEX_MoA.pdf> (10 June 2011), para. 1.

50 *Ibid.*, para. 3.

5) Ensure equitable, fair and predictable treatment for research project participants;
6) Exchange planning and other information affecting cooperative activities on a timely basis, using regular meetings of established or ad hoc joint research committees as appropriate;[51]
7) Release data and research results publicly as soon as possible, honoring the data policies of the States involved; and
8) Cooperate in promoting MSR for mutual benefit based on an overall balance of advantages.

These precepts also complement the four stated principles for MSR in Part XIII UNCLOS at Art. 240: that MSR shall be conducted exclusively for peaceful purposes, with appropriate scientific methods and means compatible with the Convention, in compliance with protection and preservation of the marine environment, and without unjustifiably interfering with other legitimate uses of the sea while being duly respected by other such uses. Finally, the first two precepts repeat common ideas derived from national Arctic strategy documents in part III of this paper. Precept 1) repeats the need for greater understanding and Precept 2) combines the notions of strengthened sovereignty and international cooperation expressed in each of the national strategy documents.

IV. Scientific Organizations

A number of scientific organizations have articulated purposes and principles relevant to providing reliable access for scientific research in the Arctic Ocean.[52] This section highlights two Arctic organizations,

---

[51] The two separate agreements contain an identical para. 3, see Science and Technology Cooperation Agreement Norway-US, see note 37; Science and Technology Cooperation Agreement Denmark-US, see note 38.

[52] See e.g. the recent MoU between Nordic polar research organizations of 26 June 2009, reported in the 5th issue of the European Polar Board Newsletter (June 2010), see <http://www.esf.org/research-areas/polar-sciences/news.html> (13 June 2011). One of the four areas of focus in the MoU is 'Infrastructure Coordination and Access', but it deals more with access to research stations:

> The cooperation and aligning of observations between European and international research stations in the Arctic and Antarctic is becoming more and more essential in terms of high quality research, monitoring and cost effectiveness of

the International Arctic Science Committee ('IASC') and the International Arctic Social Sciences Association ('IASSA'), and two that are not Arctic-specific, the International Council for Science ('ICSU') and the North Pacific Marine Science Organization ('PICES').[53]

1. International Arctic Science Committee and International Council for Science

IASC is 'a non-governmental, international scientific organization established to encourage and facilitate international consultation and co-operation in all aspects of arctic research'.[54] Following IASC's 2009 merger agreement with the Arctic Ocean Science Board ('AOSB'), the latter is now known as the Scientific Standing Committee for Marine Sciences of IASC.[55]

IASC's mission is 'to encourage, facilitate and promote leading-edge multi-disciplinary research to foster a greater scientific understanding

---

access. The European Polar Framework MOU is the proper instrument for providing Europe with high standards variable geometry platforms and clusters of facilities for large scale observations and monitoring (at 6).

53 The ICSU was formerly known as the International Council of Scientific Unions, and has been a Formal Associate of the United Nations Educational, Scientific and Cultural Organization ('UNESCO') since 1995, see e.g. Commission on the Limits of the Continental Shelf, 'Scientific and Technical Guidelines of the Commission on the Limits of the Continental Shelf' (13 May 1999) CLCS/11, Annex, List of International Organizations.

54 IASC Council Meeting, Arctic Science Summit Week (2009) draft Agenda, ANNEX 2.1.1a, BYLAWS -Draft-, see <http://arcticportal.org/uploads/Cv/hw/CvhwwsOPGTd1wu_rlgygQg/ANNEX-Council-Meeting-ASSW-2009.pdf> (13 June 2011) para 1.1.

55 Agreement between The Arctic Ocean Science Board and The International Arctic Science Committee (27 March, 2009) (unsigned version), *ibid.* ANNEX 2.1.3:

> The mission of AOSB, "to facilitate research in the Arctic Ocean and surrounding seas through the support of multinational and multidisciplinary natural science and engineering programs" will remain unchanged. It will be carried out, however, within the broader mission of IASC which is "to encourage and facilitate cooperation in all aspects of arctic research, in all countries engaged in arctic research and in all areas of the Arctic."

of the arctic region and its role in the Earth system'.[56] This linkage between the Arctic and its role in the earth system can provide an important expansion of the first precept proposed above, promoting improved scientific understanding generally through MSR. One of several strategic means IASC identifies to achieve its mission is directly related to the question of access: '[p]romoting international access to all arctic regions and the sharing of knowledge, technologies, logistics and other resources.'[57] Another is by '[s]eeking to ensure that scientific data and information from the Arctic are sustained, are freely exchanged, are made available to those living in the region, and are easily accessible for anyone seeking such information'.[58]

IASC is an International Scientific Associate of ICSU, the first non-Arctic organization mentioned above. The Scientific Committee on Antarctic Research ('SCAR') and other non-polar scientific organizations are also ICSU International Scientific Associates.[59] The polar regions are but one area of interest for ICSU, which addressed the question of access for research purposes in its April 2010 Statement on Universality of Science in the Polar Regions.[60] The Statement makes no direct reference to the legal regime of MSR, but calls on all parties conducting or influencing polar research to support '1. continued and responsible access to all areas of the Arctic and Antarctic for research purposes'. The Statement further acknowledges the need to develop and adhere to policies, procedures and regulations to ensure effective access mechanisms; a point worth including in any statement of common precepts.[61] The statement emphasizes the importance of data access, calling

---

[56] *Ibid.* ANNEX 2.1.1.a, para. 2.1.

[57] *Ibid.* ANNEX 2.1.1.a, para. 2.2.6.

[58] *Ibid.* ANNEX 2.1.1.a, para. 2.2.5.

[59] United States Department of State, Handbook of the Antarctic Treaty System, 9th ed. (2002), Chapter VI, Facilitation of International Scientific Cooperation.

[60] ISCU Statement, 'Universality of Science in the Polar Regions' (April 2010), see <http://www.icsu.org/publications/icsu-position–statements/universality-polar-regions> (13 June 2011).

[61] *Ibid.*, stating that

ICSU calls on all parties conducting or influencing polar research to support the principle of Universality of Science in general, and specifically to support:

1. continued and responsible access to all areas of the Arctic and Antarctic for research purposes;

for both 'full, open, and timely access to polar research data and information' (para. 2) and for polar scientists to continue to recognize their 'responsibilities toward sharing and stewardship of data and information' (para. 5). A substantial portion of the background material accompanying the Statement elaborates at length on data practices, with a special emphasis on continuing the work of the International Polar Year 2007-2009.

It bears emphasis that since the end of the IPY, as seen above, both IASC and ICSU have restated the importance of access to all areas or regions of the Arctic for research.

2. International Arctic Social Sciences Association

IASSA adopted Guiding Principles for the Conduct of Research in 1998.[62] IASSA is primarily an Arctic social sciences association but its Guiding Principles also apply to the natural and health sciences.[63] The

---

[...]

4. development of and adherence to policies, procedures, and regulations — and provision of resources—to ensure the long-term effectiveness of the mechanisms that are necessary to deliver 1-3;

[...]

This statement is endorsed by the Executive Board of the International Council for Science (ICSU) and its Committee on Freedom and Responsibility in the conduct of Science.

Items 2 and 3 indicated by the ellipses above read in full:

2. full, open, and timely access to polar research data and information for research and educational purposes and for sustainable development of polar regions;

3. continued development of international research capacity, coordination, and collaboration, including sharing of data and information and pooling of additional research capabilities;

[...]

62 IASSA General Assembly, 'Guiding Principles for the Conduct of Research' (23 May 1998), see <http://www.iassa.org/about-iassa/research-principles> (13 June 2011).

63 *Ibid.*, stating that '[t]hese Principles have been formulated to provide guidelines for all researchers working in the North in the social, natural and health sciences, and in the humanities' (Preface).

principles 'are intended to promote mutual respect, communication and partnerships between researchers and northern residents.'[64] The ten IASSA Guiding Principles introduce one consideration for access not articulated in any of the science and technology bilateral agreements discussed in part III, above: interaction with affected individuals and communities, including obtaining their consent and sharing research results with them. However, the first IASSA Guiding Principle does encompass the need to consult with appropriate authorities that is presumed, if not stated expressly, in the bilateral agreements above:

> 1. The researcher should consult with the appropriate regional and/or local authorities regarding planned research within their territories. Informed consent should be obtained from appropriate authorities and from any individuals involved in the research.[65]

Of the nine other IASSA principles, one complements precepts for MSR in the Arctic stated above and one adds the dimension of traditional and local knowledge. In keeping with the precept proposed in Part III, above, to make research results available to the public, Principle 3 provides:

> 3. Research results should be presented to local communities in non-technical terms and where possible translated into local languages. Copies of research reports and other relevant materials should be made available to local communities.

The fifth IASSA Principle calls for incorporating local and traditional knowledge and experience:

> 5. The researcher must respect local cultural traditions, languages, and values. Efforts should be made to incorporate local and traditional knowledge and experience and to acknowledge the principle of cultural property.

---

64 *Ibid.*

65 The IASSA Guiding Principles, *ibid.*, predate the 2007 UN Declaration on the Rights of Indigenous Peoples (UNGA Res 61/178 'United Nations Declaration on the Rights of Indigenous Peoples' (13 September 2007) GAOR 61st Session Supp 49 vol 3, 15), which provides in Art. 32(2):

> States shall consult and cooperate in good faith with the indigenous peoples concerned through their own representative institutions in order to obtain their free and informed consent prior to the approval of any project affecting their lands or territories and other resources, particularly in connection with the development, utilization or exploitation of mineral, water or other resources.

Two other features of the IASSA Guiding Principles stated in their opening paragraph can also be adapted for common precepts for MSR access in the Arctic. First the 'statement is not intended to replace other national, professional or local guidelines' (Preface). Adding 'international' to this list would be in keeping with the shared understanding of the five Arctic coastal States that they do not want to replace the rules for MSR access under Part XIII UNCLOS, but simply render them more reliable. Finally, the IASSA Principle 'that there must be continuing assessment of the principles' (*id.*) is equally applicable to any Arctic MSR access precepts: they should not be viewed as static but as subject to continuous improvement through regular reviews based on practical experience.

3. North Pacific Marine Science Organization

PICES is a treaty-based intergovernmental organization. It is an especially important model for the question of MSR access to areas under coastal State jurisdiction because the Convention that establishes PICES ('PICES Convention') creates an integral relationship between PICES' purposes and the provisions for MSR in the UNCLOS.[66] Among PICES' purposes are 'to promote and coordinate marine scientific research in order to advance scientific knowledge of the area concerned' (Art. III (a)), which is the northern North Pacific and adjacent seas especially northward of 30 degrees North (Art. II). Further PICES purposes are to promote and coordinate MSR, including 'research with respect to the ocean environment and its interactions with land and atmosphere, its role in and response to global weather and climate change [...]'[67] as well as 'to promote the collection and exchange of information and data related to marine scientific research in the area concerned' (Art. III(b)).

All three Pacific States in the Arctic – Canada, Russia and the United States – are PICES members, along with Japan, China and Korea, each

66 Convention for a North Pacific Marine Science Organization (PICES) (done 12 December 1990, entered into force 24 March 1992), see <http://www.pices.int/about/convention.aspx> (12 July 2011).

67 *Ibid.*, Art. III(a); see also WS Wooster and SF Tjossem, 'Scientific Cooperation in the North Pacific: The PICES Project', San Diego Int'l L.J. 6 (2004-2005) 191 et seq.

of which has significant research programs in the Arctic.[68] The idea for a Pacific ICES (International Council for the Exploration of the Sea) was first discussed at a Food and Agriculture Organization ('FAO') meeting in Vancouver in 1973 and took almost 20 years to be realized; the PICES Convention entered into force in 1992.[69] In the preamble, the Contracting Parties recognize that, due to the vastness of the North Pacific Ocean, 'scientific understanding of the area can be best achieved through a spirit of international scientific cooperation on a mutually beneficial basis'. This language is comparable to two of the nine precepts identified under bilateral agreements above: improved scientific understanding and mutual benefit. Further, desiring to promote 'scientific cooperation and avoid duplication of effort' the parties explicitly acknowledge 'that the activity of the organization must be based on the principles and rules of the international law of the sea applicable to marine scientific research' (*id.*).

Critically, Art. XII(1) PICES Convention provides that nothing in the Convention or activities pursuant to it 'shall prejudice or in any way affect' a members'

(a) sovereignty, sovereign rights and jurisdiction over its territorial sea, 200 nautical mile zone, or continental shelf, including its jurisdiction over marine scientific research;
(b) the rights of a Contracting Party to manage its national research programs;
(c) other international agreements, bilateral or multilateral, to which Contracting Parties are party.

Additionally, under Art. XII(2), '[n]othing in this Convention shall be construed as authorizing the Organization to regulate the activities of the Contracting Parties.' All components of Art. XIII PICES Convention are in keeping with the nine precepts identified under bilateral agreements above.

The scientific organizations briefly reviewed above offer three additional common precepts for Arctic MSR, as well as 'limiting language' for those precepts. The three additional precepts relate to access to all areas of the Arctic, availability of data, and involvement of Arctic residents and communities. The 'limiting language' can be placed as book-

---

68 PICES, 'About', see <http://www.pices.int/about/default.aspx> (13 June 2011).

69 PICES, 'History', see <http://www.pices.int/about/default.aspx> (13 June 2011).

ends on the precepts themselves and thus appears below as the introductory paragraph (which borrows heavily from the PICES Convention), and the two concluding paragraphs, respectively. For clarity's sake, the complete list of precepts, with prefatory and concluding language, appears in the next and final section of this paper.

V. Conclusion

All five Arctic coastal States recognize that reliability of access for scientists to their areas of the Arctic Ocean for Marine Scientific Research under the UN Law of the Sea Convention can be strengthened without creating any new legal mechanisms. Existing documents offer a number of common areas of agreement for how MSR access reliability can be improved and national sovereignty strengthened. A review of four sources: the UNCLOS itself, national Arctic strategy documents, existing bilateral science and technology cooperation agreements between Arctic States, and documents of scientific organizations, suggests the following statement of such common precepts.

* * *

Ten Common Precepts of Marine Scientific Research in the Arctic – A Draft

These precepts are based on the principles and rules of the international law of the sea applicable to marine scientific research ('MSR'). They complement, strengthen and support and in no way prejudice or replace those principles and rules.

States participating in MSR in the Arctic Ocean shall strive to:

1. Promote greater scientific understanding of the Arctic region and its role in the earth system through MSR;
2. Develop and adhere to policies, procedures and regulations to promote and ensure responsible access to all areas of the Arctic, including the extended continental shelf, for research purposes;
3. Exercise and strengthen national sovereignty through scientific cooperation with other States;
4. Continually improve transparency, timeliness and consistency of MSR permitting and notification processes across the Arctic;

5. Facilitate mobility of researchers according to national laws;
6. Ensure equitable, fair, timely and predictable treatment for research project participants;
7. Draw on indigenous expertise, perspectives and knowledge and, as appropriate, obtain prior informed consent from, involve, and report to local communities;
8. Exchange planning and other information affecting cooperative activities on a timely basis, using regular meetings of established or ad hoc joint research committees as appropriate;
9. Release data and research results publicly as soon as possible, honoring the data policies of the States involved; and
10. Cooperate in promoting MSR for mutual benefit based on an overall balance of advantages.

Nothing in these precepts shall in any way prejudice or affect a coastal State's sovereignty, sovereign rights and jurisdiction over MSR and its maritime zones, its rights to manage its national research programs, or any international agreements to which it is party. These precepts shall be reviewed regularly, in the spirit of continuous improvement and sustainability.

* * *

Mutual benefit and international cooperation are key to the documents analyzed throughout this paper. States and organizations do not repeat these core ideas in the variety of settings studied here merely to express lofty sentiments. These formulations and precepts are critical for sovereign States as they work to promote their neighbors' stability and their own strategic interests in a changing Arctic through better science and better scientific access to the Arctic. These draft precepts are offered to the scientific and diplomatic communities as a starting point for shaping into a tool that will promote mutual benefit and international cooperation for all who share an interest in peaceful marine scientific research in the Arctic Ocean.

Appendix I: White Paper

Marine Research Access in the Arctic Ocean: Background for Potential Guidelines in a Changing Arctic*

*by Betsy Baker*** *and Hajo Eicken****

Scientists seeking to work in the Arctic Ocean have long faced physical and geopolitical constraints on access to its waters and its continental shelf. This was true before the UN Convention on the Law of the Sea ('UNCLOS')[1] entered into force in 1994 and it is true in 2010 as the scientific community builds on accomplishments and collaborative efforts of the 2007-2009 International Polar Year ('IPY').

Two major factors affecting Arctic Ocean access have undergone substantial change in the last decade: the geopolitics of confirming rights over the extended continental shelf under the UNCLOS, and major reductions in Arctic Ocean summer ice extent, greatly increasing the area of open water accessible to surface vessels and, potentially, for resource extraction. At the same time, technological innovations, such as increasing use of autonomous sensor platforms, are beginning to transform environmental research in extreme environments. These developments render it timely to consider what adjustments and innovations can improve the procedures currently governing access to the Arctic Ocean for marine scientific research ('MSR').

---

* Note: In order to render abbreviation, capitalization and citation formats consistent with other articles in this volume, the editors have made slight modifications from the White Paper as originally posted on the IARC website in March 2010, see note 8, above, and <http://www.iarc.uaf.edu/workshops/2009/4/> (8 June 2011) (click on 'download whitepaper').

** Vermont Law School & Dartmouth College, Institute of Arctic Studies, Dickey Fellow (2009-2010).

*** University of Alaska Fairbanks, Geophysical Institute & International Arctic Research Center.

1 United Nations Convention on the Law of the Sea (concluded 10 December 1982, entered into force 16 November 1994) 1833 UNTS 397.

This White Paper has four aims:

a. To provide background on current practice for marine scientific research access to the Arctic Ocean, definitions of MSR, and the potential for more restricted access to the Arctic continental shelf;
b. To introduce IPY-related initiatives for improved research access to the Arctic;
c. To propose possible fora in which to pursue more formalized access for MSR and to outline the need for further work and communication; and
d. To initiate a discourse within the international research community that can lead to the development of plausible scenarios for future research access, and in a collaborative setting to develop consensus and promote fair access for scientific research purposes.

With these four steps, this White Paper lays the groundwork for the next phase of our project, to be covered in a subsequent document: proposing a set of guidelines that will improve the reliability of access to the Arctic Ocean for scientific research, clarify access for scientific research there after extended continental shelves have been delineated, enhance the protection of national interests for Arctic Ocean coastal States, and promote the growth of trust amongst scientists, government authorities and local residents necessary for scientific research to proceed. Ideally, the non-binding guidelines to emerge from this project can serve as the foundation for any legally binding arrangement or more formal Memoranda of Understanding ['MoU(s)'] that might be deemed necessary in the future.

I. Background on Current Practice, Definitions of MSR and the Potential for More Restricted Access to the Arctic Continental Shelf

1. Current Practice

Each of the five Arctic Ocean littoral States now follows some variation of the rules set out in Part XIII UNCLOS that require scientists to obtain coastal State permission before conducting MSR in its territorial sea, exclusive economic zone ('EEZ') or on its continental shelf. Under 'normal circumstances', consent for access to the EEZ or continental shelf may not be unreasonably delayed or denied (Art. 246 UNCLOS) if the scientists seeking permission – through the appropriate government body, e.g. the Office of Ocean and Polar Affairs, US Department of State – follow procedures established by the coastal State and otherwise comply with the requirements of Part XIII UNCLOS, e.g., that the research is to be for peaceful purposes and to increase scientific knowledge of the marine environment for the benefit of all mankind. By way of example, Norway's procedures[2] reference application forms which in turn are based on the Draft Standard Forms developed by the United Nations Division of Ocean Affairs and Law of the Sea pursuant to Part XIII UNCLOS.[3]

In April 2009 the Intergovernmental Oceanographic Commission ('IOC') of the United Nations Educational, Scientific and Cultural Organization ('UNESCO') Advisory Body of Experts on the Law of the Sea ('ABE-LOS') published survey results regarding MSR permissions requested and granted by all coastal States from 1998-2002.[4] The results for the five Arctic Ocean littoral States (all waters, not just the Arctic Ocean) appear in Table 1.

---

2 Norway, Directorate of Fisheries, Marine Scientific Research, 'Application Form and Regulations', see <http://www.fiskeridir.no/english/fisheries/marine-scientific-research> (15 June 2011) para. 9.

3 United Nations, Division for Ocean Affairs and the Law of the Sea, The Law of the Sea: Marine Scientific Research, A Guide to the Implementation of the Relevant Provisions of the United Nations Convention on the Law of the Sea (1991).

4 E Tirpak, 'Excel File Analysis of Response to IOC Questionnaire No. 3' (13 April 2009), see <http://ioc-unesco.org/index.php?option=com_oe&task=viewDocumentRecord&docID=3571> (15 June 2011) ('Tirpak [2009-a]'); E Tirpak, 'IOC Questionnaire N°3: The Practices of States in the Field of Marine Scientific Research (MSR) and Transfer of Marine Technology (TMT)' (13 April 2009), see <http://ioc-unesco.org/index.php?option=com_oe&task=viewDocumentRecord&docID=3570> (15 June 2011); E Tirpak, 'Practice of IOC Member States in the Fields of Marine Scientific Research and Transfer of Marine Technology – An Analysis of Responses to ABE-LOS Questionnaire No. 3' (1 April 2009), see <http://www.ioc-unesco.org/index.php?option=com_oe&task=viewDocumentRecord&docID=3515> (15 June 2011).

Table 1. MSR permission requests received and approved by Arctic littoral States (all waters)[5]

1998-2002 MSR Permissions*	Canada	Denmark	Norway	Russian Federation	United States
Requests received annually	103	200	68	106	70
Percent Approved	98%	95%	99%	78%	100%

*Statistics shown are for all adjoining waters of each State, not just for the Arctic Ocean. Figures for the Arctic Ocean only are not separately published.

The survey results help considerably to address the lack of basic information regarding MSR practices generally noted by scholars and scientists studying MSR permitting processes.[6] Nonetheless, anecdotal reports indicate a growing concern specific to the Arctic about declining reliability of MSR access to the Arctic Ocean.[7]

---

5 Tirpak [2009-a], *ibid.*, IOC Survey of the Practice of States with Respect to Part XIII UNCLOS.

6 FHT Wegelein, Marine Scientific Research: The Operation and Status of Research Vessels and Other Platforms in International Law (2005); JA Knauss and MH Katsouros, 'Recent Experience of the United States in Conducting Marine Scientific Research in Coastal State Exclusive Economic Zones', in TA Clingan (ed.), The Law of the Sea: What Lies Ahead? (1986) 297 et seq.

7 International Arctic Research Center ('IARC'), 'Workshop: International Scientific Collaboration and Legal Regimes in a Changing Arctic Ocean', University of Alaska Fairbanks, Alaska (21 September 2009), see <http://www.iarc.uaf.edu/workshops/2009/4/> (15 June 2011); R Macnab, O Loken and A Anand, 'The Law of the Sea and Marine Scientific Research in the Arctic Ocean', Meridian Fall/Winter (2007), 1 et seq., see also <http://www.polarcom.gc.ca/uploads/Publications/Meridian%20Newsletter/MeridianFall2007.pdf> (15 June 2011).

2. What is Marine Scientific Research?

The UNCLOS does not define 'marine scientific research'. Several States take the view that, at a minimum, hydrographic surveys are excluded from the permission requirements of Part XIII (though other regulatory regimes may apply). This view is based on the distinction between 'marine scientific research and hydrographic surveys' in Art. 21 UNCLOS, under which the coastal State may adopt laws and regulations regarding innocent passage through its territorial sea.

The US Department of State considers that activities *not* amounting to marine scientific research include:

> prospecting for and exploration of natural resources; hydrographic surveys (for enhancing the safety of navigation); military activities including military surveys; environmental monitoring and assessment of marine pollution pursuant to section 4 of Part XII of the Convention; the collection of marine meteorological data and other routine ocean observations, including through the voluntary ocean observation programs of the Joint IOC-WMO Technical Commission on Oceanography and Marine Meteorology (JCOMM) and the Argo program; and activities related to submerged wrecks or objects of an archeological and historical nature.[8]

The activities listed above and not considered to be MSR, while not subject to the consent requirements referenced in part I.1 above, are subject to other legal regimes. Roach proposes 'marine data collection' as an umbrella term without legal content 'under which to consider the various activities for which the law of the sea does provide varying regimes depending on the maritime zone involved.'[9]

It is beyond the purview of this paper to discuss in detail possible definitions of MSR. The second MSR White Paper will address the pros and cons of defining MSR more specifically for any MSR access regime that might develop for the Arctic Ocean. For this paper, we simply highlight the fact that, as sea ice diminishes and vessel traffic and indus-

---

8 US Department of State, Office of Ocean and Polar Affairs ('OPA') within the Department's Bureau of Oceans and International Environmental and Scientific Affairs ('OES'), 'Marine Scientific Research Authorizations', see <http://www.state.gov/g/oes/ocns/opa/rvc/index.htm> (15 June 2011).

9 JA Roach, 'Marine Data Collection: Methods and the Law', in MH Nordquist, TTB Koh and JN Moore (eds), Freedom of Seas, Passage Rights and the 1982 Law of the Sea Convention (2009) 171 et seq.

trial activities increase in the Arctic Ocean,[10] surveying and operational information needs will increase and will require reliance on methods such as deployment of ice-tethered, autonomous sensor packages[11] that, in other oceans, may not be considered classic hydrography. Along these lines, we hope to focus future discussion on activities that are directly relevant for or part of research campaigns – such as hydrographic surveys, environmental monitoring and assessment of marine pollution, collection of marine meteorological data and other routine ocean observations, and archeological and historical studies – and to more fully articulate the different legal regimes that may apply to MSR and other kinds of 'marine data collection' in the Arctic.[12]

3. Potential for More Restricted Access to the Central Arctic Ocean: Shelf vs. Water Column vs. 'the Area'

All five Arctic Ocean littoral States are actively seeking to confirm, or have had confirmed, their rights over their extended continental shelves under the process spelled out in Art. 76 UNCLOS. Those rights comprise the exclusive right to explore and exploit the living and non-living resources of the seabed and subfloor, and sedentary species (Art. 77 UNCLOS). The confirmation of extended continental shelf rights has the potential to further restrict research access in the Arctic Ocean. Macnab has speculated that, when the process is completed – potentially decades from now – only a small portion of the Arctic Ocean will remain open to all States for MSR under the high seas freedoms guaranteed by UNCLOS. That portion would include only the Arctic Ocean seabed and ocean floor and subsoil thereof that remain beyond the limits of national jurisdiction. However, under Art. 257 UNCLOS, seaward of the EEZ MSR would still be allowed in the water column and, presumably, on the sea ice. The Central Arctic Ocean is already surrounded by an 'unbroken band' of 200 nm EEZs.[13]

---

[10] Arctic Council, Protection of the Marine Environment Working Group ('PAME'), 'Arctic Marine Shipping Assessment 2009 Report' (26 April 2009), see <http://www.pame.is/amsa/amsa-2009-report> (15 June 2011).

[11] A Proshutinsky et al., 'An Array of Ice-based Observatories for Arctic Studies', Eos Trans. AGU 85(46) (2005) 484 et seq.

[12] Roach, see note 9.

[13] Macnab, Loken and Anand, see note 7.

What does this mean for scientists seeking permission for MSR in the Arctic Ocean? Coastal States may regulate MSR access to both the water column and the continental shelf within their exclusive economic zones, typically 200 nm from the baselines from which the breadth of their territorial sea is measured. Seaward of the EEZ all States regardless of geographic location, as well as competent international organizations, have the high seas freedom, in keeping with the UNCLOS and coastal State regulations, 'to conduct marine scientific research in the water column beyond the limits of the exclusive economic zone' (Art. 257 UNCLOS). Seaward of the EEZ, on the continental shelf – as opposed to the water column –, in principle coastal States may withhold consent only for those parts of the shelf that they have designated in advance for actual or imminent exploitation or detailed exploratory operations (Art. 246(6) UNCLOS). However, this provision offers no practical hurdle to a State designating much or all of its continental shelf as being part of this exploration/exploitation area and thus more restricted for MSR acccess. Finally, all States have the right to conduct MSR in 'the Area', which the UNCLOS defines as 'the seabed and ocean floor and subsoil thereof beyond the limits of national jurisdiction' (Art. 1(1) UNCLOS), i.e. not included or delineated as the continental shelf of the coastal State.

In confirming their extended continental shelf rights, all five States are abiding by the relevant procedures prescribed in the UNCLOS. Norway received final Recommendations from the Commission on the Limits of the Continental Shelf ('CLCS') in 2009.[14] The Russian Federation, building on its initial submission in 2001, plans to submit additional data as the Commission requested in its interim Recommendation.[15] Canada and Denmark/Greenland have until 2013 and 2014, respectively, to initiate the submission process at the CLCS. The United States will not be able to submit data to the CLCS unless it accedes to the UNCLOS.

---

14 Commission on the Limits of the Continental Shelf, 'Summary of the Recommendations of the Commission on the Limits of the Continental Shelf in regard to the Submission Made by Norway in respect of Areas in the Arctic Ocean, the Barents Sea and the Norwegian Sea on 27 November 2006' (27 March 2009), see <http://www.un.org/Depts/los/clcs_new/submissions_files/nor06/nor_rec_summ.pdf> (15 June 2011).

15 For a short summary of the CLCS Recommendation concerning Russia, see UNGA, 'Oceans and the Law of the Sea: Report of the Secretary General' (8 October 2002) UN Doc A/57/57/Add.1, paras 38-41.

Even if the limits of all five States' continental shelves are not finally determined for another two or three decades, the scientific community is well advised to prepare in advance for what will be eventual control of the Arctic littoral States over significant portions of the Arctic continental shelf.

II. The Arctic Council and IPY-related Discussion of Access to the Arctic for Research Purposes

The International Polar Year 2007-2009 generated an unprecedented level of international cooperation and coordinated activities in Arctic marine scientific research, and increased interest in access issues. A scoping report prepared for the Arctic Council, 'Maximizing the Legacy of IPY', and presented to the Arctic Council in April 2009 highlighted 'access to study areas and infrastructure' as one of the four areas for follow-up actions.[16] This italicized language regarding access is repeated verbatim in the Arctic Council's April 2009 Tromsø Declaration, a document circulated widely in policy circles, as one component the Council supports under 'continued international coordination to maximize the legacy of IPY [...].'[17]

Subsequently, the Senior Arctic Officials ('SAOs') discussed the scoping study recommendations at their November 2009 meeting in Copenhagen. The final report of that meeting[18] references two statements relevant to the scoping study and the question of study access to the Arctic (access generally; the Arctic Ocean was not specified as such):

---

[16] Arctic Council, 'Senior Arctic Officials (SAO) Report to Ministers', Tromsø, Norway (April 2009), see <http://arctic-council.org/filearchive/FINAL%20SAO%20Report%20to%20Ministers%20April%202009.pdf> (15 June 2011).

[17] Arctic Council, 'Tromsø Declaration on the Occasion of the Sixth Ministerial Meeting of The Arctic Council, Tromsø', Norway (29 April 2009), see <http://arctic-council.org/filearchive/Tromsoe%20Declaration-1.pdf> (15 June 2011).

[18] Arctic Council, Meeting of Senior Arctic Officials, 'Final Report', Copenhagen, Denmark (12-13 November 2009), see <http://arctic-council.org/file archive/SAO%20Meeting%20nov09-%20FINAL.pdf> (15 June 2011).

1.) The SAOs observed that

> Action in some recommended areas, such as enhancing access to study areas and maintaining infrastructure, needs to be pursued nationally and in the bodies that deal with these issues. Funding is a national responsibility.[19]

Notably, the SAO did not identify any such bodies.

2.) The SAOs also took the following

> Decision: SAOs agreed that the Arctic Council would continue to contribute to the legacy of IPY by asking the working groups to make use of the most up-to-date research results in ongoing assessment processes, as well as through contributions to, inter alia, SAON. SAOs agreed to support a joint Arctic Council/ATCM workshop in June 2010 in conjunction with the Oslo IPY Conference to further discuss IPY legacy issues.[20]

Close cooperation between the Arctic Council and the Antarctic Treaty Consultative Meeting on post-IPY projects, such as the June 2010 IPY meeting noted above, was suggested in the Final Report of the SAO's 2007 meeting (in Narvik, Norway) as well. There the IPY Joint Committee 6th session in October 2007 Quebec was reported to have 'asked that the Arctic Council continue to help with access to areas in the Arctic for research.'[21]

III. Possible Fora in Which to Pursue Improved Access for Arctic MSR: The Need for Further Work and Communication

The SAO observations reported above, that post-IPY follow-up on access issues is best pursued nationally and in relevant bodies dealing with such issues, indicates the need for coordinated, complementary organization on this and related issues.

Independent next steps should include:

---

19 *Ibid.*, 11.

20 *Ibid.*

21 Arctic Council, Meeting of Senior Arctic Officials, 'Final Report', Narvik, Norway (28-29 November 2007), see <http://arctic-council.org/filearchive/Narvik%20-FINAL%20Report-%2023Apr08.doc> (15 June 2011) 15.

- The Arctic research community reviewing and possibly amending or revising the issues and goals identified in this White Paper, based on discussions and feedback solicited, e.g., at scientific working group meetings or international conferences and through other appropriate channels. As one example, the Polar Research Board of the US National Research Council organized a workshop on access impediments to research in the Arctic and related issues.[22] Similar opportunities in other countries should be identified and publicized.
- Identifying bodies [e.g. Intergovernmental Oceanographic Commission, International Arctic Science Committee/Arctic Ocean Sciences Board ('IASC/AOSB'), International Arctic Buoy Program ('IABP'), International Study of Arctic Change ('ISAC'), Sustained Arctic Observing Networks ('SAON', and others] and national authorities interested in improving access to the Arctic Ocean for MSR, and appropriate contacts in each of them (respecting that some bodies will not want to engage in access-related activities that might be considered political).
- Collecting information on related initiatives for MSR access that these national entities and relevant bodies have undertaken, whether successful or unsuccessful. A study of reasons for failed initiatives may also prove useful.
- Working with experts within international bodies that have successfully developed guidelines and ensured access for marine research in other geographic settings (e.g., Joint IOC/WMO Technical Commission for Oceanography and Marine Meteorology) to explore how such models could be applied to the Arctic.
- Reviewing and potentially revising definitions of key activities related to marine scientific and operational research in light of recent technological and scientific advances.

Arctic research has a long history of strong international collaboration and coordination, with all Arctic coastal States strongly reaffirming this tradition during the recent IPY. All Arctic coastal States have much to gain from joint agreements and further initiatives towards improved scientific access in a rapidly changing Arctic. Further work along the

[22] Polar Research Board of the US National Academies, 'US Scientific Access to the Russian Arctic' (22 September 2010), see <http://dels.nas.edu/Past-Events/Scientific-Access/AUTO-5-11-31-B> (15 June 2011).

lines described above can help ensure safe operations and marine activities in Arctic Ocean environments.

Climate Change, Marine Science and Delineation of the Continental Shelf

by Vladimir Golitsyn[*]

I. Impacts of Climate Change on the Arctic

The increasingly negative impact of changes in the global climate on our planet's ecosystems and, in particular, on the functioning of the oceans, is a matter of growing concern for the community of States. This was clearly demonstrated by the tense discussions that took place prior to and during the Copenhagen round of negotiations held in 2009 within the framework of the United Nations Convention on Climate Change.[1] While climate change has a negative impact on practically all the ecosystems of our planet, the global warming associated with climate change has an especially significant adverse impact on the ecosystems of high-latitude regions where the climate is changing almost twice as fast as in lower-latitude regions. The Arctic, which is a high-latitude region, is, therefore, especially vulnerable to the effects of global warming. As pointed out in the recently released report of the Aspen Institute Commission on Arctic Climate Change, *The Shared Future*, the Arctic is among the first regions in which human-induced climate impacts are being seen and it will – in all probability – be the first region where climate change will lead to transformative ecological, economic and social change. The report further notes that the Arctic's warmer temperatures and decreases in permafrost, snow cover, glaciers and sea ice also have

---

* Judge, International Tribunal for the Law of the Sea, Hamburg, Germany.

1 United Nations Framework Convention on Climate Change (with Annexes) (adopted 9 May 1992, entered into force 21 March 1994) 1771 UNTS 107.

wide-ranging consequences for the physical and biological systems in other parts of the world.[2]

It should be emphasized that all these negative developments resulting from climate change are not caused by activities in the Arctic. They are the consequence of industrial activities associated with the growing release of greenhouse gases in other parts of the world. Consequently, they can be prevented and mitigated only through efforts undertaken at an international level, in particular, within the United Nations Framework Convention on Climate Change. However, this does not mean that developments associated with climate change should not receive urgent attention at a regional level as well. In the Arctic, for example, the further recession of pack ice will eventually result in the development on a large scale of new economic activities, including exploration and exploitation of oil and gas resources in the Arctic Ocean, as well as shipping and fishing. The prospect of such activities raises concerns regarding the proper protection of the fragile Arctic marine environment.

The emerging threats posed by the negative impacts of climate change were acknowledged with deep concern by the five coastal States bordering on the Arctic Ocean, namely Canada, Denmark, Norway, the Russian Federation and the United States of America. The Ilulissat Declaration, adopted by them on 28 May 2008 at a meeting held in Greenland, states that the Arctic Ocean stands at the threshold of significant changes and that climate change and the melting of ice have a potential impact on vulnerable ecosystems, the livelihoods of local inhabitants and indigenous communities, and the potential exploitation of natural resources.[3] Recognizing that under the circumstances a better understanding of the ecological dynamics of Arctic ecosystems is crucial for meeting these new challenges, the five Arctic coastal States committed themselves to close cooperation, including in the collection of scientific data concerning the continental shelf, the protection of the marine environment and other scientific research.

In the Ilulissat Declaration the five Arctic coastal States rejected the assumption of those who claim that non-Arctic players, such as the European Union, China and some other States, in addition to main-user interests related to the exploitation of offshore hydrocarbon re-

---

[2] The Aspen Institute, Energy and Environment Program, *The Shared Future: A Report of the Aspen Institute Commission on Arctic Climate Change* (2011) 9-10.

[3] Ilulissat Declaration (28 May 2008) (2009) 48 ILM 382.

sources in the Arctic Ocean, also have so-called non-user interests, including the protection and preservation of the marine environment and safeguarding biodiversity. This allegedly entitles them to become involved in the governance and regulation of the marine Arctic to safeguard the non-user interests in their own right on behalf of the international community.[4] The Ilulissat Declaration states in this regard that by virtue of their sovereignty, sovereign rights and jurisdiction in large areas of the Arctic Ocean the five coastal States are the ones who are in a unique position to address challenges faced in the Arctic Ocean in connection with climate change and therefore they see no need to develop a new comprehensive international legal regime to govern the Arctic Ocean.[5]

II. Importance of Cooperation in Marine Scientific Research in the Arctic Ocean

As pointed out in the report of the Aspen Institute Commission, referred to above, the Arctic remains a remote and expensive place to conduct scientific research, placing a premium on cooperation, coordination of monitoring efforts, and the sharing of research platforms.[6] Although the Ilulissat Declaration calls for cooperation in the collection of scientific data, and the Arctic States' national policies have the aim of facilitating cooperation in scientific research in the framework of the Arctic Council, it should be acknowledged that following the conclusion of the International Polar Year ('IPY') in 2009, the level of collective efforts in scientific research in the Arctic has been substantially reduced and therefore a concerted effort is required to build on the legacy of the IPY.[7]

Although there are various definitions of the Arctic, it is generally accepted that the Arctic region can be defined as the area north of the Arctic Circle (66° 32' N), which is the approximate limit of the midnight sun and the polar night. There are eight Arctic States that have

---

4 T Koivurova, EJ Molenaar and DL VanderZwaag, 'Canada, the EU, and Arctic Ocean Governance: A Tangled and Shifting Seascape and Future Directions', *Journal of Transnational Law and Policy* 18 (2009) 247 et seq. (253).

5 Ilulissat Declaration, see note 3.

6 The Aspen Institute, see note 2, 29.

7 *Ibid.*, 9.

land territories north of the Arctic Circle namely: Canada, Denmark (Greenland), Finland, Iceland, Norway, the Russian Federation, Sweden and the United States. They are all members of the Arctic Council – a central coordinating body for activities in the Arctic. The largest part of the Arctic is occupied by the Arctic Ocean surrounded, as pointed out in the Ilulissat Declaration, by the territories of the five Arctic States bordering the Arctic Ocean.

The Arctic Ocean is the smallest and shallowest of the world's major oceans. The conduct of scientific research in the Arctic Ocean is governed by the provisions of the United Nations Convention on the Law of the Sea ('UNCLOS').[8] All Arctic States, with the exception of the United States, are parties to this Convention. It was President Reagan who decided not to sign the Convention. However, in his Ocean Policy Statement in 1983 he announced that the United States accepted, and would act in accordance with, the Convention's balance of interests relating to traditional uses of the oceans – everything but deep seabed mining – and instructed the government to abide by, or as the case may be, to enjoy the rights accorded by, the other provisions, and to encourage other countries to do likewise.[9] It is worthy of note that in the Ilulissat Declaration the five coastal Arctic States confirmed their commitment to the legal framework of the law of the sea which, according to the Declaration, provides for important rights and obligations concerning the delineation of the outer limits of the continental shelf, the protection of the marine environment, including ice-covered areas, freedom of navigation, marine scientific research, and other uses of the sea.[10]

It may be assumed that conduct of scientific research in maritime areas, including the exclusive economic zone ('EEZ'), constitutes one of the traditional uses of the oceans and that all Arctic States, even the United States, are bound by the rules of the law of the sea governing such research as reflected in the relevant provisions of UNCLOS. However, as will be shown later, the situation is much more complicated in the case

---

[8] United Nations Convention on the Law of the Sea (concluded 10 December 1982, entered into force 16 November 1994) 1833 UNTS 397.

[9] *Accession to the 1982 Law of the Sea Convention and Ratification of the 1994 Agreement Amending Part XI of the Law of the Sea Convention*, Written Testimony of JD Negroponte, Deputy Secretary US Department of State, before the Foreign Relations Committee, United States Senate on 27 September 2007, Senate Treaty Document 103-39 (3).

[10] Ilulissat Declaration, see note 3.

of the extended continental shelf, the outer limits of which are subject to verification by the Commission on the Limits of the Continental Shelf established under the Convention.

III. International Legal Regime Governing Marine Scientific Research in the Arctic Ocean

UNCLOS has a special chapter, Part XIII, devoted to marine scientific research. It contains elaborate provisions regarding the promotion of marine scientific research, general principles governing its conduct, promotion of cooperation, publication and dissemination of information and knowledge resulting from marine scientific research. However, under UNCLOS, marine scientific research is currently unrestricted only on the high seas. The Convention states in Art. 87(1)(f) that on the high seas all States have freedom of scientific research, subject to Parts VI ('Continental Shelf') and XIII ('Marine Scientific Research'). It further states in Arts 256 and 257 respectively that all States have the right to conduct marine scientific research in the Area, in conformity with the provisions of Part XI, and in the water column beyond the limits of the exclusive economic zone.

Pursuant to UNCLOS, in maritime areas where the coastal State exercises its sovereignty or has sovereign rights, rules governing marine scientific research rest on the principle of consent of the coastal State, which is mandatory in respect of the territorial sea and implied in the case of the exclusive economic zone and the continental shelf.

1. Territorial Sea – Express Consent

In the case of the territorial sea, under Art. 245 the coastal State, in the exercise of its sovereignty, has the exclusive right to regulate and authorize marine scientific research in its territorial sea and such research therein can be conducted only with the express consent of and under the conditions set forth by the coastal State. Consequently, in maritime areas of the Arctic Ocean which constitute the territorial sea of the five Arctic coastal States, marine scientific research can be conducted only with the express consent of those States.

2. EEZ and Continental Shelf Within 200 nm – Implied Consent

In the case of the exclusive economic zone and the continental shelf, pursuant to Art. 246(1) and (2), the coastal State in the exercise of its jurisdiction has the right to regulate and authorize marine scientific research and such research can be conducted in these areas only with the consent of the coastal State. UNCLOS at the same time stipulates in para. 3 of the same Article that the coastal State, in normal circumstances, is supposed to grant its consent for marine scientific research projects by other States in its exclusive economic zone or on the continental shelf in order to increase scientific knowledge of the marine environment for the benefit of all mankind. To this end, the coastal State is required to establish rules and procedures ensuring that such consent will not be delayed or denied unreasonably.

UNCLOS imposes in Art. 248 an obligation on States which intend to undertake marine scientific research in the exclusive economic zone or on the continental shelf to provide the coastal State, not less than six months in advance of the expected starting date of the project, with a full description of it including, *inter alia*, the nature and objectives of the project, the method and means to be used, including name, tonnage, type and class of vessels and a description of scientific equipment, precise geographical areas in which the project is to be conducted, the expected date of first appearance and final departure of the research vessels, or deployment of the equipment and its removal, as appropriate, and the extent to which it is considered that the coastal State should be able to participate or to be represented in the project.

With reference to implied consent, UNCLOS provides in Art. 252 that foreign States may proceed with a marine scientific research project six months after the date upon which the information required pursuant to Art. 248 was provided to the coastal State unless within four months of the receipt of the communication containing such information the coastal State has informed the State or organization conducting the research that it has withheld its consent under the provisions of Art. 246.

While stating, as noted above, that the coastal State is, in normal circumstances, supposed to grant its consent for marine scientific research projects by other States, the Convention also stipulates in Art. 246(5), that coastal States may in their discretion withhold their consent to the conduct of a marine scientific research project of another State if that project, *inter alia*, is of direct significance for the exploration and exploitation of natural resources, whether living or non-living, involves drilling into the continental shelf or involves the construction, operation or use of artificial islands, installations and structures referred to in Arts 60 and 80.

3. Extended Continental Shelf – Designation of Areas for Exploitation or Exploratory Operations

However, according to UNCLOS the coastal State has less flexibility in restricting marine scientific research of other States on the extended part of its continental shelf, in other words on the continental shelf beyond 200 nm of its coast. The Convention provides in Art. 246(6) that coastal States may not exercise their discretion to withhold consent in respect of marine scientific research projects to be undertaken on the continental shelf beyond 200 nm. However, this prohibition to withhold consent does not apply to specific areas which coastal States may at any time publicly designate as areas in which exploitation or detailed exploratory operations focused on those areas are occurring or will occur within a reasonable period of time. Coastal States are required to give reasonable notice of the designation of such areas, as well as any modifications thereto, but shall not be obliged to give details of the operations therein.

4. Summary

It follows from the foregoing that in the Arctic Ocean, as in other parts of the world, marine scientific research under UNCLOS can be undertaken in the territorial sea only with the express consent of the coastal State. In the exclusive economic zone and on the continental shelf within 200 nm such consent is implied but can be withheld by the coastal State in its discretion, in particular, if such research is of direct significance for the exploration and exploitation of natural resources. As any marine scientific research, with a bit of imagination, may be linked to natural resources, the coastal State can withhold its consent practically for any marine scientific project of another State in its exclusive economic zone or on the continental shelf within 200 nm. In the case of the extended continental shelf beyond 200 nm, at first glance the coastal State does not have any such discretion. However since UNCLOS authorizes the coastal State to designate any area of that shelf as being reserved for exploitation or exploratory operations without giving details of such operations, the coastal State does, in effect, have the means to close substantial areas of its extended continental shelf to marine scientific research by other States.

In summary, it may be concluded that the conduct of marine scientific research in the Arctic Ocean depends to a great extent on the good will of the five Arctic States bordering the Arctic Ocean and that their consent in one form or another is required for such research in maritime areas where these States have sovereignty or exercise sovereign rights. Consequently, for a potential marine scientific research project operator it is of vital importance to know the legal status of the marine areas where the planned project is to be conducted. While the situation is clear with respect to the exclusive economic zone and the continental shelf within 200 nm, the picture is still rather murky as regards the extended continental shelf.

IV. Limits of the Extended Continental Shelf of the Arctic States in the Arctic Ocean

1. Introductory Remarks

At present there is a lot of uncertainty regarding the outer limits of the continental shelf in the Arctic Ocean. Only Norway has been able, so far, to define the outer limits of its continental shelf in the Arctic Ocean. In 2006 it made a submission on the issue to the Commission on the Limits of the Continental Shelf and in 2009 the Commission issued its recommendations on the submission.[11] This allowed Norway to finalize the establishment of the outer limits of its continental shelf in accordance with Art. 76(8) UNCLOS. The Russian Federation was the first State to make a submission to the Commission on the Limits of the

[11] Commission on the Limits of the Continental Shelf, 'Continental Shelf Submission of Norway in respect of Areas in the Arctic Ocean, the Barents Sea and the Norwegian Sea: Executive Summary' (December 2006), see <http://www.un.org/Depts/los/clcs_new/submissions_files/nor06/nor_exec_sum.pdf> (2 August 2011); Commission on the Limits of the Continental Shelf, 'Summary of the Recommendations of the Commission on the Limits of the Continental Shelf in regard to the Submission Made by Norway in respect of Areas in the Arctic Ocean, the Barents Sea and the Norwegian Sea on 27 November 2006' (27 March 2009), see <http://www.un.org/Depts/los/clcs_new/submissions_files/nor06/nor_rec_summ.pdf> (2 August 2011).

Continental Shelf in 2001.[12] However, the Commission decided in 2002 that the information provided to it by the Russian Federation was insufficient and recommended that the Russian Federation makes a revised submission to the Commission with regard to the Central Arctic Ocean. Canada and Denmark have yet to make their submissions to the Commission. It is anticipated that they will be submitted to the Commission in 2013 and 2014 respectively. It is difficult to say with any certainty whether the United States will eventually become a party to UNCLOS.

It should be emphasized that the sovereign rights of the coastal State over its continental shelf are unique in their nature and are different from sovereign rights exercised in the exclusive economic zone. The continental shelf constitutes the natural prolongation of the coastal State's land territory. According to Art. 77(2) UNCLOS, these rights are exclusive in the sense that if the coastal State does not explore the continental shelf or exploit its natural resources, no one may undertake these activities without the express consent of the coastal State. Pursuant to Art. 76(8) UNCLOS the limits of the continental shelf are also established by a coastal State on the basis of recommendations of the Commission on the Limits of the Continental Shelf and this act by the coastal State is final and binding. The role of the Commission is purely advisory. Its task is to verify the accuracy of the determination by the coastal State in accordance with the provisions of Art. 76 UNCLOS of the limits of its continental shelf and to make recommendation thereon.

Lack of clarity regarding the outer limits of the continental shelf in the Arctic Ocean raises the question of whether, with the exception of Norway, any other Arctic State can at this stage designate, pursuant to Art. 246(6) any part of the seabed of the Arctic Ocean beyond 200 nm as an area of its continental shelf in which exploitation or exploratory operations are occurring or will take place. This uncertainty has direct implications for marine scientific research projects if such projects are planned to be undertaken in areas where some of the Arctic States may have overlapping claims over potential extended continental shelves.

---

[12] Russian Federation, Continental Shelf Submission, Executive Summary, attached to Commission on the Limits of the Continental Shelf, 'Receipt of the Submission Made by the Russian Federation to the Commission on the Limits of the Continental Shelf' (20 December 2001) CLCS.01.2001.LOS (Continental Shelf Notification), see <http://www.un.org/Depts/los/clcs_new/submissions_files/submission_rus.htm> (2 August 2011).

It is worth noting that in accordance with Art. 76(3) UNCLOS, oceanic ridges do not constitute part of the continental shelf of the coastal State. Pursuant to para. 6 of that Article, submarine ridges are included in the continental shelf, but the outer limit of the continental shelf in that case cannot exceed 350 nm. However, in accordance with Art. 76(6), the 350 nm limitation does not apply to submarine elevations that are natural components of the continental margin, such as its plateau, rises, caps, banks and spurs.

Available information allows the following picture to be drawn with regard to positions taken or which could be taken by the four Arctic States on their extended continental shelves in the Arctic Ocean.

2. Norway

As noted above, Norway has already finalized the establishment of the outer limits of its continental shelf on the basis of the recommendations of the Commission in accordance with Art. 76(8) UNCLOS.

3. The Russian Federation

According to the Executive Summary of the Russian submission to the Commission on the Limits of the Continental Shelf, the Alpha, Mendeleev and Lomonosov Ridges are natural components of the continental margin and therefore are submarine elevations to which the 350 nm limitation does not apply.[13] Several States submitted their comments with regard to the submission by the Russian Federation. However, only three of them are of relevance to this presentation.

In its comments on the Executive Summary of the Russian submission forwarded on 28 February 2002 to the United Nations Secretariat for circulation to the States Members of the United Nations and members of the Commission, the United States argued that the Alpha-Mendeleev Ridge System is a volcanic feature of oceanic origin that has been formed on, and occurs only within the area of, the oceanic crust that underlies the Amerasia Subbasin of the deep Ocean Basin and therefore *is not part of any State's continental shelf* (emphasis added). It further argued that the Lomonosov Ridge is a freestanding feature in the deep,

---

13 *Ibid.*

oceanic part of the Arctic Ocean Basin, and *not a natural prolongation of the continental margins of either Russia or any other State* (emphasis added).[14]

In its note verbale to the United Nations Secretariat of 18 January 2002 concerning the Russian submission to the Commission, Canada stated that it is not in a position to determine whether it agrees with the submission without the provision of further supporting data to analyze and that Canada's inability to comment should not be interpreted as either agreement or acquiescence by Canada to the Russian Federation's submission. Canada also stressed in the note verbale that the Russian Federation's submission and any recommendations by the Commission in response are without prejudice to the question of delimitation of the continental shelf between Canada and the Russian Federation.[15]

Denmark, in its comments on the Russian submission, forwarded to the United Nations Secretariat in the form of a note verbale on 4 February 2002, also stated that it is not able to form an opinion on the submission because a qualified assessment would require more specific data. However, such absence of opinion does not imply Denmark's agreement or acquiescence to the Russian Federation's submission. Denmark further stated that it is not in a position to evaluate the possible impact of an extended Russian continental shelf beyond 200 nm on the extended shelf appurtenant to Greenland, and is therefore unable to state that the Russian claim would not be met by overlapping Danish/Greenland claims to continental shelf areas beyond 200 nm in the Arctic. Denmark stressed in the note verbale that in accordance with UNCLOS, including its Annex II, the actions of the Commission shall not prejudice matters relating to delimitation of boundaries between States in opposite or adjacent coasts and consequently, the Russian Federation's submission

---

14 United States, 'Note to the Secretary-General of the United Nations' (28 February 2002) attached to UN Secretary General, 'United States: Notification Regarding the Submission Made by the Russian Federation to the Commission on the Limits of the Continental Shelf' (18 March 2002) CLCS. 01.2001.LOS/USA, see <http://www.un.org/Depts/los/clcs_new/submissions_files/submission_rus.htm> (2 August 2011).

15 Canada, 'Note to the Secretary-General of the United Nations' (18 January 2002) attached to UN Secretary General, 'Canada: Notification Regarding the Submission Made by the Russian Federation to the Commission on the Limits of the Continental Shelf' (26 February 2002) CLCS.01.2001.LOS/CAN, see <http://www.un.org/Depts/los/clcs_new/submissions_files/submission_rus.htm> (2 August 2011).

and the Commission's recommendations are without prejudice to the delimitation of the continental shelf between Denmark/Greenland and the Russian Federation.[16]

4. Canada

Canada has not yet made a submission to the Commission regarding its extended continental shelf in the Arctic Ocean and is expected to do so not earlier than 2013. Canada produced a map based on a desktop study which indicates that Canada's margin areas in the eastern Arctic are based on the Alpha, Mendeleev and Lomonosov Ridges.[17] Consequently, as pointed out by Professor McDorman, it appears that Canada's view mirrors that of the Russian Federation that these ridges are natural components of the continental shelf, albeit the Canadian continental shelf and not the Russian continental shelf, and that these features are neither submarine nor oceanic ridges. This is also confirmed by Monique Allain in her article on Canada's claim to the Arctic. She notes that according to a report of the Canadian Department of Foreign Affairs and International Trade, the early results suggest that the Alpha-Mendeleev Ridges in the eastern Arctic are attached to the Canadian landmass.[18] She also refers to information on the natural resources of Canada provided by Marc St. Onge who states that the Lomonosov Ridge is connected to the Canadian landmass.[19] Thus, there appears to

---

16 Denmark, 'Note to the Secretary-General of the United Nations' (4 February 2002) attached to UN Secretary General, 'Denmark: Notification Regarding the Submission Made by the Russian Federation to the Commission on the Limits of the Continental Shelf' (26 February 2002) CLCS.01.2001. LOS/DNK, see <http://www.un.org/Depts/los/clcs_new/submissions_files/submission_rus.htm> (2 August 2011).

17 TL McDorman, 'The Outer Continental Shelf in the Arctic Ocean: Legal Framework and Legal Developments', in D Vidas (ed.), *Law, Technology and Science for Oceans in Globalization* (2010) 499 et seq. (511).

18 MA Allain, 'Canada's Claim to the Arctic: A Study in Overlapping Claims to the Outer Continental Shelf', *Journal of Maritime Law and Commerce* 42 (2011) 1 et seq. (11).

19 *Ibid.*, 11; Presentation of M St-Onge, Senior Research Scientist, Regional Geology, Department of Natural Resources, at Meeting 28 of the Standing Committee on National Defence, Canada, 2nd Session, 40th Parliament (16 June 2009).

be a potential disagreement between the Russian Federation and Canada regarding the status of the three aforementioned ridges and consequently the extent of Canadian and Russian continental shelves in the Arctic Ocean.

5. Denmark

Denmark has not made its submission to the Commission either and is expected to do this in 2014. According to Professor McDorman, a report on the results of survey work jointly done by Canada and Denmark in the Arctic demonstrates that the Lomonosov Ridge is attached to the North America and Greenland plates.[20] So it may be expected that Denmark will also claim in its submission to the Commission that the Lomonosov Ridge constitutes a natural prolongation of Greenland's continental margin which will be in conflict with the statement on the status of this ridge in the Russian submission.

6. The United States

The United States is not a party to UNCLOS. John Negroponte, Deputy Secretary of State, in his presentation to the US Senate Foreign Relations Committee, stated that the United States has one of the largest continental shelves in the world and in the Arctic it could run as far as 600 nm from the coastline.[21] The US holds the view that the Chukchi plateau and its component elevations north of Alaska fit the category of submarine elevations under Art. 76(6) UNCLOS, and, therefore, are not subject of the 350 nm limitation applicable to submarine ridges.[22] As pointed out by Professor McDorman and Monique Allain, there is likely to be an overlap between the Canadian and American claims to the continental shelf in the western Arctic.[23]

Thus, while claiming that the Alpha, Mendeleev and Lomonosov Ridges are not natural components of the continental margin of any

---

[20] McDorman, see note 17, 512.

[21] Negroponte, see note 9.

[22] McDorman, see note 17, 514.

[23] Allain, see note 18, 12 and 15.

State in the Arctic Ocean, the US, in effect, bases its own position on the same premise as that of the Russian Federation, Canada and Denmark.

V. Conclusion

The Commission on the Limits of the Continental Shelf will be in a very difficult situation if all three States, namely the Russian Federation, Canada and Denmark, insist in their submissions that the respective ridges are natural prolongations of their land territories and constitute parts of the continental margin. It is questionable whether the Commission can actually decide on this issue in light of the provisions of Art. 9 Annex II UNCLOS, mentioned by both Canada and Denmark in their notes verbales on the Russian submission. This Article states that the actions of the Commission shall not prejudice matters relating to delimitation of boundaries between States with opposite or adjacent coasts.

Interaction between the Commission and some of these States may also be complicated by the provisions of Art. 8 Annex II UNCLOS which provides that in the case of disagreement by the coastal State with the recommendations of the Commission, the coastal State may, within a reasonable time, make a revised or new submission to the Commission. As pointed out by some commentators, since the Convention does not specify how many times the coastal State can disagree with the recommendations by UNCLOS, this may result in a ping-pong exchange of communications between the Commission and the coastal State concerned with no clear answer to the question of how long it could last. It could lead to an awkward situation where in the case of two or three conflicting claims, the Commission sides with one Arctic State and the other State or two States disagree with the recommendations of the Commission triggering a ping-pong exchange of communications as referred to above.

The matter is further complicated by the fact that UNCLOS does not define what should be understood by submarine ridges or submarine elevations that are natural components of the continental margin. The ridge provision of Art. 76 UNCLOS is a compromise and therefore has ambiguous legal and scientific wording. According to Symonds and Brekke, members of the Commission, '[i]t is now well recognised that one of the most contentious and difficult aspects of applying the definition of the continental shelf contained within article 76 [...] relates to the way it handles ridge-like seafloor highs.'[24] They point out in this regard that much of the difficulty in interpreting the 'ridge' provisions of the Convention stems from improvements in science as it is now becoming apparent that seafloor features are 'more complex and diverse than was envisaged when article 76 was being negotiated and drafted'[25] and that '[c]rustal composition and type is of importance to scientists as seafloor highs have not only morphological expressions, but have varying geological characteristics and origins.'[26]

There may be various options for addressing this complex situation. One of them could be for the Russian Federation, Canada and Denmark to make a joint submission to the Commission. The three Arctic States could also proceed with separate submissions but reach a common understanding concerning the geological nature of the Alpha, Mendeleev and Lomonosov Ridges. Such an understanding could be reflected in a joint statement conveying to the Commission their shared view regarding the geological nature of the Alpha, Mendeleev and Lomonosov Ridges. It could also take a form of agreed language on the subject that would be incorporated into their individual submissions. To achieve this, the three Arctic States should closely cooperate in sharing data and strive to agree on common interpretations. Once the Commission acknowledges that it is satisfied with the scientific data presented to it by the three Arctic States, then they will proceed with the delimitation of their overlapping extended continental shelves. Monique Allain argues, for example, that it would be in the best interest of Canada to combine its data with the other Arctic States and coordinate it as well so that they all have the same interpretations and apply the provisions of Art. 76 in the same way.[27]

Monique Allain also mentions another possible way of avoiding deadlock between the Arctic States: reaching a boundary agreement before proceeding with submissions to the Commission.[28] However, given the

---

[24] PA Symonds and H Brekke, 'A Scientific Overview of Ridges Related to Article 76 of the UN Convention of the Law of the Sea', in MH Nordquist, JN Moore and TH Heidar (eds), *Legal and Scientific Aspects of Continental Shelf Limits* (2004) 141 et seq. (141).

[25] *Ibid.*, 158.

[26] Allain, see note 18, 33.

[27] *Ibid.*, 19.

[28] *Ibid.*, 19.

rigid time frame for submissions established by UNCLOS, as interpreted by the Meetings of States Parties, it is highly unlikely that the Arctic States will have sufficient time to finalize the delimitation process.

The United States position, as reflected in its note verbale on the Russian Federation's submission, does not appear sustainable. The United States cannot on the one hand base its own claim to the continental shelf on the same premise as that of Russian Federation, Canada and Denmark and on the other assert that the Alpha, Mendeleev and Lomonosov Ridges are not natural components of the continental margin of any State in the Arctic Ocean. As the United States, at least at this stage, is not a party to UNCLOS, its assertion of its right to the continental shelf in the Arctic Ocean is not subject to scrutiny by the Commission. So, the United States has a clear advantage over the other Arctic States. It may be assumed with sufficient certainty that the other Arctic States will not favorably view a situation where the United States can proceed to establish the limits of its extended continental shelf in the Arctic Ocean while their submissions are awaiting consideration or have resulted in recommendations by the Commission which might not be agreeable to some or all of them.

In concluding, it may be said that although in light of the negative impact of climate change on the ecosystems of the Arctic, there is an urgent need for enhanced marine scientific research in the Arctic Ocean, remaining uncertainty regarding the extended continental shelf over which the Arctic States exercise sovereign rights undoubtedly complicates the conduct of scientific research projects in this region.

From the DAMOCLES to ACCESS Projects (Sixth & Seventh EU Framework Programmes 2005-2015) IAOOS – An Advanced Arctic Ocean Observing System (2011-2019)

by Jean-Claude Gascard

I. Introduction

The Arctic domain is primarily composed of an ocean surrounded by continents. We are aware of the very important role the World Ocean has in general on the earth's climate. The Arctic Ocean plays a very significant role in climatology because of powerful feedback mechanisms such as the albedo, the main cause of global warming amplification in Arctic regions. There are many positive feedback mechanisms contributing to amplifying climate changes in the Arctic. For instance, the shift in wind regime is accelerating sea ice motion and increasing sea ice deformation, fracturing and ridging.[1] Therefore it contributes to produc-

---

* Doctor, Université Pierre et Marie Curie/Centre national de la recherche scientifique, Paris, France; coordinator of the DAMOCLES and ACCESS projects.

DAMOCLES: Developing Arctic Modeling and Observing Capabilities for Long-term Environmental Studies (2005-2010).

ACCESS: Arctic Climate Change, Economy and Society (2011-2015).

IAOOS: Ice, Atmosphere, Ocean Observing System (2011-2019).

[1] A Lammert et al., 'Comparison of Three Weather Prediction Models with Buoy and Aircraft Measurements under Cyclone Conditions in Fram Strait', *Tellus A* 62(4) (2010) 361 et seq.

ing more open waters (leads and polynyas) and consequently to diminishing the albedo and increasing the absorption of solar radiation by the upper ocean that is immediately converted into heat, melting more sea ice.[2] The comparison of observations from the Surface Heat Budget of the Arctic Ocean project ('SHEBA 1997-1998') with model outputs has shown that model behavior in terms of cloud and ice cover was rather good during the winter period, but this was not the case during the summer when sea ice breaks off.[3]

In fact, global warming in Arctic regions impacts the three components of the Arctic cryosphere: permafrost, sea ice and glaciers. The most spectacular change involves the drastic reduction of the minimum summer sea ice extent, which enhances permafrost thawing and glacier calving. This extreme reduction of Arctic sea ice cover has a huge impact on the earth's radiative budget and fluxes at the air-surface interface.[4] It is responsible for a warming of the upper ocean and of the lower troposphere, a change of atmospheric circulation (polar vortex and storm tracks), an increase of cold air outbreaks (which increase heat flux release from the ocean and low cloud formation) and sudden stratospheric warming events, an acceleration of Greenland ice melting and sea-level rise and an increase of permafrost thawing releasing large amounts of greenhouse gases ('GHG') into the atmosphere.

The remarkable observations made during the Fourth International Polar Year ('IPY') in 2007-2008 further permitted the detailing of first order changes in the Arctic sea ice and the atmospheric circulation in the entire Arctic domain and beyond. Because of its much greater thermal inertia, the ocean appeared much more robust and more stable than the atmosphere and the Arctic sea ice. But for how long will that last? The shallow cold halocline is inhibiting heat transfer from below. But how long will the Arctic halocline resist the erosion by wind stress and atmospheric forcing acting on larger and larger open ocean areas that are no longer protected by sea ice?

---

2 C Lüpkes et al., 'Influence of Leads in Sea Ice on the Temperature of the Atmospheric Boundary Layer during Polar Night', *Geophysical Research Letters* 35 (2008) L03805.

3 K Wyser et al., 'An Evaluation of Arctic Cloud and Radiation Processes during the SHEBA Year: Simulation Results from Eight Arctic Regional Climate Models', *Climate Dynamics* 30 (2008) 203 et seq.

4 E Jacobson and T Vihma, 2010: 'Atmospheric Moisture Budget in the Arctic Based on ERA-40 Reanalysis', *International Journal of Climatology* 30 (2010) 2175 et seq. (online).

II. Developing Arctic Modeling and Observing Capabilities for Long-term Environmental Studies (2005-2010)

Developing Arctic Modelling and Observing Capabilities
for Long-term Environmental Studies

DAMOCLES was an integrated ice-atmosphere-ocean monitoring and forecasting project designed for observing, understanding and quantifying climate changes in the Arctic. It was designed to evaluate and improve global and regional climate forecasting models based on validation, assimilation and integration of observed data. The ultimate goal was to lengthen the lead-time of extreme climate changes predicted to occur in the Arctic within this century according to the Arctic Climate Impact Assessment ('ACIA') and thus to improve the ability of society to mitigate its impacts. The DAMOCLES consortium represented 48 institutions from 11 European countries including Russia.

The specific objectives of DAMOCLES were

- *to determine* the processes responsible for present variability and changes in the Arctic climate system;
- to improve our capabilities to *predict* Arctic climate changes, and in particular extreme climate events;
- *to design* optimal components of a long-term integrated monitoring and forecasting system for the Arctic Ocean;
- *to assess* impacts of an extreme climate event such as the disappearance of the Arctic perennial sea ice.

The thickness of the Arctic sea ice has decreased by a factor of 2. While it was more than 3 m thick on average 20 to 30 years ago, the sea ice is now less than 2 m thick on average. The summer retreat is also very spectacular. In September 2007 the total sea ice extent remaining at the end of the summer reached approximately 4 million km^2 compared to 8 million km^2 20 years ago. This halving of thickness and extent equates to a 75% sea ice mass loss. This is huge. But there is more. The speed of the transpolar sea ice drift has also increased by a factor of 2. The French schooner Tara took about 500 days to drift from the Laptev Sea to Fram Strait from September 2006 to January 2008. In comparison, the Norwegian ship Fram took three years (approx. 1000 days) to accomplish a similar drift 114 years earlier.[5] The age of Arctic sea ice has also changed drastically. The Arctic multi-year ice floes (perennial sea ice) are disappearing and the first-year ice floes are invading most of the Arctic domain. This tendency has increased very significantly during the past 10 years and has contributed to significantly decreasing the albedo.

---

[5] J-C Gascard et al., 'Exploring Arctic Transpolar Drift during Dramatic Sea Ice Retreat', *EOS Trans. AGU* 89(3) (2008) 21 et seq.

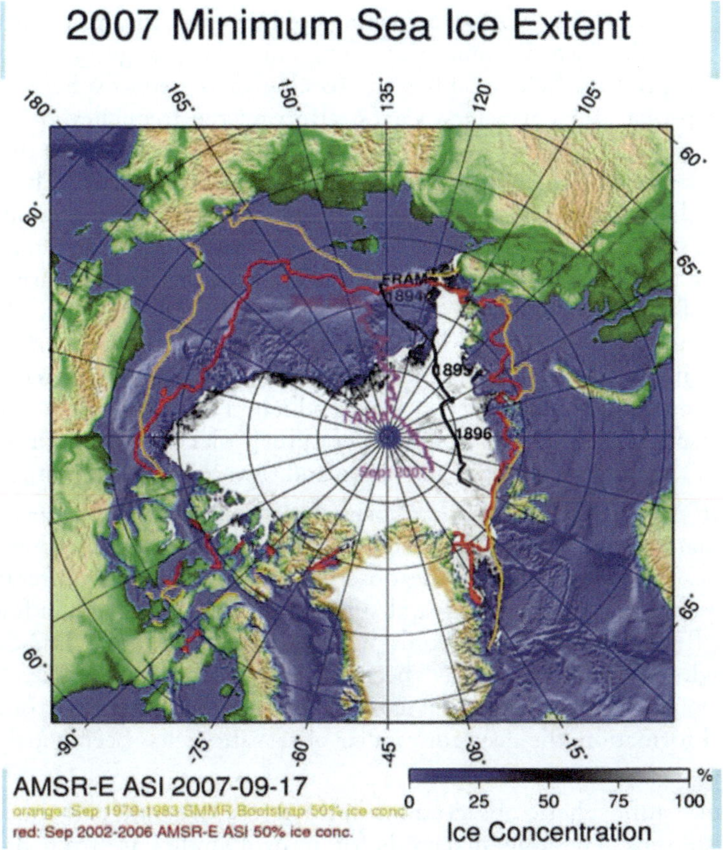

This map was created by the Polar View consortium at the University of Bremen in Germany. It represents the extreme minimum sea ice extent reached on 17 September 2007

The Arctic atmosphere has also changed considerably. The dominant mode of variability of atmospheric pressure at sea level is no longer the Arctic Oscillation (characterized by a monopole with a single extremum), but rather a dipolar anomaly (with two extrema). A vast low pressure system extends all over the north of Eurasia in contrast with a high pressure system spreading from Alaska to Greenland north of America. The mean surface temperatures in the Arctic region are higher during the autumn and winter seasons. The cumulative number of freezing degree-days during the whole freezing period from September to May the following year has decreased by as much as 1500° C over the past 30 years. This corresponds to a decrease of 1 m of sea ice. The polar lows advecting heat and moisture from low to high latitudes and cold air outbreaks ('CAO') advecting cold air from the Arctic down to mid latitudes have become more frequent. The change in the tropospheric pressure field modifies the forcing of the stratosphere by upward propagating planetary waves, affecting the frequency of sudden stratospheric warmings ('SSW') that are highly correlated with the Arctic Oscillation but maybe also with the dipolar anomaly.[6] The 1990s were characterized by a decrease in SSW occurrence and a cooling of the stratospheric polar vortex. The reverse situation has been observed during the 2000s with an increase in SSW occurrence and a warming of the polar vortex. A change in polar stratospheric temperature also affects the rate of recovery of the ozone layer. The warmer surface impacts the frequency of Arctic haze episodes, tropospheric meteorology and the amplitude of planetary waves which in turn directly modify the stratospheric circulation and temperature [Polar Stratospheric Cloud ('PSC') frequency, chlorine activation and thus ozone depletion].

The IPY projects related to atmospheric studies provided a very valuable data base for studying the aerosol and trace gas distributions in the Arctic. However, most of the observations resulted from aircraft experiments during the spring and summer periods, with only a few aircraft flights over the pole (only 3 or 4 flights of the US DC-8). Although important results have been obtained on the link between transport processes and chemical composition and between aerosol and cloud formation, the need for regular observations has been widely recognized.

Regular atmospheric observations in the Arctic are only performed at a few stations. The longest records for aerosol studies are from Barrow, Alaska (156 W, 71 N), Alert, Canada (62 W, 82 N) and Zeppelin, Svalbard (20 E, 78 N). Sulfate and black carbon trends have been discussed by Hirdman using numerical models[7] and show the importance of the transport and emission changes from Eastern Asia. Because of its remoteness, the Arctic troposphere was long believed to be extremely clean. However, it is now known that the Arctic is directly affected by pollution transport from all surrounding countries, and in summer, when forest fires are prevalent in the boreal region (as recently in Rus-

---

6 RG Graversen et al., 'Warm Winds from the Pacific Caused Extensive Arctic Sea-Ice Melt in Summer 2007', *Climate Dynamics* 36(11-12) (2010) 2103 et seq.

7 D Hirdman et al., 'Long-term Trends of Black Carbon and Sulphate Aerosol in the Arctic: Changes in Atmospheric Transport and Source Region Emissions', *Atmospheric Chemistry and Physics* 10 (2010) 9351 et seq.

sia) they are a strong high-latitude source of black carbon.[8] We still lack a measurement network extending to eastern longitudes and the pole itself to characterize this transport pathway and its evolution over a full annual cycle. Regarding vertical profiling, the US Atmospheric Radiation Measurement ('ARM') facility in Barrow (Alaska) is the most advanced. It is equipped with light detection and ranging ('lidar') (microplulse and ceilometers) combined with a Cimel sun-photometer, radar and other infrared ('IR') sensors to profile the lower layers.[9] Its observations do not include the use of a depolarization channel as used during SHEBA for the analysis of mixed phase clouds.[10]

With the advent of active sensors (lidar and radar), space observations can now be used which permit the analysis of the change in cloudiness and to some extent aerosol transport along with the sea ice evolution. Data from these space missions (CALIPSO or CloudSat, flying in the AQUA satellite constellation) can be easily accessed from the French archive centre ICARE in Lille. Additional observations from other polar orbiters such as the Infrared Atmospheric Sounding Interferometer ('IASI') now permit vertical soundings of state parameters and chemical species. Maps of the carbon monoxide ('CO') IASI distribution over the pole are now produced regularly and can be used for clustering the air masses according to emissions sources (forest fires, anthropogenic emissions).

One of the most important achievements of the 4th IPY concerned automatic Ice-Tethered Profilers ('ITP') taking daily vertical profiles of temperature and salinity in the upper 1 km of the Arctic Ocean (deep basin). About 15 profilers drifted across the Arctic Ocean deep basin during each of the two IPY years (2007-2008). The data set collected from these profilers is unique. Several thousand profiles were taken all year long during the 4th IPY, in comparison with just a few hundred profiles before. Ice Mass Balance ('IMB') platforms deployed jointly with a few Conductivity, Temperature and Depth ('CTD') profilers

---

8 D Lavoue et al., 'Modeling of Carbonaceous Particles Emitted by Boreal and Temperate Wildfires at Northern Latitudes', *Journal of Geophysical Research* 105 (2000) 26871 et seq.

9 See US Department of Energy, ARM Climate Research Facility, 'Instruments: NSA Barrow Facility', see <http://www.arm.gov/sites/nsa/C1/instru ments> (15 July 2011).

10 MD Intrieri et al., 'An Annual Cycle of Arctic Cloud Characteristics Observed by Radar and Lidar at SHEBA', *Journal of Geophysical Resources* 107(C10) (2002) 8030.

provided another unique data set regarding sea ice thickness variability over time. DAMOCLES, a pilot project for IPY under the Sixth Framework Programme of the European Union, was an opportunity for promoting modern advanced technology using underwater acoustics for sound fixing and ranging and tracking sea gliders under sea ice, for measuring sea ice draft with upward-looking sonars and operating acoustic tomography in complex regions such as the Fram Strait. The extensive and appropriate use of modern technology is the backbone for a modern Arctic Ocean observing system.

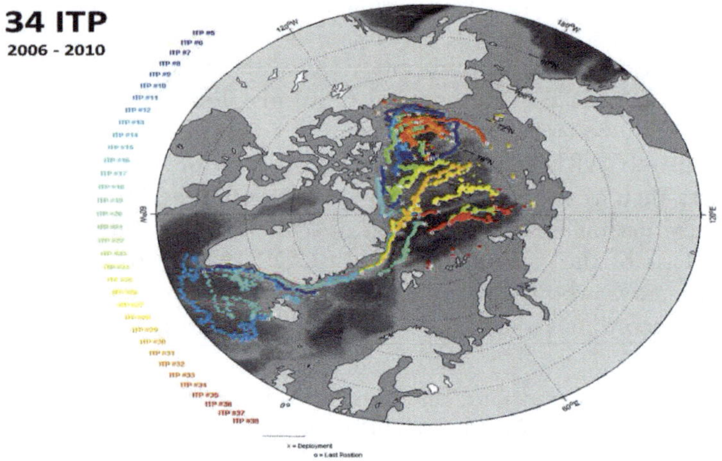

The Ice-Tethered Profiler data were collected and made available by the Ice-Tethered Profiler Programme based at the Woods Hole Oceanographic Institution[11]

In summary

- During the 4th IPY (2007-2009) northern hemisphere winters, the circulation of the Arctic middle atmosphere was disturbed by major warmings that resulted in record-breaking winter temperatures.
- In both years, the stratospheric vortex was disrupted, being displaced in 2007-2009 and split in 2009-2010.[12]

[11] See Woods Hole Oceanographic Institution, 'Ice-Tethered Profiler: An Autonomous Instrument for Sustained Observation of the Arctic Ocean', see <http://www.whoi.edu/itp> (20 July 2011).

- There has been an increase in the frequency of midwinter warmings in recent years and the variations in the past two winters are widely considered to be exceptional.
- Arctic Highs in Siberia, that develop in the lower troposphere during the cold season, influence the phase and magnitude of the dominant northern hemisphere teleconnection pattern, the Arctic Oscillation, including stratospheric warmings.[13]
- From 2002, extensive regions in the Arctic have had surface air temperature ('SAT') anomalies of greater than 3° C during late autumn. These temperatures contribute to gradients in the 1000-500 hPa thickness field influencing sub-Arctic winds.[14]
- There is a strong need for improved understanding of Arctic regional feedbacks and their impacts on decadal-scale climate variability.[15]
- The main parameters characterizing Arctic sea ice changed drastically during the past 30 years: extent decreased by a factor of 2 at the end of the summer, thickness decreased by a factor of 2, drift speed increased by a factor of 2, age decreased and first-year ice ('FYI') is now dominant as in the Antarctic Ocean.
- The extreme sea ice summer minimum extents in 2005 and 2007 were characterized by *a quasi disappearance of FYI and second-year ice ('SYI')* resulting in no replenishment of multi-year ice ('MYI').

---

[12] R Collins et al., 'The Arctic Middle Atmosphere in the Earth System: IPY Observations and Recent Model Results', Presentation at the State of the Arctic Conference *At the Forefront of Global Change,* Miami, Florida (16-19 March 2010).

[13] J Cohen and J Jones, 'The Importance of Arctic Highs to the Winter Climate of the High and Mid-Latitudes of the Northern Hemisphere', Presentation at the State of the Arctic Conference *At the Forefront of Global Change,* Miami, Florida (16-19 March 2010).

[14] JE Overland and M Wang, 'Arctic Sea Ice Behaving Strangely Impacts Mid-Latitudes', Presentation at the State of the Arctic Conference *At the Forefront of Global Change,* Miami, Florida (16-19 March 2010).

[15] K Dethloff et al., 'Arctic Feedbacks and Atmospheric Teleconnection Patterns', Presentation at the State of the Arctic Conference *At the Forefront of Global Change,* Miami, Florida (16-19 March 2010).

- In 2008 & 2009, FYI & SYI spread widely at the expense of *MYI that reduced drastically.*
- Thinning of sea ice started 30 years ago, well before any change was reported in the other main Arctic sea ice characteristics.
- A thinner sea ice regime exhibits less predictability than a thicker sea ice regime[16] so that uncertainty in the timing of sea ice loss over the next decades will remain high. Less thick ice will probably be formed during milder winters. More sea ice is likely to melt during longer summers affecting sea ice extent in the summer.
- There is a strong need to dedicate a lot of attention to the evolution of the Arctic sea ice thickness distribution which is a sensitive element of the Arctic climate system. We are looking forward to Cryosat 2 in addition to new *in situ* observations and advanced numerical modeling.

II. Arctic Climate Change, Economy and Society (2011-2015)

ACCESS
Arctic Climate Change
Economy and Society

ACCESS is a European project selected in response to the first joint call 'the Ocean of Tomorrow' of theSeventh EU Framework Programme Research and Development. It started on 1 March 2011 and will end on

---

16 MM Holland, 'Is the Loss of Perennial Arctic Sea Ice Reversible?', Presentation at the State of the Arctic Conference *At the Forefront of Global Change,* Miami, Florida (16-19 March 2010).

1 March 2015. Coordinated by the University Pierre & Marie Curie, Paris, France, and involving 27 partners from 10 European countries including Russia, the main objective of ACCESS is to assess climate impacts on marine transportation (including tourism), fisheries, marine mammals and the extraction of oil and gas in the Arctic Ocean. ACCESS is also highly concerned with Arctic governance and options for policy-makers.

Context and Objectives: Arctic climate change is going to have strong impacts on both marine ecosystems and human activities in the Arctic, which in turn has important socio-economic implications for Europe and the World. ACCESS will propose Arctic climate change scenarios and evaluate their impacts on marine transportation (including tourism), fisheries, marine mammals and the extraction of hydrocarbons in the Arctic for the next decades with particular attention to environmental sensitivities and sustainability. ACCESS is engaged in integrating Arctic climate change and socio-economic impacts and identifying Arctic governance options. ACCESS will also engage in close cooperation with indigenous people. An open and inclusive forum will give the opportunity to all stakeholders interested in the ACCESS consortium activities and cross-cutting issues, to interact with ACCESS partners.

ACCESS is composed of 5 integrated working groups. These groups perform the interdisciplinary research needed to address societal, economic, ecosystem and policy consequences of current and projected climate change impacts in the Arctic Ocean by

- understanding the complex workings of the ocean-ice-atmosphere system within the Arctic Ocean through a combination of monitoring and modeling;
- assessing the opening of marine transportation in the Arctic Ocean North of Europe and Siberia, through the Canadian Archipelago and the North Pole in the context of climate change;
- examining Arctic fisheries, aquaculture and livelihoods in the context of climate change;
- foreseeing the development of Arctic offshore oil and gas activities with respect to the harsh and fragile environment in the context of climate change scenarios;
- assessing the interplay of Arctic institutions, governance strategies and policy options with regard to Arctic States, indigenous peoples and global civil society in the context of climate change.

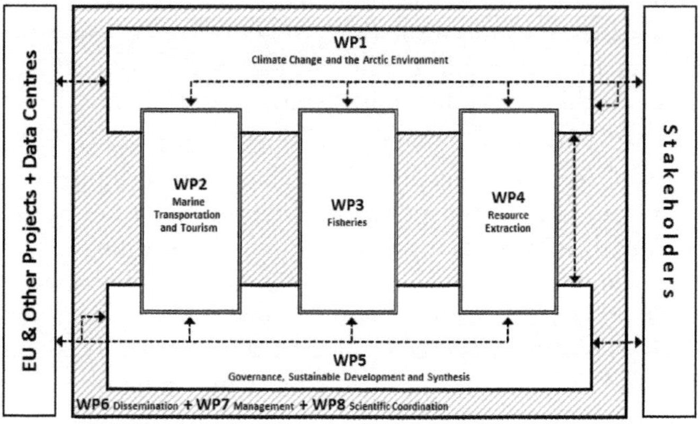

To ensure international dissemination of ACCESS activities, specific links will be set up with internationally renowned organizations such as the Arctic Information Centre at the University of Lapland in Rovaniemi (Finland). ACCESS will establish close links with international organizations overseeing international research in the Arctic and specifically with Arctic Council working groups and the International Arctic Science Committee ('IASC').

Communication within ACCESS is led by the Université Pierre et Marie Curie ('UPMC') and involves collaboration with all ACCESS partners, the steering committee, the coordinator, the manager of ACCESS and the European Commission. The project results will be disseminated by various means such as quarterly newsletters and will be distributed electronically via a website (www.access-eu.org).

III. IAOOS – An Advanced Arctic Ocean Observing System

An advanced Arctic Ocean Observing System ('IAOOS') will rely on a fully *integrated atmosphere-ice-ocean ensemble* for collecting and transmitting in *real time* key parameters related to coherent *features* (weather patterns and atmospheric circulation, ice floes and sea ice thickness distribution, large-scale ocean circulation and mesoscale ocean eddies) and internal *structures* (atmospheric inversion layer, ocean halocline & thermocline, sea ice pressure ridge & leads etc...) typical of the three domains and the *fluxes at the interfaces* between them including solar radiation. Such a system should provide basic information to estimate heat fluxes by *advection, radiation, convection and diffusion* within the three domains. It will consist of a combination of *in situ measurements and remote sensing*. Vertical profiles throughout the troposphere (10-15 km), the sea ice (a few meters) and the upper ocean (1 km) will be essential. Being at the junction between atmosphere, sea ice and ocean, *sea-level observations* will be a key component for measuring precisely sea ice thickness and snow depth, surface air temperature and pressure, temperature profiles across the snow and ice, and sea surface temperature. One of the most critical issues will concern *the time and space resolution* related to the broad range of variability that characterizes air-ice-ocean interactions that strongly depend on relevant *processes* governing heat and momentum transfer within and between the three domains. We will need to cover local, regional and pan-Arctic scales as well as weather and climatology. For that reason the operational lifetime for such a system must be *long* (several years or decades). The system will have to cover the whole Arctic Ocean, the shallow shelves and the deep basins and also the bordering lands (coastal areas).

Regarding an advanced Arctic Ocean Observing System, one approach was very successfully experimented during the International Arctic Buoy Project ('IABP') using long-term sea ice drifting platforms meandering across the entire Arctic Ocean over the past 30 years. A second approach was experimented during SHEBA, when an icebreaker (the Canadian icebreaker Des Groseillers) drifted in the Beaufort Gyre for one year (1997-1998) to provide a base for taking intensive soundings in the lower atmosphere, sea ice and the upper ocean. More recently DAMOCLES took advantage of advanced technology for implementing a modern IABP-SHEBA combined strategy for observing the Arctic Ocean at the pan-Arctic scale during two consecutive years (IPY 2007-2008).

The system will have to operate over several years in an autonomous mode with automatic long-lived drifting stations deployed all over the deep part of the Arctic Ocean and this will constitute the Lagrangian array. In addition there will be a need to monitor ocean gateways in Fram Strait, Bering Strait and Nares Strait, as well as the shallow Arctic shelves that represent half the size of the Arctic Ocean including the

Barents Sea and the Canadian Archipelago and the continental margin, all these regions being strongly influenced by seasonal sea ice (ice-free in summer, ice-covered in winter). This second ensemble will constitute the Eulerian array since it will mainly be instrumented with subsurface moorings. The complementarity and adaptability of the two systems will be essential to address main questions and resolve main issues regarding Arctic sciences. The two systems will have to be long-lived (several years) and be autonomous and reliable. The major differences between the Eulerian and Lagrangian arrays will be (1) the absence of atmospheric measurements in the case of the Eulerian array except those provided by shore weather stations such as Barrow (United States), Eureka (Canada), Ny-Ålesund (Norway) and Tiksi (Russia), located at the periphery of the Arctic Ocean, which will complement the atmospheric measurements taken with the offshore Lagrangian array, (2) the real time data transmission in the case of the Lagrangian array that will be highly appreciated for improving Numerical Weather Prediction ('NWP') in the Arctic domain. In the following, some elements are given for such a system, its basic elements for sounding the atmosphere, sea ice and the ocean simultaneously and how it will function and deliver information in real time.

1. The Large-Scale Lagrangian Array System (IAOOS)

This array will be based on 15 autonomous platforms operating at any given time in the Arctic Ocean for a period of five years. During IPY the most successful field work experiment concerned the deployment of an array of ITP across the entire Arctic Ocean over a long period of time (> 1 year). These ITP are drifting platforms equipped with profilers recording temperature in the upper ocean and sea ice (but nothing yet in the atmosphere except surface air temperature and atmospheric pressure). This idea is not new. The 30 year-old IABP took advantage of platforms drifting on sea ice documenting the ice motion in addition to surface air temperature and pressure. But IABP never used profilers in the ocean, in the atmosphere and throughout the sea ice. IAOOS is the first attempt to collect simultaneously vertical profiles throughout the three domains.

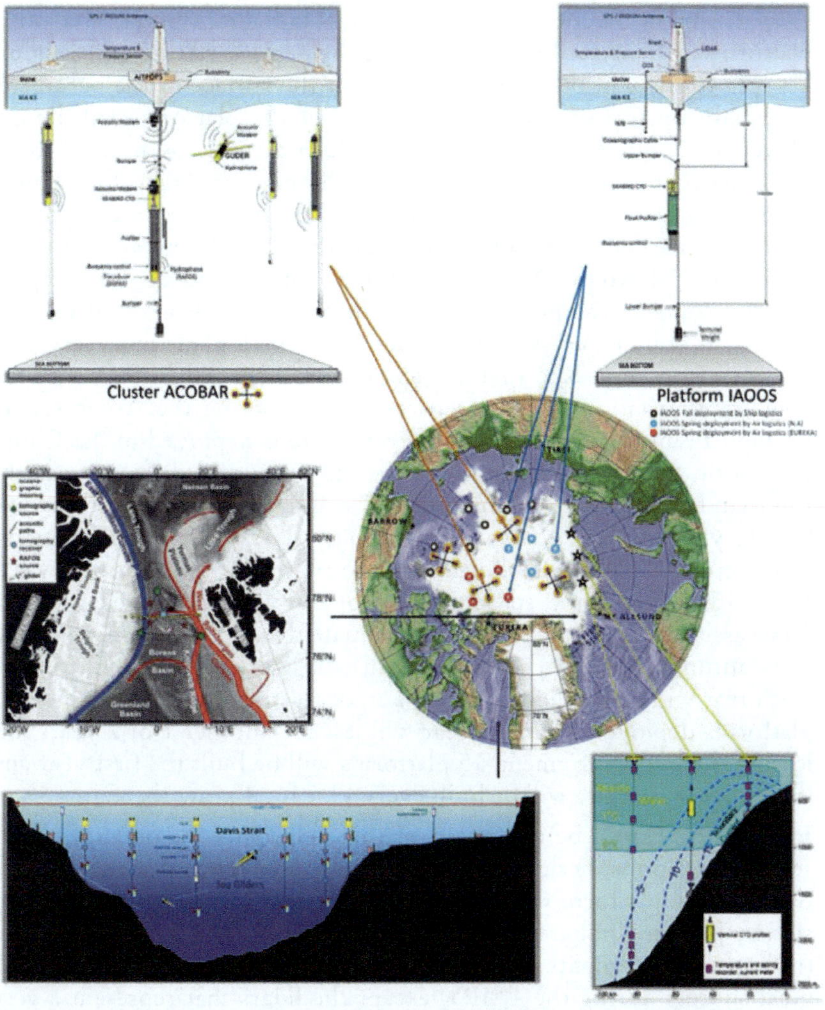

This figure represents the main elements of an advanced Arctic Ice Atmosphere Ocean Observing System including some Lagrangian components (ACOBAR and IAOOS) on the top part and some Eulerian components [Fram Strait, Davis Strait and Nansen and Amundsen Basins Observational System ('NABOS')] on the bottom part of the figure. For simplicity and clarity similar instrumentation in Bering Strait and the Canadian Archipelago is not shown

These platforms, deployed on sea ice and in open water, will drift according to sea ice motions and currents mainly imposed by the Arctic transpolar drift and the Beaufort Gyre. Some of the platforms will drift away from the Arctic Ocean, exiting through Fram Strait, and other platforms will inevitably be destroyed by sea ice rafting and ridging. The platforms will be designed to float at the surface of the ocean and to remain on top of sea ice floes. None of the platforms will be deployed in exclusive economic zones (200 nm from the nearest coastline). In order to maintain the network of 15 platforms distributed more or less evenly over the Arctic Ocean, it will be necessary to compensate for the sea ice drift that will entrain platforms in different areas (including exiting the Arctic Ocean through Fram Strait) and also to compensate for some of the platforms being destroyed by sea ice rafting and ridging. It will be necessary to replace 7 out of 15 platforms every year in order to compensate first sea ice drift and second instrument loss. Some of the stations will be retrieved when exiting the Arctic Ocean through Fram Strait. Most of the seven stations deployed in the Chukchi Sea on the Pacific side close to Bering Strait will drift towards the Eurasian basin close to Fram Strait and will have to be replaced the following year by new stations being deployed at their initial positions. Most of the platforms deployed near the North Pole will drift towards Fram Strait and some stations will hopefully be recovered in Fram Strait assuming the conditions are adequate during spring, summer and early autumn (not during the dark and cold winter season). In total 55 platforms will be needed for an experiment lasting 6 to 7 years since the platforms deployed in the 5th year will last for another 1 or 2 years following the last deployment. 15 platforms will be built the first year and successively 10 more will be built every year for 4 years thereafter.

Each platform will be composed of three elements for oceanographic, sea ice and atmospheric soundings (vertical profiles as described previously). Each platform will transmit data in near real time to a receiving station [Institut polaire français ('IPEV'), Brest, France] via satellites (iridium). All elements of this equipment were tested in the field with great success during the 4th IPY, except the lidars that represent a very innovative component of the new equipment. The Optical Depth Sensor ('ODS') had been tested for space missions on planet Mars and the constraints in the harsh Arctic environment are not as stringent as they are on Mars (except the need for protection from snow). The data will be transmitted according to existing standards and norms (ARGO format for oceanographic data, GTS format for meteorological data). The data will be transferred on a daily basis to well established data centers [Coriolis for ARGO, ICARE for the atmospheric data, the National Snow and Ice Data Center ('NSIDC') for the sea ice and snow data, the European Centre for Medium-Range Weather Forecasts ('ECMWF') and Meteo France for weather information such as surface air temperature and pressure via the GTS system].

The equipment proposed for the atmosphere is based on a combination of satellite and ground-based observations. As shown by previous field experiments, namely SHEBA, as well as space observations, mixed-phase clouds occur frequently in the Arctic (from 25% in winter to 95% in summer), and their impact on the surface radiation budget is large. This cannot be observed from space, as water droplets are not detected by radar when they are small, and lidar cannot penetrate through lower layers. Observations from the surface are thus needed. It is the objective of IAOOS to perform atmospheric observations over the central Arctic region covered by iced ocean, where no station has yet been established using combined autonomous (and unattended) microlidar and optical depth sensors. Only limited records are available, for example the microlidar observations performed at Ny-Ålesund in March-April 2007.[17]

In the ocean we are targeting the first 1000 m beneath the surface in order to document precisely the surface mixed layer, the halocline and the Atlantic and/or Pacific water masses advected into the Arctic Ocean via Fram Strait and Bering Strait respectively. In the sea ice we need to document a few meters of thick ice in order to infer the temperature profiles through the ice layer and the sea ice thickness as a function of time in order to monitor sea ice melting and freezing and sea ice deformation. In the atmosphere we do not currently have any system able to profile the entire troposphere and up to the stratosphere except from satellites. But profiles obtained in the lower troposphere from satellites are subject to a lot of errors and bias. So we need automatic profilers operated from the ground and looking upward.

Lidar solid-state technology is currently the only technology able to operate in a remote harsh environment and in a completely automatic way. This is a technical challenge because of the environment and the limited available power. The great innovation of this proposal would be for the first time to operate three automatic profilers collecting data in the upper 1000 m of the ocean, through a few meters of sea ice and across the troposphere from a single platform drifting on sea ice (or in water) and transmitting the data in real time via iridium to the satellite (see figure 2). The drifting platform will be fitted with a Global Posi-

---

[17] A Hoffmann et al., 'Ground-based Lidar Measurements from Ny-Ålesund during ASTAR 2007', *Atmospheric Chemistry and Physics* 9 (2009) 9059 et seq.

tioning System ('GPS') making it possible to locate it to a great degree of accuracy over the whole long period of time.

a. Ocean [Polar Ocean Profiling System ('POPS')]

During IPY two systems based on different technology were tested successfully: the ITP system developed at Woods Hole (United States) and the POPS system developed by France and Canada on the basis of ARGO float technology. We will adopt this second version entirely built by the French company NKE.

The system is composed of a surface buoy unit capable of floating at the surface of an ocean free of ice as well as being deployed on sea ice. This surface unit contains the GPS and iridium transmitters for geo-localization and real time data transmission to dedicated satellites. It also contains the processor for data acquisition and the lithium battery supply for one year of operations (at least). An 800 m long cable is attached to the buoy underneath and loaded with a 50 kg deadweight at the very end in order to keep the cable as vertical as possible even during strong sea ice drift entraining the surface buoy and cable. Along this 800 m long cable an ARGO-like float equipped with a CTD will scan up and down from the surface down to 800 m depth and up again at any given pace, taking vertical profiles of temperature and salinity once or twice per day. At the end of each profile, the data are immediately transmitted by iridium to satellites and to land. These profiles are very important for keeping us informed about the ocean mixed layer depth, the depth and strength of the halocline, the Atlantic layer and/or the Pacific layer under the halocline. These are fundamental observations allowing us to compute the heat flux from the ocean to the ice or to the atmosphere.

b. Sea Ice (Ice Mass Balance)

To measure ice mass balance (sea ice thickness with 2 cm accuracy) a novel system was developed recently during DAMOCLES and IPY whereby the sea ice mass balance was monitored through a series of tightly spaced thermistors with heater elements.

By utilizing developments in addressable digital temperature technology, a system that reduces the number of cables to just three regardless of the number of thermistors has been designed. Thus, there is no theoretical limit to the number of sensors on a chain. The novel twist to the system was the inclusion of a heater element with each temperature sensor. Each sensor is periodically heated and by monitoring the thermal response, the medium in which the sensor is embedded (air, snow, ice, water) can be identified.

The system has several distinct advantages: (i) it can continuously monitor changes in atmospheric, snow, ice and water conditions at any period upward of 1 second; (ii) the data is transmitted in real time via the iridium network; (iii) sampling frequency can be changed, even after deployment; (iv) it is cheap and robust; and (v) can be deployed in minutes by non-specialists.

c. Atmosphere (Lidar and Optical Depth Sensor)

As previously detailed, it is necessary to use new instruments in order to better document the vertical structure and properties of clouds, haze and aerosols. The unique advantage of drifting stations deployed in the Arctic would be the provision of information of highly variable local processes conditioning radiation fluxes through aerosols, Arctic haze and clouds in the troposphere and PSCs in the winter stratosphere. To this end we aim to install combined lidars and OD Sensors in the developed network.

The lidar will be based on the unattended ceilometer system as already implemented at Barrow station. However it has to be reduced in size and modified to operate with low power consumption. The design will thus have to be modified to fit into a very small temperature-controlled package. Cimel in France has considerable expertise in autonomous instruments [Cimel is the world leader in sun-photometers, the basic instrument of the Aerosol Robotic Network ('AERONET') network] and has developed microlidars for profiling. It is proposed here to create a dual channel diode based backscatter microlidar (this will permit analysis of the full waveform signal and not only the cloud base or top altitude as for a ceilometer) including depolarization. Data inversion will be usefully aided by the determination of the optical depth from ODS for the exploitation phase.

An ODS instrument that would be able to carry out the measurements proposed here, well adapted to the stringent constraints imposed by the fact that the drifting stations will be unattended for one year, has already been designed for the observation of very similar measurements (aerosol and low cloud optical depth, high cloud thickness and altitude) on small stations on the surface of the planet Mars. The atmospheric optical depth ('AOD') is derived from the observation of sun or moon scattered light at two wavelengths – blue and near infrared – at the zenith, while the altitude of the clouds is provided by the variation of a color index, the ratio of the two channels, with the solar zenith angle at twilight. The capacity of the instrument for providing this information has been fully demonstrated by long-term comparison with the measurements of co-located AERONET sun-photometers in Burkina Faso for studying Saharan dust episodes and cirrus and sub-cirrus clouds near the tropopause.

The development of an unattended micro-lidar system, operating in the Arctic over two years, is a real challenge. To overcome the expected problems a thermal model of the buoy is to be developed by the Institut national des sciences de l'univers/Division technique ('INSU/DT') using their expertise in stratospheric balloon-borne experiments to identify critical operating parameters for and realization constraints on the system to be manufactured by industry.

2. The Small-Scale Lagrangian Array: the ACOBAR Cluster

The IAOOS is specially designed to address large-scale air, sea ice and ocean interactions. It is not designed for small-scale processes occurring in the atmosphere, sea ice and ocean. However, we do need specially designed experiments in dedicated regions to address key topical questions and issues such as the boundary layer processes in the lower atmosphere and the upper ocean, sea ice mechanics, the atmospheric inversion layer, the ocean halocline, double diffusion, brines and frazil ice formation, deep convection etc.…

The so-called Acoustic Technology for Observing the interior of the Arctic Ocean ('ACOBAR') cluster is one specially designed experiment addressing upper ocean phenomena such as mixed layer processes, sea ice-ocean interactions (heat and salt transfer), frazil ice and brine formation, ocean halocline formation and variability, double diffusion and staircase microstructure, Pacific and Atlantic waters influencing the upper ocean, and solar radiation influencing the surface ocean mixed layer. ACOBAR is an EU project funded under the Seventh Framework Programme coordinated by the Nansen Environmental and Remote Sensing Center ('NERSC') (located in Bergen, Norway; project coordinators: PI Stein Sandven and Hanne Sagen). ACOBAR is a follow on DAMOCLES-like project comprising an advanced ocean observing system mainly based on underwater acoustic systems for long range fixing and ranging (SOFAR/RAFOS), acoustic modems for high-rate data transmission at short range (kms), upward looking sonar ('ULS') for measuring sea ice drafts from floats drifting at shallow depths, acoustic tomography inferring temperature field from inversion techniques, and sea-gliders operating under sea ice.

The ACOBAR cluster, as represented in the upper right part of the figure, is mainly composed of four Acoustic Ice Tethered Platforms ('AITP') drifting with sea ice and used mainly to navigate sea gliders under sea ice.

3. The Eulerian Component of IAOOS

The Lagrangian array of platforms drifting with sea ice is principally complemented by a set of bottom moored arrays deployed in gateways such as Fram Strait, Bering Strait, Nares and Davis Straits, along the continental slope (the NABOS array), on Arctic shelves (Barents Sea, Kara Sea, East Siberian Sea) and in the Canadian Archipelago (see the map).

a. Arctic Gateways

There are major differences between the main Arctic gateways and the optimum systems adapted to each of them are consequently just as diverse. It is very important to keep all of these sub-systems in operation over a long period of time. Advances in technology should enable some significant innovations such as the use of sea gliders across major straits. As Bering Strait is much shallower and narrower than Fram Strait and Davis Strait, it will accommodate bottom moored instrumentation rather than sea gliders. The same is true for the Canadian Archipelago. What is really important is the possibility to acquire data in near real time and, as for the Lagrangian system, to acquire data simultaneously regarding the atmosphere and sea ice as well.

b. Arctic Shelf Break

Because of the large extent of Arctic shelf seas north of Europe and Siberia, the shelf break separating these shallow seas from the Arctic Ocean deep basin plays a major role in driving and controlling most of the exchanges between the shelf seas and the deep basins of the Arctic

Ocean. It is also in this region that NABOS is concentrating all efforts by deploying long-term moorings along and across the continental slope north of Eurasia. This is a very active and dynamic region controlling the circulation of major water masses entering the Arctic Ocean from either the Pacific or the Atlantic Ocean.

c. Arctic Shelves

Arctic shelves represent half of the total surface of the Arctic Ocean. Due to the shallowness of the shelves (< 200 m depth), bottom topography controls major phenomena such as brine-enriched shelf waters that are more likely one of the main sources for the cold halocline extending over all the Arctic Ocean. On the Canadian side the halocline is strongly influenced by Pacific waters, in contrast with Atlantic waters in the Eurasian basin, that circulate much deeper and have less influence on the halocline that remains much colder. Brines are formed during sea ice formation and are the coldest, the saltiest and the densest waters we can find around the Arctic.

4. The Remote Sensing Component of IAOOS

The great advantage of satellite borne instruments for remote sensing observations in the Arctic domain is the Pan Arctic coverage obtained on a daily basis for long-term applications (years), days and nights at all seasons.

Remote sensing from satellites plays a specific role in the Arctic for observing sea ice in particular. Advanced scanning radiometers such as the Advanced Microwave Scanning Radiometer for the Earth Observing System ('AMSR-E') are capable of providing a remarkable ensemble of information regarding sea ice concentration. Cryosat was launched recently to provide sea ice thickness information from freeboard measurements. European Remote Sensing ('ERS') scatterometers provided extremely useful information regarding sea ice types and ages. Radarsat provided high resolution of sea ice dynamics and sea ice motion. GRACE measured gravity fields and thereby provided indications about freshwater variability.

Data from space missions such as CALIPSO or CloudSat, flying in the AQUA satellite constellation, can be easily accessed from the French archive centre ICARE in Lille (France). Additional observations from other polar orbiters such as IASI now permit vertical soundings of state parameters and chemical species. Maps of the CO IASI distribution over the pole are now produced regularly and can be used for clustering air masses according to emission sources (forest fires, anthropogenic emissions). Data are available from the French data centre ETHER at the Institut Pierre Simon Laplace ('IPSL')/UPMC.

All satellite borne instruments need *in situ* calibration and validation. So there is a close connection between remote sensing and *in situ* measurements. For instance Cryosat measuring sea ice thickness from free-board measurements will need *in situ* IMB detecting total sea ice thickness or ULS measuring ice draft from underneath. ULS can be mounted on moored instruments, drifters and submarines or autonomous underwater vehicles ('AUVs'). There are also several airborne instruments such as electromagnetic sensor ('EM') measuring sea ice thickness from slow aircrafts and helicopters flying at low altitudes or lasers for measuring the upper topography of sea ice.

IV. General Comments

- Climate change studies require full access to the Arctic domain whilst respecting regional jurisdictions.
- Climate change studies and Arctic marine science research require transparency and unlimited information & data sharing.
- There is a strong need for a fully integrated international (worldwide) body dedicated to Arctic scientific research. That would require some intense coordination involving IASC & the International Arctic Social Sciences Association ('IASSA'), the World Meteorological Organization ('WMO') & the International Council for Science ('ICSU'), the Intergovernmental Oceanographic Commission ('IOC') & the United Nations Educational, Scientific and Cultural Organization ('UNESCO').
- It has to be kept in mind that future Arctic observing systems will have to bridge gaps between academic institutions working on fundamental issues such as climate change and all institutions and private organizations involved in human activities, socio-economic affairs and Arctic governance.
- The Arctic is part of the worldwide common heritage of humankind. It deserves full protection.

Internationally Coordinated, Cooperative Arctic Marine Science during the Fourth International Polar Year: Lessons for Future Arctic Ocean Science Agreements

by Hajo Eicken[*]

I. Introduction

The Arctic Ocean region has undergone major changes over the past decade, many of them linked to a transformation of the sea ice cover from predominantly perennial to predominantly seasonal ice.[1] In parallel, and partly as a result of milder ice conditions, resource development and ship-based tourism have driven up Arctic maritime traffic and fostered geopolitical discourse at the national and international level concerning future uses of the Arctic Ocean and associated international agreements and regulatory regimes.[2] As illustrated by the Arctic Council's recent search and rescue agreement,[3] tracking of maritime activity

---

[*] International Arctic Research Center & Geophysical Institute, University of Alaska Fairbanks, Fairbanks, AK 99775-7320, USA.

[1] J Maslanik et al., 'Distribution and Trends in Arctic Sea Ice Age through Spring 2011', *Geophysical Research Letters* 38 (2011) L13502.

[2] LW Brigham, 'The Fast-Changing Maritime Arctic', *Proceedings of the U.S. Naval Institute* 136 (2010) 55 et seq.; PA Berkman and OR Young, 'Governance and Environmental Change in the Arctic Ocean', *Science* 324 (2009) 339 et seq.; AL Lovecraft and H Eicken (eds), *North by 2020: Perspectives on Alaska's Changing Social-Ecological Systems* (2011) 681 et seq.

[3] Arctic Council, 'Agreement on Cooperation on Aeronautical and Maritime Search and Rescue in the Arctic' (May 2011), see <http://arctic-council.

as well as effective regulation and emergency response require data on environmental and other hazards at high spatial and temporal density. At present, it is largely an emerging network of Arctic observing systems that is providing what little information is available to effectively govern and manage different maritime uses and responses,[4] with different entities, including academia, federal agencies and international programs contributing to these efforts. However, given the scope of such an endeavor once fully implemented and its inherently transboundary nature in an Arctic Ocean with highly dynamic sea ice and surface waters, future uses and stewardship of the Arctic Ocean will be closely tied to internationally coordinated, collaborative marine scientific research and operational environmental observations.

Such activities may be impacted by pending and future claims that seek to confirm jurisdiction of Arctic coastal States over the extended continental shelves under the provisions of the United Nations Convention of the Law of the Seas ('UNCLOS').[5] This development warrants an examination into current practice and potential recommendations concerning international scientific research and long-term observing programs that may play an important role as the Arctic Ocean continues on its trajectory of major environmental and socio-economic change. The Fourth International Polar Year ('IPY', March 2007 through March 2009) as a major international collaborative effort underway at a time of punctuated change, such as the record summer sea ice minimum of 2007,[6] provides an opportunity for such a review and analysis of current practice and implications for future collaborative regimes and scientific access in the Arctic. This brief contribution focuses on studies of the sea ice cover as a helpful illustration of broader issues concerning best practices and potential recommendations for Arctic maritime research and operations. Specifically, Section II below presents a brief overview of selected efforts underway during and beyond the IPY to help illustrate key requirements for successful international cooperative efforts. Section III examines the traits of successful frameworks and best practices

---

org/filearchive/Arctic_SAR_Agreement_EN_FINAL_for_signature_21-Apr-2011.pdf> (24 June 2011).

4 H Eicken et al., 'Environmental Security in Arctic Ice-covered Seas: From Strategy to Tactics of Hazard Identification and Emergency Response', *Marine Technology Society Journal* 45 (2011) 37 et seq.

5 United Nations Convention on the Law of the Sea (concluded 10 December 1982, entered into force 16 November 1994) 1833 UNTS 397.

6 Maslanik et al., see note 1.

to foster cooperative scientific approaches that can bear on present-day and future maritime activities as well as improved understanding of the fundamentals of Arctic environmental change.

II. The 4th IPY as a Testbed of Internationally Coordinated, Cooperative Research: Examples From Sea Ice Research

The activities of the 4th IPY were overseen by the International Council for Science ('ICSU') and the World Meteorological Organization ('WMO') and fully endorsed and supported by the nations represented in these bodies, including the Arctic coastal States.[7] In many respects, the IPY served as a testbed for technologically or conceptually advanced approaches to the study of the polar regions, with a focus in the Arctic on environmental and socio-economic change, including its impact on and tracking by Arctic residents.[8] Out of a total of roughly 90 endorsed Arctic IPY projects, more than 20 included a major (if not central) sea ice component. Since these projects cover five of the seven IPY research themes, they can serve to illustrate overarching challenges and opportunities in pursuing internationally coordinated, cooperative research.

1. A Nascent, Internationally Coordinated Arctic Observing System

The 4th IPY provided a major push towards the implementation of a pan-Arctic, networked and internationally coordinated observing system on land and in particular in the Arctic Ocean. Arctic coastal States invested heavily in observing system infrastructure, including Canada's ArcticNet program with its dedicated research icebreaker, the *Amundsen*, the European Union's Developing Arctic Modeling and Observing Capabilities for Long-Term Environmental Studies ('DAMOCLES') program with major deployment of advanced sensor technology throughout the Eurasian Arctic, and Russia's investment into coastal observing stations such as the Tiksi Observatory, a cooperative effort between Russian and US research and operational agencies. In the US,

---

[7] I Krupnik et al. (eds), *Understanding Earth's Polar Challenges: International Polar Year* 2007-2008 (2011) 87 et seq.

[8] Krupnik et al., *ibid.*, 357 et seq.

the National Science Foundation ('NSF'), along with support from partners such as the National Oceanic and Atmospheric Administration ('NOAA'), made major investments into the Arctic Observing Network ('AON'), an IPY initiative under the auspices of the US interagency Study of Environmental Arctic Change ('SEARCH').[9] The AON illustrates two important aspects of modern environmental observing systems in the service of society.[10]

First, it was implemented as an effort driven by the scientific community with international coordination of deployment of sensors and resources from the very start, such as through coordination with DAMOCLES[11] and other international programs.[12] This was achieved through the long history of existing international ties and collaborative frameworks which had prepared the ground for the intensive observations conducted during the IPY, with the density of observations greatly increased (see example shown in Figure 1). This type of bottom-up collaboration has been highly effective in combining infrastructure and resources internationally (including substantial support by non-Arctic nations in Asia and Europe) in order to sustain observations in challenging and remote Arctic environments to benefit all nations.

Second, the AON is one of the first large-scale international programs that has included a stipulation enforced by the funding agency (NSF) for immediate release of data collected through the observing network without any of the traditional embargo period – typically between 1 and 3 years – that has commonly limited initial data access to the prin-

---

9 Arctic Observing Network (ed.), *Arctic Observing Network (AON) Program Status Report – 2009* (2010); MO Jeffries et al. (eds), 'Arctic Observing Network: Toward a U.S. Contribution to Pan-Arctic Observing', *Arctic Research of the United States* 21 (2007) 3 et seq.

10 J Calder et al., 'Community White Paper: An Integrated International Approach to Arctic Ocean Observations for Society', in J Hall, DE Harrison and D Stammer (eds), *Proceedings of OceanObs'09: Sustained Ocean Observations and Information for Society*, Vol. 2 (2009).

11 DAMOCLES and SEARCH formed a joint project ('S4D') to this end, with activities including joint workshops and model intercomparison projects.

12 B Dickson and E Fahrbach, 'Observing Our Northern Seas during the IPY: What Was Achieved, What Have We Learned, Where Do We Go From Here? A Report of the Arctic Ocean Sciences Board and of the IPY Joint Committee' (2010), see <http://iasc.arcticportal.org/index.php/home/groups/working-groups/marineaosb/publications> (1 August 2011).

cipal investigators for a given set of measurements.[13] This approach was based on recommendations made by a Data Working Group assembled under SEARCH that recognized the importance of such open data policies to encourage collaboration and discovery in a rapidly changing Arctic, requiring quick access to data in order to effectively track and respond to major changes such as the 2007 record ice minimum. Furthermore, this data policy was also seen as an important step in addressing the information needs of Arctic residents and other decision-makers responding to Arctic change.

2. Cooperative Activities to Improve Understanding and Prediction of Sea Ice: The Arctic Sea Ice Outlook

The rapidity and magnitude of Arctic environmental changes has challenged the scientific community and government agencies to analyze, digest and respond on time scales much shorter than the traditional cycle of research planning, data analysis, review and publication of results. This development was highlighted by the record summer sea ice minimum extent of 2007,[14] almost one quarter below the previous record set in 2005 that had immediate impacts ranging from major effects on marine ecosystems to increases in ship-based tourism.[15] In response to this challenge, and under the leadership of the SEARCH and DAMOCLES programs, the international sea ice observation and modeling community created a forum for the synthesis, discussion and review of seasonal ice predictions, the SEARCH Arctic Sea Ice Outlook.[16] Each year since 2008, between May and October, around 20 of the leading sea ice research and modeling groups have prepared and updated a monthly prediction of pan-Arctic and regional September minimum ice extent,

---

[13] AON data are available through the portal maintained by the Cooperative Arctic Data and Information Service; see <http://aoncadis.org> (1 August 2011).

[14] For the period 1979 through 2011 for which consistent, reliable satellite records of ice extent are available; Maslanik et al., see note 1.

[15] LW Cooper et al., 'Rapid Seasonal Sea-Ice Retreat in the Arctic Could be Impacting Pacific Walrus (*Odobenus rosmarus divergens*) Recruitment', *Aquatic Mammals* 32 (2006) 98 et seq.; Brigham, see note 2.

[16] J Overland et al., 'International Arctic Sea Ice Monitoring Program Continues Into Second Summer', *Eos Trans AGU* 90(37) (2009) 321 et seq.; see also <http://www.arcus.org/search/siwo> (1 August 2011).

along with a detailed explanation of the approach taken, its uncertainties and indications on what type of observations would help improve these projections.

The Outlook has spawned a number of research projects, increased insight into the degree of predictability of different sea ice processes and advanced the evaluation and development of prediction models. However, in the context of internationally coordinated research an important outcome has been to serve as a testbed for improved coordination of observations that can help reduce uncertainties in tracking and predicting seasonal ice evolution. This has been achieved in part by creating an informal, internet-based forum for collaboration and exchange of information between operators and research scientists, currently representing all Arctic coastal States and many non-Arctic nations. As an example of this approach, consider how in 2009 uncertainty about the ice evolution in the North American Arctic and discussion of the long-term outlook provided by the US National Ice Center prompted near real time transmission of information about the state of the ice cover via satellite phone by the DAMOCLES lead scientist onboard the Chinese research vessel *Xuelong*, operating in the Canadian Basin.

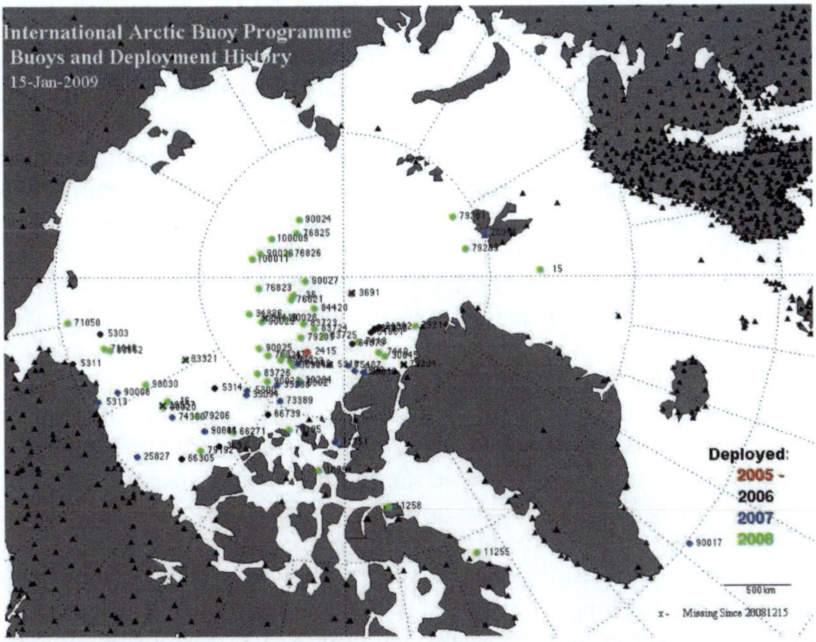

Fig. 1. Distribution of sea ice buoys in the Arctic Ocean on 15 January 2009 towards the end of the IPY. Note the surge of deployments associated with the IPY in 2007 and 2008 (shown in blue and green color). The comparative lack of buoys in the Eurasian Arctic is a result of the lack of perennial sea ice in this region with reductions in summer ice extent[17]

3. Improving Weather and Sea Ice Prediction, Understanding of Change and Emergency Response Through the International Arctic Buoy Program ('IABP')

The IABP is a highly successful example of how the pooling of resources between different countries and between research and operational agencies can provide information that is now critical to improved weather forecasts for the northern hemisphere and Arctic maritime operations, as well as for understanding large-scale Arctic environmental variability and change.[18] IABP data are collected in support of the

---

17 Image courtesy of IABP, see <http://iabp.apl.washington.edu/> (1 August 2011).

18 IG Rigor and JM Wallace, 'Responses of Sea Ice to the Arctic Oscillation', *Journal of Climate* 15 (2002) 2648 et seq.; see also <http://iabp.apl.washington.

WMO, the World Climate Research Program and the World Weather Watch Program. Buoy resources are pooled among the IABP partners which come from both Arctic coastal States and non-Arctic nations, and deployed by research vessels, operational agencies and on occasion by residents in Arctic communities. Buoys are deployed throughout the Arctic and the operating principles of the IABP govern deployment from or drift into the respective partner countries' exclusive economic zones ['EEZ(s)'].[19] As evident from Figure 1, the IPY has resulted in a major boost in the number of deployed buoys, but also in the number of countries involved in deployment and the sophistication of sensor packages. Thus many buoys are now capable of measuring ice surface and bottom melt and thickness, improving understanding and prediction of seasonal ice retreat at the regional level.[20] Typically, all the data from buoy sensors (position, surface air pressure, air temperature and potentially other variables such as ice thickness) are transmitted in near real time, including data feeds into the WMO operational networks, which have been shown to substantially improve weather forecasts and assessment of impacts of sea ice change within and outside of the Arctic.[21] Due to the dynamic nature of sea ice and uncertainties in predicting ice drift over periods of weeks, let alone years, it is not possible to anticipate the trajectories of these buoys. Thus, deriving the full benefits from such a hybrid operational and research sensor network also requires international agreements on access over the continental shelves.

4. Technological Advances Transform Marine Science and Improve Emergency Preparedness and Response

Over the past decade, a number of substantial advances have begun to transform research in the ice-covered Arctic Ocean. This includes major

edu> (1 August 2011); J Inoue et al., 'Impact of Observations from Arctic Drifting Buoys on the Reanalysis of Surface Fields', *Geophysical Research Letters* 36 (2009) L08501.

[19] IABP operating principles are laid out the in the following document accessible online: IABP, 'Operating Principles' (Revised May 2007), see <http://iabp.apl.washington.edu/overview_principles.html> (1 August 2011).

[20] C Polashenski et al., 'Seasonal Ice Mass-Balance Buoys: Adapting Tools to the Changing Arctic', *Annals of Glaciology* 52 (2011) 18 et seq.

[21] Inoue et al., see note 18.

advances in the technology and data transmission from ice-tethered platforms,[22] with increasing coverage in the Arctic[23] as well as increasing use of autonomous underwater and aerial vehicles. In addition to the mature and comprehensive set of instruments monitoring Arctic sea ice from space,[24] airborne measurements in particular of ice thickness, snow depth and other ice characteristics have now reached a stage where airborne surveys can provide data that may enter directly into seasonal forecasts of ice evolution and provide updates on the state of the Arctic sea ice cover.[25] In some Arctic locations, industrial activity such as offshore oil and gas exploration and development have resulted in substantially increased observation and data density employing state-of-the-art methods.

Hence, at present and in the near-term foreseeable future the infrastructure and resources available to operational agencies and others involved in emergency preparedness and response are likely to consist to a significant degree of a mix of marine scientific research assets described above, along with instrumentation deployed by industry, rather than dedicated operational monitoring networks. The scientific community increasingly acknowledges the research opportunities and societal mandates of use-inspired research.[26] Since there is substantial overlap in the methodology and instrumentation employed by the scientific community and operators, it can be postulated that any future Arctic observing system is likely to be a hybrid construct, overseen by a consortium of

---

[22] R Krishfield et al., 'Automated Ice-tethered Profilers for Seawater Observations Under Pack Ice in All Seasons', *Journal of Atmospheric and Oceanic Technology* 25 (2008) 2091 et seq.

[23] More than one tenth of the buoys shown in Figure 1 for January 2009 possess capabilities that extend to measurements in the ice or ocean underneath; in 2011 this fraction is at roughly one fifth of all drifters.

[24] RA Massom, 'Principal Uses of Remote Sensing in Sea Ice Field Research', in H Eicken et al. (eds), *Field Techniques For Sea Ice Research* (2009) 405 et seq.

[25] C Haas et al., 'Synoptic Airborne Thickness Surveys Reveal State of Arctic Sea Ice Cover', *Geophysical Research Letters* 37 (2010) L09501.

[26] For sea ice, parallel and potential conflicting uses of ice-covered regions by different stakeholders can be tracked and parsed effectively in the context of ice uses or services delivered by sea ice; see H Eicken et al., 'Sea-ice System Services: A Framework to Help Identify and Meet Information Needs Relevant for Arctic Observing Networks', *Arctic* 62 (2009) 119 et seq. (132).

stakeholders and with blurred boundaries between operational and academic research components.

III. Building Frameworks and Best Practices to Foster Internationally Coordinated, Cooperative Arctic Marine Science

The advances made during the 4th IPY as a testbed of state-of-the-art marine science (as outlined in Section II), offer important lessons for the implementation and support of internationally coordinated, collaborative research. As emphasized by several contributors at the conference on 'Arctic Science, International Law and Climate Change – Legal Aspects of Marine Science in the Arctic Ocean' in March 2011, international coordination and collaboration in marine science builds on successful international science partnerships, but also requires consistent and transparent application of existing rules and regulations governing Arctic marine science within different countries' EEZs or territorial waters. In this context, the scientific community can help advance cooperation by working towards internationally agreed-upon best-practices and (non-binding) statements to sustain and improve access and collaboration within the legal frameworks and international agreements governing such observation programs. Based on an evaluation of IPY outcomes (focusing on the sea ice themes) four principal sets of conclusions and recommendations have been derived.

1. International Collaboration

The expansion of Arctic observing network components during the IPY was driven partly by urgent information needs of decision-makers and stakeholders. For example, an increasing number of surface drifters (Figure 1) is now deployed by industry to improve understanding of sea ice circulation in regions of planned oil and gas development.[27] At the same time, the urgency of some of the questions emerging from the observed reductions in summer ice extent have promoted community-based observations and participation by Arctic residents in research activities, including deployment of surface drifters.

---

[27] Eicken et al. [2011], see note 4.

The increasing involvement of such groups in defining the aims of an observing system is guiding an expansion of sea ice research into the operational, applied sector. With a burst of activity during the IPY, the focus is now on consolidation and optimization of a pan-Arctic observing system through international collaboration and coordination. Initiatives under the International Arctic Science Committee ('IASC'), such as the International Study of Arctic Change, the World Climate Research Program ('WCRP'), or the Arctic Council, such as the Sustained Arctic Observing Network initiative, are working towards this goal and may be able to provide important perspectives on how international agreements that are part of the legal regime can work in concert with such efforts at consolidation. Of particular value at the present time would be a survey of past and ongoing cooperative observation efforts that have successfully addressed key challenges with respect to international coordination, cooperation and scientific access. Developing best-practices documents and guidance based on such efforts will help achieve broader success.

2. Lessons Offered by IABP

The remoteness and harsh environment of the Arctic have fueled advances in autonomous sensor systems and remote sensing as important elements of an observing network. The IABP is an excellent example of how an existing collaborative framework can help maximize the benefits derived from technological advances – within the context of international agreements and legal regimes governing marine research. IABP is particularly relevant because it is collaboration between research scientists and government agencies that provides operational data in near real time. Similar to other networks that have undergone a comparable evolution (e.g., the Argo profiling float network[28]), IABP may thus help ensure that as sensor capabilities expand and a more mobile, less predictable ice cover may disperse sensors more widely, buoy deployment can continue to serve the information needs of Arctic nations. In particular, since ice drift of sensors such as those shown in Figure 1 cannot be controlled and only poorly predicted on timescales of the lifetime of such buoys, a key point to resolve along the lines of current practice

[28] A Mateos and M Gorina-Ysern, 'Climate Change and Guidelines for Argo Profiling Float Deployment on the High Seas', *ASIL Insight* 14 (2010), see <http://www.asil.org/insights100408.cfm> (2 August 2011).

with Argo concerns the need to inform or notify a partner to the IABP agreement (equivalent to informing International Oceanographic Commission Member States in the case of the Argo network[29]). Such questions are of increasing importance because of increases in the drift velocity and improvements in sensor technology and buoy design that increase the lifetime of such drifting sensors. At the same time, with data on the location of drifters typically available in near real time, the act of 'information' in regards to drifter position takes on a different meaning as potentially all the data collected by such buoys is accessible online, with no embargo or withholding of data.

3. Open Data/Access: A Proposal for Coupling Data and Scientific Research Access

In light of the developments outlined in the previous section, the IPY has also helped advance the concept of rapid - ideally real time - access to all data collected within the context of an evolving Arctic observing system. In the US, the National Science Foundation as a major supporter of the AON, reflected decision-maker and scientific community interests (as expressed in the SEARCH Data Policy[30]) in stipulating that data from the AON cannot be embargoed and has to be made available immediately after collection. Due to the size of the AON with more than 50 projects, this open-access policy has had a significant impact, and ensures that long-term Arctic observations collected through a research network can also help serve increasingly important operational needs. In the context of UNCLOS, operational research (typically characterized by unrestricted data release and real time integration into operational networks such as those overseen by the WMO) is not subject to the permission scheme required for 'marine scientific research', an undefined term of art in the Convention. This exception implies recognition of the broader value of such data in service of mankind or a larger group of nations. Here, it is argued that the idea of coupling data access and scientific access (Open Data/Access or 'OD/A'), as implicit in the UNCLOS regime and agreements governing programs such as IABP, is a potentially helpful concept that can inform the development

---

29 Mateos and Gorina-Ysern, *ibid.*

30 SEARCH Data Policy, see <http://www.arcus.org/search/searchscience/data.php> (2 August 2011).

of best practices and guide the international coordination and harmonization of observing network activities.

In contrast with Argo, where similar questions have surfaced, the IABP has the great advantage that data can be made available any time and potentially at high sampling and transfer rates, since the buoys are surface-based and can hence communicate information without the restrictions of the submerged Argo drifters which have to surface to relay data. The International Polar Decade, as a WMO endorsed endeavor, provides an ideal framework to build on the IPY and explore viable ways to implement OD/A guidelines in different settings.

4. A Case in Point: Safety of Maritime Operations

Maritime traffic and coastal and offshore oil and gas development are on the rise in several Arctic nations, often in locations where ice drift and currents may quickly disperse pollutants beyond a country's borders. As a result, emergency preparedness, prevention and response ('EPPR') are of increasing importance, as underscored by the Arctic Council's recent agreement on search and rescue.[31] Effective EPPR activities will have to draw on data from pertinent observing systems, with high demands placed on data availability and spatio-temporal resolution during an emergency.[32] Hence, such information needs may turn into a powerful driver outside of the research community towards collaborative, internationally coordinated activities governed by OD/A practices outlined above. In some regions, such as the Canadian Mackenzie Shelf and the shelf off Sakhalin, where oil and gas exploration and development is moving forward, one might argue that key principles of OD/A are already being followed. Thus, sensor deployment and long-term observations by scientific research groups is integrated into multilateral industry partnerships. By agreeing to the coordination practices of the overarching program, improved access has been obtained in exchange for collaborative contributions to industry program goals.[33]

---

[31] Arctic Council, 'Agreement on Cooperation on Aeronautical and Maritime Search and Rescue in the Arctic', see note 3.

[32] Eicken et al. [2011], see note 4.

[33] For example, sea ice and ocean measurements by Japanese research groups have been facilitated through such agreements with industry; e.g., K

IV. Conclusions

The comparative lack of observing assets in a remote Arctic Ocean combined with technological advances and increasing recognition of the value of near real time open data access by the scientific community has resulted in increasing recognition by all major stakeholders that cooperative, coordinated approaches towards environmental observations are in the best interests of all involved. Furthermore, the boundaries between fundamental and operational research have been blurred by the increasing recognition of the value of use-inspired science on the part of the scientific community and major advances in sensor technology that allow scientifically and operationally relevant measurements to be taken by the same set of instruments. Finally, the ever-increasing degree of international collaboration in science and operations, along with new partnerships being built between industry, Arctic communities, agencies and academia has prepared the ground for a major review and potentially substantial advance in the practice and theory of scientific access in an Arctic Ocean that is undergoing substantial, rapid change. At this point, the scientific community needs to develop and articulate a vision for coordinated and cooperative international marine science that maintains and ideally improves EEZ and extended continental shelf access,[34] while working with legal scholars and the relevant stakeholders to explore pragmatic approaches that ensure improved understanding of and effective responses to a changing Arctic Ocean. In this setting, IASC as a consultative body to the Arctic Council may help advance the dialog and provide a framework for emerging best practices that meets the requirements and information needs of all involved.

Shirasawa et al., 'The Thickness of Coastal Fast Ice in the Sea of Okhotsk', *Cold Regions Science and Technology* 42 (2005) 25 et seq.

[34] On this topic, see also the contribution by B Baker, 'Common Precepts of Marine Scientific Research Access in the Arctic', in this volume; B Baker and H Eicken, 'Marine Research Access in the Arctic Ocean: Background for Potential Guidelines in a Changing Arctic', unpublished White Paper (10 March 2010), see <http://www.iarc.uaf.edu/workshops/2009/4/> (8 June 2011) (click on 'download whitepaper'); the paper is also attached as an Appendix to the contribution of B Baker, in this volume.

Inuit, Circumpolar Law & Politics, Resource Development

(Speech)

by Udloriak Hanson[*]

My name is Udloriak Hanson. I was born and raised in Iqaluit, Nunavut, Canada's Arctic. I am Special Advisor to Mary Simon, President of Inuit Tapiriit Kanatami ('ITK'), the national organization representing the Inuit of Canada. Let me begin by thanking those of you who put a human face on the Arctic. Lars-Otto Reiersen's[1] presentation highlighted people/Inuit. It made a real connection between marine research, science and people, as did Hajo Eicken's.[2]

This may not be the forum to propose this, but I urge you to consider the need for more social sciences in the Arctic. You have to ask yourselves – will my research create a better place for the people that live there? How? 'how?' is where I will focus my comments. Inuit have had a long history with scientists: some good and some bad. We want to move from being subjects and bystanders to participants, contributors and beneficiaries of science. It is so important to make your science applicable to the Arctic communities.

Here is your first Inuktitut lesson of the day. The name used by Canadian Inuit to describe the Inuit homeland within Arctic Canada is *Inuit Nunangat.*

---

* Special Advisor to the President of Inuit Tapiriit Kanatami, Canada.

1 See the contribution by L-O Reiersen and S Wilson, 'The Arctic – the Sentinel for Environmental Processes and Effects', in this volume.

2 See the contribution by H Eicken, 'Internationally Coordinated, Cooperative Arctic Marine Science during the Fourth International Polar Year: Lessons for Future Arctic Ocean Science Agreements', in this volume.

I thank the conference organizers for inviting me to speak at this very important conference. The agenda offers a most impressive combination of topics, speakers and participants. It is an honor to be here. While the conference has legal, scientific and research dimensions, I should note, at the outset, that I am not a lawyer, I am not a scientist and I am not a researcher, but by virtue of being Inuit, and an employee of a national Inuit organization, I hope that I can positively contribute to this discussion.

Inuit have a strong respect for the *rule* of law and the *role* of law. While we now share core liberal democratic institutions and values with other Canadians, we must not forget that law in the Arctic preceded our contact with outsiders. As a hunting and gathering society inhabiting Arctic lands and water through both historic and pre-historic times, we developed our own forms of customary law. Many aspects of customary law continue to guide our behavior, whether out on the land or within communities and family settings.

Newcomers to *Inuit Nunangat* brought with them European based laws, primarily unwritten Anglo-Canadian common law and the various statutes and regulations that come from the Parliament of Canada and provincial and territorial legislatures. By the middle of the last century, colonization from outside had reduced us to the margins of power and enforced a new kind of law in our own homeland. In the last 40 years – about one working lifetime – we have confronted the legacy of colonization. We have used the tools at our disposal to regain a share of control over our land, our waters, and our lives. The crucial part of this effort has been the renewal of confidence in ourselves – and in the intrinsic value of our customs, language, and beliefs.

This renewal has involved making calculated and creative use of southern law to challenge our disempowerment within the terms of the very laws applied to us from the outside. We cited common law aboriginal title to confront hydro-electric and mining developments proceeding without our permission in litigation. We relied on common law aboriginal rights to negotiate and conclude five comprehensive land claims agreements that govern the top third of Canada stretching from the Alaska border to Labrador.

We insisted those land claims agreements supply us with powers in the realms of both property law – fee simple title, royalty payments and water rights – and public law – joint Inuit/government management boards, regional government institutions and the entire new territory of Nunavut. When promises made by the Crown in those agreements have not been fulfilled, we have brought litigation in the courts to enforce them, making use of their contractual, statutory, and constitutional force. With land claims agreements, the negotiations phase often seems to be all about architecture; sadly, the implementation phase often seems to be all about minimizing. There is now a landmark lawsuit by Nunavut Inuit before the Nunavut Court of Justice in response to the failure of the Crown to implement, fairly and fully, the promises made in the 1993 Nunavut Land Claims Agreement.[3]

For lawyers, this case is worth watching. As you would expect, we do not limit our efforts to the executive and judicial branches. We appear regularly at parliamentary committees, and make concrete proposals for both policy reform and for legislative amendments. Inuit, through their organizations, are working to prevent the further colonization and appropriation of Inuit knowledge in the areas of research, science and policy by asserting and sharing an Inuit perspective and translating Inuit knowledge so that it can be more easily understood by others. We believe and recognize that true Inuit knowledge stewardship requires the consideration, support and fostering of distinct local worldviews, perspectives, ways of life, language and culture, rather than the simple incorporation of Inuit knowledge into mainstream methods and systems of western science. Knowledge stewardship must be led by Inuit and their representative organizations in collaboration with others. While there remain significant challenges in expanding the research base to incorporate Inuit knowledge, we are providing solutions to ensure inclusion of Inuit in key research, science and policy areas that affect our lives and homeland.

And we have done everything we can, within limited resources, to communicate closely and candidly with fellow citizens in Canada and with the wider global community. Inuit have worked very hard, and, I think, been quite successful, in what could best be described as 'organizing ourselves'. Our community-based organizations are the building blocks of our regional organizations, which in turn form the basis of national organizations, and our national organizations work together at the international level through the Inuit Circumpolar Council.

The unity of Inuit as a people, and the profound consequences that Inuit unity has in relation to international and domestic law and politics for the circumpolar area, is a key starting point in the Circumpolar

---

[3] Nunavut Land Claims Agreement (signed 25 May 1993, ratification legislation entered into force 10 June 1993) <http://www.tunngavik.com/category/publications/nunavut-land-claims-agreement/> (9 August 2011).

Inuit Declaration on Sovereignty in the Arctic adopted in April 2009.[4] This Declaration has changed the geo-political landscape. It is a central and increasingly important reference point in legal and policy debates about the future of the circumpolar area. Partnership with Inuit means that Inuit have the right to reject resource development ideas that are bad for Inuit, bad for the Arctic, or bad for the planet. And partnership with Inuit means that Inuit have the right to attach conditions to Arctic development strategies, policies, and projects – including research projects.

Three questions may flow from this: First, why do Inuit, for the purposes of our fundamental well-being, need to attach conditions? Second, why does the world need to pay attention to those conditions? And, third, what outcomes could a balancing of Inuit and Arctic State interests deliver in *Inuit Nunangat*, and, perhaps, by extension, to other parts of the circumpolar world? I would like to address each of these things very briefly.

Why do Inuit need to attach conditions in the first place? Let me give you some context. The relationship between Inuit and the outside world, primarily European and European-descended States and peoples, is a long and complex one. It may have started on a footing of an approximate equality of power, and the motivation of mutually beneficial commercial exchange. But it became a very one-sided arrangement in the 19th and 20th centuries. And we are still working through some of the negative effects of that colonization. Effects ranging from coerced relocations to and between communities and residential schools abuse, to the personal, family and inter-generational crises brought about by disregard and disrespect for traditional Inuit knowledge, culture and values. We are recovering, we have made progress and we have hope but, as anyone knows who looks at some basic socio-economic indicators of how Inuit are doing – employment, educational achievement, life expectancy, suicide, family violence –, we still have a long way to go. History, of course, is never simple. But some of its lessons are both stark and powerful. We are determined to set conditions on Arctic resource development because the primary responsibility for doing so rests with us. As stewards of the land, the only satisfactory assurance

4 Inuit Circumpolar Council, 'A Circumpolar Inuit Declaration on Sovereignty in the Arctic' (April 2009), see <http://www.inuit.org/fileadmin/user_upload/File/declarations/ICC_Sovereignty_Declaration_2009_pages.pdf> (9 August 2011).

that it will be done rests with us. That is the core, the consequence, and the burden of self-determination. Let me move to my second question.

Why does the world need to pay attention to our conditions? There are two sorts of reasons: practical reasons and moral reasons. The practical reasons turn on the reality that, as one of the indigenous peoples of the world, Inuit have made maximum use of the evolving principles and processes of domestic and international law to secure recognition of the fundamental rights and interests that flow from our being the first and majority inhabitants of *Inuit Nunangat.* In the domestic arena, we have secured our rights and interests through a variety of means and instruments: constitutional recognition, statutory recognition, modern treaties, inter-jurisdictional agreements, self-government agreements and new territorial and regional governments.

The list is as long as our opportunities and imaginations have taken us and we have not finished yet. These domestic arrangements do not add up to our having unqualified control of our traditional homeland. They do add up to a formidable amount of power sharing between Inuit and the four Arctic States we inhabit. They do add up to a kind of 'tripwire' that will make it very difficult for outsiders to operate in the mixed jurisdictional and political realities of our world, without having Inuit on-side. In some cases, having Inuit on-side will take the form of active Inuit partnership in all commercial dimensions of specific developments. In some cases, having Inuit on-side will take the form of free, prior and informed Inuit consent. Whatever the form and diversity of Inuit participation, the willingness of Inuit to support major projects is now a critical factor at every level, from the front end to the head offices of the investment bankers.

The strength of our domestic power-sharing arrangements is reinforced by the growing status and profile of indigenous rights as a fundamental part of human rights at the international level. For all sorts of reasons, respect for legal rights and interests makes practical sense. Respect for the well-being of indigenous peoples also commands compelling moral authority. The history of the globe is littered with indigenous peoples cast crudely to one side in the race for lands and resources. Building a peaceable and sustainable planet gives all of us ample incentive – and ample responsibility – to turn to more collaborative, constructive and respectful approaches.

Finally, what outcomes could a balancing of Inuit and Arctic State interests deliver in Inuit Nunangat *and, perhaps, by extension, to other parts of the circumpolar world?* We are not just the traditional hunters and gatherers of *Inuit Nunangat.* We are also the frontline environmental watchdogs and police officers. This does not make us hostile to new forms of development, or locked in a kind of paralyzing nostalgia for days of old. It does however make us a critical force in ensuring that the development of Arctic resources is done in ways that are measured, informed, transparent and accountable. And that makes sustaining the well-being and cultural continuity of Inuit necessary and central considerations.

Scientific and traditional Inuit knowledge each has an important role to play in such an approach but research scientists must pay particular attention to a number of key Inuit concerns. After all the scientific data is most useful when attached to the rich context and culture to which it applies. The burden is on those who propose to conduct scientific research in *Inuit Nunangat* to explain and rationalize their research proposals to Inuit communities and regions, paying particular attention to Inuit sensitivities in relation to sustaining healthy wildlife and marine populations and a healthy natural environment. The successful effort last year by Inuit from the Baffin region in using the courts to block a program of seismic research in Lancaster Sound demonstrates that lack of adequate consultation often has bad results.

In order to facilitate exchanges between Inuit and the scientific research community, ITK has recently established the Inuit *Qaujisarvingat*: the Inuit Knowledge Centre. The Centre will focus efforts to ensure an increasingly active role for Inuit in research that leads to the generation of innovative knowledge for improved research, science, and policy decision making within a Canadian, circumpolar and global context.

ITK is also working hard to make sure that collaborative networks that involve academic and industry scientists pay growing attention to Inuit involvement and priorities. Expanding Inuit research capacity must be on the agenda of all those with serious, long term commitment to Arctic research. So, too, must be mobilizing research and related education and training budgets to provide economic opportunities in Inuit communities and regions that are very much suffering from economic and social well-being gaps in comparison with outsiders.

In closing, I would urge all of you to continue to pay attention to where Inuit figure in the development of the Arctic, including natural resource development projects on land and offshore, and the land and marine research and infrastructure that underpin such development. And, as you would suspect, I would also urge you to go about whatever it is you do – setting priorities, focusing your research, advising colleagues, communicating to the public – in ways that are respectful and accommodating of the central place of Inuit in the history, current reality, and future of the Arctic. Any rational and enduring Arctic policy, whether international or domestic, must put enhancing and sustaining the well-being of Inuit at its heart. I invite you to embrace these things as a defining and long-term dimension of your work on Arctic issues. We might not be at top of mind for many of you, but I hope this presentation provides you with a sense that we (Inuit) have the legal and moral authorities to have the last word.

Thank for your attention. Taima. Qujanamiik.

Conclusions of the Chair

*by Susanne Wasum-Rainer**

The dedicated and learned discussions during our conference have proven again how sensitive and important the Arctic region and the issue of the Arctic is. Some diverging views were expressed but also a considerable number of converging ones. There was, among other, broad consensus on the pivotal importance of the Arctic region for the world climate and for the need to protect its fragile natural environment. Freedom of marine scientific research is an essential element in this regard. That is why we had chosen this subject as the main focus of the Second Berlin Arctic Conference. I have tried to summarize the main points of our discussion and drawn some Conclusions of the Chair. These Conclusions may facilitate the continuation of the discussions and contribute to the finding of the best solutions for the problems at stake.[1]

---

* Director General for Legal Affairs, Legal Advisor, German Foreign Office.

[1] See also Papers from the International Conference at the German Federal Foreign Office on New Chances and New Responsibilities in the Arctic Region (11 – 13 March 2009), in G Witschel et al. (eds.), *New Chances and New Responsibilities in the Arctic Region* (2010).

For further background information see:

Arctic Climate Impact Assessment (ACIA), *Impacts of a Warming Arctic* (2004);

H Corell, 'Chairman's Conclusions', in Nordic Council of Ministers, *Common Concern for the Arctic* (2008) 16 et seq.;

L Nowlan, 'Arctic Legal Regime for Environmental Protection', *IUCN Policy and Law Paper* 44 (2001) 1 et seq.;

A Proelss and T Müller, 'The Legal Regime of the Arctic Ocean', *ZaöRV* 68 (2008) 651 et seq.;

1. The Arctic is of pivotal importance for the world climate. The effects of climate change can be seen globally and in the Arctic itself. There are fundamental changes in the Arctic ice level with regard to quality, extent and thickness due to global warming and other environmental developments. These include carbon pollution and ocean acidification. The causes of climate change, originate mainly from outside the Arctic area and, accordingly, non-Arctic actors are major targets affected by climate change. However, this may change if and when economic activities in the Arctic increase.

2. The freedom of marine scientific research as enshrined in the International Law of the Sea is a core requirement. The United Convention on the Law of the Sea (UNCLOS) provides the basic legal regime for marine scientific research in the Arctic Ocean and the legal balance between national interests and common interests in this regard. International scientific cooperation could be considered a common Arctic issue.

3. Balancing the interests of Arctic coastal States and the international community needs to take place within the framework of UNCLOS. Discussions about a special regime for scientific cooperation in the Arctic have not yet led to conclusive results. At the same time, the UNCLOS regime may be open to further development in the future.

4. International cooperation is vital for Arctic governance. Exchange of data would amplify the beneficial aspects of marine scientific research in the Arctic.

5. The exploitation of the newly accessible Arctic must be conducted in a sustainable way. Economic prospects must be balanced against environmental needs.

6. Large areas of the Arctic Ocean will continue to be areas of High Seas, where freedom of marine scientific research applies.

7. All relevant actors proceed on the basis of the International Law of the Sea, in particular UNCLOS. The international rules are supplemented by various domestic laws and regulations. Domestic procedures should be simplified and best practices identified in order to support marine scientific research. Harmonization of existing permission procedures would be welcome. In particular, 'one-stop' procedures would be helpful.

I Winkelmann, 'Arktische Ressourcen nutzen und arktische Vielfalt schützen: Quadratur des Kreises?', in UE Simonis et al. (eds), *Jahrbuch für Ökologie 2009* (2008) 38 et seq.

8. It is recommended that guidelines be developed to help coastal States apply the rules of the Convention more homogeneously. Such guidelines should privilege

- marine scientific research undertaken in cooperation, including cooperation without discrimination between scientists from coastal States and States not bordering the Arctic.
- marine scientific research enhancing knowledge about environmental matters. In this respect, one could borrow from the provisions of and practice under the Antarctic Treaty.

9. Research undertaken in the Arctic needs to take into account the legitimate interests of a multitude of stakeholders: indigenous peoples, states, international organizations, researchers and economic actors alike.

10. Climate change affects the foundations of entire indigenous cultural systems. The international community should, therefore, improve cooperation with indigenous peoples in order to reflect more fully their unique attachment to the Arctic.

11. The pertinent UNCLOS provisions concerning marine scientific research, in particular Art. 246, grant the coastal States considerable leeway to interpret and establish whether a research project is resource-oriented or not. Apart from that, different legal regimes apply to research of the Continental Shelf and the High Seas above it. Formally, the scientific research regime for the Outer Continental Shelf differs from that for the Continental Shelf.

12. UNCLOS uses scientific terms in a legal context. The legal usage of these terms may differ from accepted scientific terminology. 'Nature does not accept legal boundaries and distinctions'. This highlights the need for interdisciplinary cooperation between natural scientists and lawyers in fora such as the Berlin Arctic conferences.

13. International scientific cooperation in the Arctic Ocean is a reality. Cooperation also extends to the context of claims of extended continental shelves. This cooperation benefits the sharing of logistics and resources. Joint evaluation of data helps all actors to better understand the Arctic Ocean.

14. The freedom of marine scientific research should be upheld and maintained. There are different views as to the concrete extent of the application of the principle of the common heritage of mankind. Common interests are an evolving body of international law. International scientific cooperation could be considered a common Arctic issue.

15. There is a need to address the interaction between UNCLOS and other international agreements. Further research and discussion is required to understand the legal interrelationship between different regimes.

16. A degree of uncertainty will remain about the exact extent of continental shelves beyond 200 nm, as the Commission on the Limits of the Continental Shelf will still need considerable time to complete its work. Different views exist about the exercise of coastal State jurisdiction prior to recommendations on the outer limits of the continental shelf by the Commission. However, this question should not burden future marine scientific research.

17. In the 'Area', the International Seabed Authority is a vehicle for the dissemination of results of marine scientific research. The Authority is a forum for exchange of scientific results and thus demonstrates the general thrust of the Convention towards international cooperation.

18. There is an ongoing discussion about the feasibility of complementing the binding universally agreed rules from UNCLOS with non-binding precepts. A combination of both would be beneficial for coastal States as well as scientists.

19. Difficulties in application processes for research in the Arctic hinder the development of research-related technology. The industry is cooperating with indigenous peoples, taking into account their unique position in the Arctic.

20. Understanding the Arctic climate system requires fully integrated atmospheric and solar radiation, sea ice and ocean science, including observations and modelling activities.

21. The fundamental issue for the improvement of conditions for marine scientific research is access. Access depends upon mutual trust. The Arctic Ocean should remain an area of excellent international scientific collaboration and cooperation.

Annex 1

Second International Arctic Forum 'The Arctic – Territory of Dialogue' Arkhangelsk, 22 – 23 September 2010

(Speech)

by Vladimir Putin[*]

Ladies and gentlemen, colleagues and esteemed guests,

First of all, I'd like to thank our guests for coming here today to discuss Arctic issues with us. I'm very glad to welcome all participants of this forum, "The Arctic – Territory of Dialogue." This forum was initiated by the Russian Geographical Society (RGS), whose history is bound up with the start of research efforts in the northern latitudes. Established in 1845, the RGS immediately made the Arctic its top priority. Suffice it to recall the Arctic expeditions headed by society members Georgy Sedov, Georgy Brusilov and Vladimir Rusanov, its leading role in the first International Polar Year in 1881, the creation of the world's first network of meteorological polar stations, and the idea for an ice-breaker fleet, which was suggested by another outstanding member of the Russian Geographical Society, Admiral Makarov.

It is great that the society's glorious Arctic traditions live on, as this forum bears out. The first forum held in Moscow exactly a year ago sparked a great deal of interest among all those who care about Arctic problems, exploration of Russia's Far North, its environment, ethnography, historical heritage and the building of trust and partnership in this region. This is exactly why we decided to hold it every year.

The forum is moving increasingly closer to the Arctic. This time we met in Arkhangelsk. It was primarily the forum's agenda – the region's transport infrastructure – that determined the choice of the venue. This is a major foundation for the region's harmonious and stable development and of the cooperation among Arctic states. The city of Ark-

---

* Prime Minister of Russia.

hangelsk, the ancient capital of the Russian North and the homeland of Arctic trailblazers, is a very befitting and even symbolic place. It is here, on the banks of the Northern Dvina River, that Russia built its fleet, particularly its Arctic fleet. It built the first major dockyards in the 17th century. The Arkhangelsk Region is also the homeland of the great son of the Russian people Mikhail Lomonosov, a scientist and educator who predicted the vast importance of the North for Russia and the rest of civilisation.

Legendary expeditions to study northern regions embarked from this point. The Russian-Swedish expedition in 1878 laid the foundation for the navigation of the Northern Sea Route and became the first joint Arctic project.

The development of the Arctic, primarily its waterways, is very closely linked with Arkhangelsk. The Northern Sea Route (NSR) occupies a special place here. We are planning to turn it into a key commercial route of global importance. I'd like to emphasize that we see its future as an international transport artery capable of competing with traditional sea routes in cost of services, safety and quality.

The shortest way between the biggest markets of Europe and the Asia-Pacific region lies through the Arctic. It is almost one-third shorter than the traditional southern route and presents a great opportunity to reduce shipping costs. By using it, states and private companies will gain tangible economic benefits.

I think that transportation – new sea and air corridors – can become one of the breakthrough projects uniting Arctic nations. It will allow us to make returns on our investments and test universal mechanisms of cooperation.

Returning to our plans, I'd like to recall that we have already carried out major test runs of hydrocarbon shipping along the NSR this year. Transit pilotage of vessels is also gaining momentum. According to tentative estimate, shipping may reach 700,000 tonnes this year, and this is only the beginning.

Russia plans to carry out a series of measures to develop the NSR. At yesterday's government meeting we discussed a draft law designed to regulate all NSR navigation issues. We hope the Duma will pass it before the end of this year.

Developing modern infrastructure along the Northern Sea Route is a major objective. We are launching a comprehensive transport project designed to ensure the dynamic development and exploration of our northern territories, resolve vital economic and social challenges and create new production lines and jobs. We are planning to expand existing ports and build new ones, for instance the Port of Varandei by the Yugorsky Shar Strait and the Sabetta Port on the Yamal Peninsula. The NSR and its major harbours will be integrated with other modes of transport. We are also planning to upgrade river, car and railway routes and communications, northern airfields, airports and polar aviation. We are going to considerably expand our ice-breaker fleet. Today we have 10 ice-breakers. We intend to build another three all-purpose nuclear-powered icebreakers and six diesel-electric ones before 2020. We have allotted 38 billion roubles for this purpose until 2014.

We will continue working to develop systems of communication, navigation and hydrography in the Arctic, primarily with the use of our GLONASS global positioning system. Our experts are already developing a multi-purpose Arctic satellite system that will monitor the environment of the Far North. In addition, specialists are working on a project to create the "North Pole" ice-resistant observatory platform. We will pay special attention to transport safety.

Participants in the Arctic Council ministerial meeting last May signed the first legally binding pan-Arctic document – the Agreement on Cooperation on Aeronautical and Maritime Search and Rescue in the Arctic. Under this agreement, we are building a system of warning, monitoring and response for natural and man-made disasters in our Arctic zone. Under this programme we will build 10 all-purpose rescue centres in the Far North by 2015.

Russia will continue playing an active role in developing and consolidating the international legal foundation for the Arctic, in particular, the agreement on oil pollution prevention and control, which is currently under development. This entirely new field of international cooperation is extremely important.

Climate change, which is gradually increasing the navigation period, and technical progress are paving the way to new, still unexplored areas of the Far North, where economic activity is likely to grow.

Today, we had a video conference with one of such growth point. I'm referring to the Prirazlomnoye deposit on the shelf of the Pechora Sea. We have already installed one of the world's largest hydrocarbon platforms there. Russia is starting to develop the Arctic shelf and opening a new chapter in the history of Arctic exploration. Very soon it will contain pages on the commissioning of the Shtokman deposit in the Barents Sea and the development of resources in the Kara Sea and on the Yamal Peninsula. You know about Rosneft's agreement on long-term strategic cooperation with the American company ExxonMobil, one of the world's leaders in this field. They plan to build a special centre in St Petersburg to provide scientific and technical support for shelf projects, including the study of environmental issues.

All our plans will be carried out in compliance with the toughest environmental standards. A careful, civilised attitude to nature is a requirement of all development programmes. Active economic development of the Arctic will be beneficial only if we maintain a rational balance between economic interests and environmental protection for the long term, not just for 10, 15 or 20 years. I mentioned the Prirazlomnoye deposit, where oil production is expected to last for at least 25 years and, hence, environmental support must be provided for this entire period. The Shtokman deposit is expected to last for 50 years. This is why only long-term environmental monitoring can help us achieve the balance I mentioned. Russia's position is borne out by its participation in the Arctic Council's first collective fund, an instrument of financial support for environmental initiatives, including those aimed at dealing with Arctic problem zones.

For our part, we have already launched a general clean-up operation in the Far North and the Russian Arctic as promised. One of the first projects is clearing Franz Josef Land of barrels with waste oil. We have allocated 2.3 billion roubles from the federal budget to this end until 2015. We will do the same on Wrangel Island and Russian villages on Spitsbergen. We will also conduct a comprehensive analysis of the environment in another seven major Arctic zones.

Implementation of these proposals will not only improve the Arctic environment but also allow us to develop unique technology for reclaiming polluted territories. Let me repeat that environmental protection should become a key theme of our activities in the Far North because for all its severity, the Arctic has the most fragile ecosystem on our planet. The price of a negligent, careless attitude towards the Arctic is very high and the consequences disastrous.

Our goal is to use all our resources to study the current state of the Arctic in detail and to develop effective instruments for reducing already inflicted damage and preventing new risks. I'd like to hope that the current forum will become a venue for discussing and finding solutions to all the problems I have mentioned. Therefore, I propose that we devote the next forum to environmental protection.

In conclusion, I'd like to wish you every success in your work. Thank you for your attention.

Closing Remarks:

Colleagues and friends, I want to thank our guests once again for expressing the positions of their countries on the development of the Arctic region as well as their own views on the importance of the Arctic, its prospects and the outlook for cooperation in this region. I have to say that I am very pleased that this discussion platform has been created by the Russian Geographical Society. We will definitely continue supporting everything related to the research and development of the Arctic.

As I said, the climate is changing and new development opportunities are emerging; this is good and, at the same time, cause for some alarm. Obviously, intensive economic activity often leads to sad consequences. But humankind has already accumulated significant experience of working without damaging the environment. If we all act together – actively, carefully, abiding by international standards, listening to expert opinions and finding compromises – I am certain that we will be able to organise our work in the Arctic in such a way that it will contribute to all nations living there and, in fact, to all of humankind. At the same time, we will act so as to preserve the region for future generations, to the extent that humans can preserve anything in its original form; what I mean is that our planet is a living organism, and everything on it is changing.

Again, I want to assure experts dealing with these issues that we will assist you in your work through the Russian Geographical Society, we will create the necessary conditions for your work and will take your results into account in our practical activities. Thank you very much.

Annex 2

United Nations **A/65/912**

General Assembly

Distr.: General
22 July 2011
Original: English

Sixty-fifth session
Agenda item 75
Responsibility of States for internationally wrongful acts

Letter dated 18 July 2011 from the Permanent Representative of Germany to the United Nations addressed to the Secretary-General

Allow me to transmit herewith the conclusions of the Chair of the International Conference on Arctic Science, International Law and Climate Protection (see annex). The Conference was held in Berlin on 17 and 18 March 2011, and was organized by the Foreign Ministries of the Federal Republic of Germany and Finland. A number of prestigious academic institutions from Finland, Germany, the Russian Federation and the United States of America supported the Conference.

Since, in the view of my Government, the dramatic changes in the Arctic matter globally, concerted monitoring and research is pivotal to understanding and hopefully mitigating the effects of climate change.

I therefore would be grateful if you could circulate the present letter and its annex as a document of the General Assembly, under agenda item 75.

(*Signed*) Peter **Wittig**
Ambassador
Permanent Representative of Germany
to the United Nations

Annex to the letter dated 18 July 2011 from the Permanent Representative of Germany to the United Nations addressed to the Secretary-General

Conclusions of the Chair of the International Conference on Arctic Science, International Law and Climate Protection: Legal Aspects of Marine Science in the Arctic Ocean*

1. The Arctic is of pivotal importance for the world climate. The effects of climate change can be seen globally and in the Arctic itself. There are fundamental changes in the Arctic ice level with regard to quality, extent and thickness due to global warming and other environmental developments. These include carbon pollution and ocean acidification. The causes of climate change, originate mainly from outside the Arctic area and, accordingly, non-Arctic actors are major targets affected by climate change. However, this may change if and when economic activities in the Arctic increase.

2. The freedom of marine scientific research as enshrined in the International Law of the Sea is a core requirement. The United Convention on the Law of the Sea (UNCLOS) provides the basic legal regime for marine scientific research in the Arctic Ocean and the legal balance between national interests and common interests in this regard. International scientific cooperation could be considered a common Arctic issue.

3. Balancing the interests of Arctic coastal States and the international community needs to take place within the framework of UNCLOS. Discussions about a special regime for scientific cooperation in the Arctic have not yet led to conclusive results. At the same time, the UNCLOS regime may be open to further development in the future.

---

* The Conference was organized by the German Federal Foreign Office in cooperation with the Ministry of Foreign Affairs of Finland. Support was provided by the Alfred Wegener Institute fornPolar and Marine Research, Bremerhaven, Germany; the Arctic and Antarctic Research Institute, St. Petersburg, Russian Federation; the Arctic Centre, University of Lapland, Rovaniemi, Finland; the International Arctic Research Centre, Fairbanks, Alaska, United States; and the Max Planck Institute for Comparative Public Law and International Law, Heidelberg, Germany.

4. International cooperation is vital for Arctic governance. Exchange of data would amplify the beneficial aspects of marine scientific research in the Arctic.

5. The exploitation of the newly accessible Arctic must be conducted in a sustainable way. Economic prospects must be balanced against environmental needs.

6. Large areas of the Arctic Ocean will continue to be areas of High Seas, where freedom of marine scientific research applies.

7. All relevant actors proceed on the basis of the International Law of the Sea, in particular UNCLOS. The international rules are supplemented by various domestic laws and regulations. Domestic procedures should be simplified and best practices identified in order to support marine scientific research. Harmonization of existing permission procedures would be welcome. In particular, 'one-stop' procedures would be helpful.

8. It is recommended that guidelines be developed to help coastal States apply the rules of the Convention more homogeneously. Such guidelines should privilege

- marine scientific research undertaken in cooperation, including cooperation without discrimination between scientists from coastal States and States not bordering the Arctic.
- marine scientific research enhancing knowledge about environmenttal matters. In this respect, one could borrow from the provisions of and practice under the Antarctic Treaty.

9. Research undertaken in the Arctic needs to take into account the legitimate interests of a multitude of stakeholders: indigenous peoples, states, international organizations, researchers and economic actors alike.

10. Climate change affects the foundations of entire indigenous cultural systems. The international community should, therefore, improve cooperation with indigenous peoples in order to reflect more fully their unique attachment to the Arctic.

11. The pertinent UNCLOS provisions concerning marine scientific research, in particular Art. 246, grant the coastal States considerable leeway to interpret and establish whether a research project is resource-oriented or not. Apart from that, different legal regimes apply to research of the Continental Shelf and the High Seas above it. Formally, the scientific research regime for the Outer Continental Shelf differs from that for the Continental Shelf.

12. UNCLOS uses scientific terms in a legal context. The legal usage of these terms may differ from accepted scientific terminology. 'Nature does not accept legal boundaries and distinctions'. This highlights the need for interdisciplinary cooperation between natural scientists and lawyers in fora such as the Berlin Arctic conferences.

13. International scientific cooperation in the Arctic Ocean is a reality. Cooperation also extends to the context of claims of extended continental shelves. This cooperation benefits the sharing of logistics and resources. Joint evaluation of data helps all actors to better understand the Arctic Ocean.

14. The freedom of marine scientific research should be upheld and maintained. There are different views as to the concrete extent of the application of the principle of the common heritage of mankind. Common interests are an evolving body of international law. International scientific cooperation could be considered a common Arctic issue.

15. There is a need to address the interaction between UNCLOS and other international agreements. Further research and discussion is required to understand the legal interrelationship between different regimes.

16. A degree of uncertainty will remain about the exact extent of continental shelves beyond 200 nm, as the Commission on the Limits of the Continental Shelf will still need considerable time to complete its work. Different views exist about the exercise of coastal State jurisdiction prior to recommendations on the outer limits of the continental shelf by the Commission. However, this question should not burden future marine scientific research.

17. In the 'Area', the International Seabed Authority is a vehicle for the dissemination of results of marine scientific research. The Authority is a forum for exchange of scientific results and thus demonstrates the general thrust of the Convention towards international cooperation.

18. There is an ongoing discussion about the feasibility of complementing the binding universally agreed rules from UNCLOS with non-binding precepts. A combination of both would be beneficial for coastal States as well as scientists.

19. Difficulties in application processes for research in the Arctic hinder the development of research-related technology. The industry is cooperating with indigenous peoples, taking into account their unique position in the Arctic.

20. Understanding the Arctic climate system requires fully integrated atmospheric and solar radiation, sea ice and ocean science, including observations and modelling activities.

21. The fundamental issue for the improvement of conditions for marine scientific research is access. Access depends upon mutual trust. The Arctic Ocean should remain an area of excellent international scientific collaboration and cooperation.

Annex 3

EUROPEAN PARLIAMENT 2009 - 2014

Plenary sitting

A7-0377/2010

16.12.2010

Report

on a sustainable EU policy for the High North
(2009/2214(INI))
Committee on Foreign Affairs
Rapporteur: Michael Gahler

Contents

Motion for a European Parliament Resolution 325
Explanatory Statement 345
Result of Final Vote in Committee 352

Motion for a European Parliament Resolution

on a sustainable EU policy for the High North
(2009/2214(INI))

The European Parliament,

– having regard to the United Nations Convention on the Law of the Sea (UNCLOS), concluded on 10 December 1982 and in force since 16 November 1994,
– having regard to the United Nations Commission on the Limits of the Continental Shelf,
– having regard to the United Nations Framework Convention on Climate Change (UNFCCC) and the Convention on Biological Diversity (CBD),
– having regard to the United Nations Declaration on the Rights of Indigenous Peoples of 13 September 2007,
– having regard to the Declaration on the Establishment of the Arctic Council (AC), signed on 19 September 1996,
– having regard to the Treaty on European Union, the Treaty on the Functioning of the European Union and in particular to Part Four thereof and to the European Economic Area (EEA) Agreement,
– having regard to the Declaration on the Cooperation in the Barents Euro-Arctic Region, signed in Kirkenes on 11 January 1993,
– having regard to the Commission Communication of 20 November 2008 on the European Union and the Arctic Region (COM(2008)0763),
– having regard to its resolution of 9 October 2008 on Arctic governance[1],
– having regard to the Council conclusions on Arctic issues of 8 December 2009[2] and on the European Union and the Arctic region of 8 December 2008[3],

---

1 OJ C 9 E, 15.1.2010, p. 41.

2 2985th Foreign Affairs Council meeting.

- having regard to the Ilulissat Declaration adopted on 28 May 2008 at the Arctic Ocean Conference,
- having regard to the Treaty between Norway, the United States of America, Denmark, France, Italy, Japan, the Netherlands, Great Britain, Ireland, the British Overseas Dominions and Sweden concerning Spitsbergen/Svalbard of 9 February 1920,
- having regard to the Northern Dimension policy and its Partnerships as well as the EU-Russia Common Spaces,
- having regard to the EU-Greenland Partnership Agreement, 2007-2012,
- having regard to the EU's Fifth, Sixth and Seventh Framework Programmes for Research and Technological Development,
- having regard to International Labour Organisation Convention 169 adopted on 27 June 1989,
- having regard to the Nordic Sami Convention of November 2005,
- having regard to United Nations General Assembly Declaration 61/295 of 13 September 2007on the Rights of Indigenous Peoples,
- having regard to Council resolutions 6/12 of 28 September 2007, 6/36 of 14 December 2007, 9/7 of 24 September 2008, 12/13 of 1 October 2009 and 15/7 of 5 October 2010,
- having regard to Finland's strategy for the Arctic Region adopted on 4 June 2010,
- having regard to the opinion of the Foreign Affairs Committee of the Swedish Parliament on Commission Communication COM(2008)0763[4],
- having regard to the joint Danish and Greenlandic strategy for the Arctic at a time of transition of May 2008,
- having regard to the Norwegian Government's Strategy for the High North of 2007, and its follow-up of March 2009,
- having regard to the Nordregio Report 2009:2, 'Strong, Specific and Promising – Towards a Vision for the Northern Sparsely Populated Areas in 2020',

---

[3] 2914th Council meeting.

[4] 2009/10:UU4.

- having regard to the Nordic Council of Ministers' Arctic Co-operation Programme 2009-2011, the Barents Euro-Arctic Council (BEAC) programme and the AC Chairmanship programme,
- having regard to the Canadian Northern Strategy of August 2009 and the follow-up statement on Canada's Arctic Foreign Policy of 20 August 2010,
- having regard to the Canadian Act to amend the Arctic Waters Pollution Prevention Act of August 2009,
- having regard to the Russian national security strategy until 2020 of May 2009,
- having regard to the American National Security Presidential Directive and Homeland Security Presidential Directive of 9 January 2009,
- having regard to the USA's Responsible Arctic Energy Development Act of 2010,
- having regard to the USA's Arctic Oil Spill Research and Prevention Act of 2009,
- having regard to the USA's Arctic Marine Shipping Assessment Implementation Act of 2009,
- having regard to the Monaco Declaration of November 2008,
- having regard to the final statement adopted at the First Northern Dimension Parliamentary Forum in Brussels on 26 September 2009,
- having regard to the Conference Statement of the Ninth Conference of Parliamentarians of the Arctic Region of 15 September 2010,
- having regard to NATO's upcoming new Strategic Concept, which will be approved by Heads of State and Government at the Lisbon Summit in November 2010, and its implications vis-à-vis the security prospects in the Arctic region, particularly the military aspects of the High North,
- having regard to Rule 48 of its Rules of Procedure,
- having regard to the report of the Committee on Foreign Affairs (A7-0377/2010),

A. whereas the Commission communication constitutes a formal first step towards responding to the European Parliament's call for the formulation of an EU Arctic policy; whereas the Council Conclusions on Arctic Issues should be recognised as a further step in the definition of an EU policy on the Arctic,

B. whereas the European Parliament has been an active participant in the work of the Standing Committee of Arctic Parliamentarians through its Delegation for relations with Switzerland, Iceland and Norway for a period of some two decades, culminating in the hosting of the full Conference of the Parliamentarians of the Arctic in Brussels in September 2010,

C. whereas Denmark, Finland and Sweden are Arctic countries and both Finland and Sweden are partially located within the Arctic Circle; whereas the EU's only indigenous people, the Sami people, live in the Arctic regions of Finland and Sweden as well as Norway and Russia,

D. whereas Iceland's application to join the EU will increase the need for the EU to take account of the Arctic region in its geopolitical perspective,

E. whereas Norway, a reliable partner, is associated with the EU through the EEA Agreement,

F. whereas there has been a longstanding engagement of the EU in the Arctic by way of its involvement in the Common Northern Dimension Policy with Russia, Norway and Iceland, including its Arctic Window, in the Barents cooperation and particularly in the Barents Euro-Arctic Council, the implications of the strategic partnerships with Canada, the United States and Russia and its participation as an active ad hoc observer in the AC,

G. whereas the gradual formulation of an EU policy on the Arctic should be based on the recognition of the existing international, multilateral and bilateral legal frameworks such as the comprehensive set of rules laid down in UNCLOS and several sectoral, bilateral and multilateral agreements which already regulate certain issues of importance to the Arctic,

H. whereas the EU and its Member States make a major contribution to research in the Arctic and whereas EU programmes, including the current Seventh Framework Programme, support major research projects in the region,

I. whereas it is estimated that about a fifth of the world's undiscovered hydrocarbon resources are located in the Arctic region, although more extensive research is needed to establish more accurately how much gas and oil there is in the region and how economically viable it would be to exploit these reserves,

J. whereas there is also strong global interest in other Arctic renewable and non-renewable resources such as minerals, forests, fish and pristine landscapes for tourism,

K. whereas the growing interest in the Arctic region of other non-Arctic actors such as China, illustrated by China's commissioning of a first icebreaker, their allocation of funding to polar research and not least the applications by South Korea, China, Italy, the EU, Japan and Singapore for status as permanent observers at the AC, indicates a different geopolitical appreciation of the Arctic on a larger scale,

L. whereas the recently established self-government in Greenland with regard to relevant policy areas including environmental legislation and resources and the recent update of the EU-Greenland Partnership Agreement has led to an increased interest in the exploration and exploitation of resources in Greenland and on its Continental shelf,

M. whereas the effects of climate change mainly originating from outside the Arctic and the globalisation of the world economy will impact the region; whereas in particular the retreat of the sea ice, as well as the potential for resources and the possible use of new technologies, is likely to produce unforeseeable environmental effects and repercussions in other parts of the planet as well as an increase in shipping in particular between Europe, Asia and North America, in exploration and exploitation of natural resources, namely gas, oil and other minerals but also natural resources such as fish, and exploitation of marine genetic resources, increased mining and logging activities and increased tourism and research activities; whereas those effects will produce new challenges but also new opportunities in the Arctic and elsewhere,

N. whereas climate change is managed by monitoring, mitigation and adaptation methods; whereas the promotion of sustainable development in using natural resources and in building new infrastructures is managed by strategic planning processes.

I. The EU and the Arctic

1. Recalls that three EU Member States – Denmark, Finland and Sweden – are Arctic States; acknowledges that the EU has no Arctic Ocean coastline so far; reaffirms the legitimate interest of the EU and other third countries as stakeholders by virtue of their rights and obligations under international law, its commitment to environmental, climate and other policies and its funding, research activities and economic interests, including shipping and exploitation of natural resources; moreover recalls that the EU has large Arctic land areas in Finland and Sweden that are inhabited by the only indigenous population group in Europe, the Sami;
2. Takes into account that through its Northern Member States and candidate countries the EU is affected by Arctic policies and likewise has an impact on Arctic policies, and recognises the ongoing work in the several partnerships of the Northern Dimension, a common policy of the EU with Russia, Norway and Iceland;
3. Underlines that certain policies that are relevant to the Arctic are exclusive Union competences, such as the conservation of marine biological resources under the common fisheries policy, others partly shared with Member States;
4. Highlights that the EU is committed to devising its policy responses in the Arctic on the basis of the best available scientific knowledge and understanding of the processes affecting the Arctic, and is accordingly already devoting sizeable research efforts to generating sound scientific evidence to support policy-making;
5. Conscious of the need to protect the fragile environment of the Arctic, underlines the importance of overall stability and peace in the region; stresses that the EU should pursue policies that ensure that measures to address environmental concerns take into account the interests of the inhabitants of the Arctic region, including its indigenous peoples, in protecting and developing the region; stresses the similarity in approach, analysis and priorities between the Commission Communication and policy documents in the Arctic States; stresses the need to engage in policies that respect the interest in sustainable management and use of the land-based and marine, non-renewable and

renewable natural resources of the Arctic region, which in turn provide important resources for Europe and are a major source of income to the inhabitants of the region;

6. Highlights the fact that a future accession of Iceland to the EU would transform the Union into an Arctic coastal entity, while noting that Iceland's status as a candidate country for accession to the EU underlines the need for a coordinated Arctic policy at EU level and represents a strategic opportunity for the EU to assume a more active role and contribute to multilateral governance in the Arctic region; considers that Iceland's accession to the EU would further consolidate the EU's presence in the Arctic Council;

7. Emphasises the importance of interacting with Arctic communities and supporting capacity-building programmes in order to improve the quality of life of indigenous and local communities in the region and gain more understanding of the living conditions and cultures of these communities; calls on the EU to promote a stronger dialogue with the indigenous peoples and the Arctic local inhabitants;

8. Stresses the need for a united, coordinated EU policy on the Arctic region, in which both the EU's priorities and the potential challenges and a strategy are clearly defined;

New world transport routes

9. Underlines the major importance of the safety and security of new world trade routes through the sea in the Arctic, in particular for the EU and its Member States' economies, these countries controlling 40% of world commercial shipping; welcomes the work in the International Maritime Organisation (IMO) on a mandatory Polar Code for shipping and the work in the Working Groups of the AC, particularly the Taskforce on Search and Rescue (SAR); underlines that the EU and its Member States should actively uphold the freedom of the seas and the right to free passage through international waterways;

10. Stresses the importance of developing new railway and transport corridors in the Barents Euro-Arctic Transport Area (Beata) to facilitate the growing need for international trade, mining and other economic development, as well as aviation

connections in the High North; draws attention in this regard to the new Northern Dimension Partnership on Transport and Logistics;

11. Suggests that important non-Arctic shipping nations using the Arctic Ocean should be included in the results of the Search and Rescue Work Initiative of the AC; therefore recommends that the Commission and the Council, together with the European Maritime Safety Agency (EMSA), coordinate EU and Member States' policies in that particular field in the IMO, the AC and other organisations;

12. Points out that in spite of the efforts on a mandatory Polar Code for shipping a faster solution to the issue of safety of Arctic shipping might be found through coordination and harmonisation of national legislation and calls on EMSA to concern itself to the maximum with Arctic shipping;

13. Welcomes other cooperation initiatives on secure and safe shipping in the Arctic and on better access to the various Northern sea routes; emphasises that this concerns not only commercial traffic but also a large and increasing volume of tourist shipping carrying EU citizens; calls for more research on the effect that climate change has on Arctic navigation and shipping routes; equally calls for assessments of the impact of the increase in navigation and commercial activities, including offshore activities, on the Arctic environment and its inhabitants;

14. Calls on the States in the region to ensure that any current transport routes – and those that may emerge in the future – are open to international shipping and to refrain from introducing any unilateral arbitrary burdens, be they financial or administrative, that could hinder shipping in the Arctic, other than internationally agreed measures aimed at increasing security or protection of the environment;

Natural resources

15. Is conscious of the need for resources for a growing world population and recognises the increase in interest in them as well as the sovereign rights under international law of the Arctic States; recommends any party involved to take steps to en-

sure the highest possible safety, social and environmental standards in exploration and exploitation of the natural resources;

16. Highlights the fact that the Environmental Impact Assessment (EIA) as well as strategic and social impact assessment processes will be central tools in the management of concrete projects and programmes in the Arctic; draws attention to Directive 2001/42/EC[5] on Strategic Environmental Assessment (SEA) and to the fact that Finland, Sweden and Norway have ratified the UNECE Convention on EIA in a Transboundary Context (Espoo Convention), which will provide a good basis for the active promotion of impact assessment procedures in the Arctic; refers in this regard also to the Bergen Statement issued by the Ministerial Meeting of the OSPAR Commission of 23 and 24 September 2010;

17. Calls on the States in the region to resolve any current or future conflicts over access to natural resources in the Arctic in the way of constructive dialogue, possibly within the AC, which constitutes a good forum for such discussion; underlines the role of the UN Commission on the Limits of the Continental Shelf (CLCS) in finding solutions for conflicts between Arctic States over delimitation of their exclusive economic zones;

18. Points in particular to the responsibility of the Arctic States to ensure that oil companies that plan to engage in offshore oil drilling within their respective maritime borders have the necessary safety technology and expertise in place and are financially prepared to prevent and respond to oil rig disasters and oil spills; notes that the extreme weather conditions and the high ecological fragility of the Arctic region render it necessary for relevant oil companies to develop special expertise in preventing and handling oil spills in the region;

19. Welcomes the new delimitation agreement[6] between Norway and Russia, in particular the expressed will to engage in closer cooperation regarding the joint management of resources, and the continued joint management of fish stocks, in the Barents Sea, including in terms of sustainability; regards in particular the bilateral cooperation between Norway and Russia as a

---

5 OJ L 197, 21.7.2001, p. 30.

6 Signed on 15 September 2010.

showcase for joint application of the highest available technical standards in the field of environmental protection while prospecting for oil and gas in the Barents Sea; points out in particular the importance of the contentious development of new technologies especially developed for the Arctic environment, such as sub-seabed installation technology;

20. Is conscious of the different interpretations of the Svalbard/Spitsbergen Treaty with regard to its applicability to the continental shelf and the maritime zones of Svalbard/Spitsbergen, and, given the relatively good accessibility of resources in the continental shelf, would welcome an agreement on the legal status of the shelf acknowledging the legal rights and duties of the costal shelf states; is confident that any disputes which may arise will be dealt with in a constructive way;

21. Recalls the position of the EU as a main consumer of Arctic natural resources, as well as the involvement of European economic actors; requests the Commission to further engage in fostering cooperation and technology transfer to ensure the highest standards and adequate administrative procedures, to establish a sound scientific basis for future trends and governance needs for Arctic resources, such as fisheries, mining, forestry and tourism, and to make full use of the EU competences to regulate in this regard; as economic activities in the Arctic will increase, calls upon the EU to promote the principles of sustainable development therein;

22. Insists that before any new commercial fisheries are opened in the Arctic region, reliable and precautionary scientific stock assessments must be conducted in order to determine levels of fishing that will conserve the targeted fish stocks and not lead to depletion of other species or to serious damage to the marine environment, and that any fishing on the high seas must be regulated by a Regional Fisheries Management Organisation that respects scientific advice and has a robust control and surveillance programme to ensure compliance with management measures, while fishing within Exclusive Economic Zones (EEZ) must meet the same standards;

23. Considers that the creation and enforcement of marine protected areas of sufficient size and diversity are an important tool in the conservation of the marine environment;

Climate change and pollution effects on the Arctic

24. Acknowledges that the EU, like other developed areas of the world, contributes substantially to climate change and hence bears special responsibility and must play a leading role in combating climate change;

25. Acknowledges that the best protection for the Arctic is a long-term and ambitious global climate agreement, but realises that the rapid warming of the Arctic makes it necessary, in addition, to work on possible further short-term measures to limit Arctic warming;

26. Regards the Arctic as a sensitive region where the effects of climate change are especially visible, having serious repercussions on other regions in the world; supports therefore the Council Conclusions on increased cooperation with the UNFCCC and the Sustaining Arctic Observing Networks (SAON) and the efforts to realise the Svalbard Integrated Observation System (SIOS) and the Arctic components of the European Multidisciplinary Seafloor Observatory (EMSO), since those initiatives ensure a unique European contribution to understanding climate and environment change in the Arctic region;

27. Recognises the disproportionately large Arctic warming impact caused by black carbon emissions from the EU and other regions in the northern hemisphere, and stresses the need for inclusion of black carbon emissions in the relevant UNECE and EU regulatory framework, such as the Convention on Long-Range Transboundary Air Pollution and the National Emissions Ceilings Directive;

28. Welcomes the ban on the use and carriage of heavy fuel oil on vessels operating in the Antarctic Area, approved by the IMO's Marine Environment Protection Committee (MEPC), which is due to enter into force on 1 August 2011; stresses that a similar ban might be appropriate in Arctic waters to reduce risks to the environment in case of accidents;

29. Supports increased cooperation with Arctic and non-Arctic states on developing the Sustaining Arctic Observing Networks (SAON) and encourages the European Environmental Agency to continue its valuable work and to promote cooporation through the European Environment Information and Ob-

servation Network (Eionet) using the guiding principles of the Shared Environmental Information System (SEIS);

30. Stresses the important role the EU and the circumpolar nations have to play in the reduction of pollution in the Arctic region caused by long-range transport, e.g. shipping; highlights in this respect the importance of the implementation of European legislation such as Regulation (EC) No 1907/2006[7]; points out that the climatic changes in the Arctic will have a major impact on coastal regions in Europe and elsewhere and on climate-dependent sectors in Europe such as agriculture and fisheries, renewable energy, reindeer herding, hunting, tourism and transport;

Sustainable socioeconomic development

31. Recognises that the effects of the melting ice and milder temperatures are also creating opportunities for economic development in the Arctic region; acknowledges the wish of the governments of the Arctic region with sovereign rights and responsibilities to continue to pursue sustainable economic development while at the same time protecting the very sensitive nature of the Arctic ecosystems, taking into account their experience in using and developing the resources of the region in a sustainable way; recommends applying ecosystem-based management principles to consolidate ecological scientific knowledge with social values and needs;

32. Underlines that it is important for the EU together with representatives of the regions in the area to discuss the importance of the Structural Funds for development and cooperation in order to face the future global challenges with a view to progress and to be able to seize the development potential of the area;

33. Is of the opinion that in order to identify the specific potential of each locality and to develop adequate settlement strategies with respect to regional differences, an inclusive process with the assistance of the national and EU levels is needed; believes that partnerships and dialogue between the levels of authority

---

7 OJ L 136, 29.5.2007, p. 3.

concerned ensures that the policies can be implemented at the most effective level;

34. Notes the special position and recognises the rights of the indigenous peoples of the Arctic and points in particular to the legal and political situation of the indigenous peoples in the Arctic States and in their representation in the Arctic Council; calls for greater involvement of indigenous people in policy-making; stresses the need to adopt special measures to safeguard the culture, language and land rights of indigenous peoples in the way defined in ILO Convention 169; calls for a regular dialogue between the indigenous peoples' representatives and the EU institutions and further calls on the EU to take into account the special needs of sparsely populated peripheral areas in terms of regional development, livelihoods and education; underlines the importance of supporting activities promoting the culture, language and customs of indigenous peoples;

35. Notes that the economies of the indigenous peoples rely to a high extent on sustainable use of natural resources and therefore that the reduction of climate change and its effects and the right of the indigenous peoples to an unpolluted natural environment are also questions of human rights;

36. Welcomes the work of the UN Special Rapporteur on the situation of human rights and fundamental freedoms of indigenous people and that of the UN Expert Mechanism on the Rights of Indigenous Peoples;

37. Welcomes the successful completion by the Expert Mechanism of its progress report on the study on indigenous peoples and the right to participate in decision-making;

38. Encourages the Arctic Member States to engage in negotiations leading to a new ratified Nordic Sami Convention;

39. Urges the EU to promote actively the culture and language rights of Finno-Ugric people living in Northern Russia;

40. Takes note of the recent legal developments regarding the EU's ban on seal products, in particular the action brought for annulment of Regulation (EC) No 1007/2009[8] (Case T-18/10, Inuit Tapiriit Kanatami v Parliament and the Council) pending

---

8 OJ L 286, 31.10.2009, p. 36.

before the General Court; notes the consultation procedure under the auspices of the World Trade Organisation (WTO) requested by Canada and Norway; expresses its hope that disagreements between the parties can be overcome following the rulings of the ECJ and the result of the WTO procedures;

41. Is aware of the increasing interest in the exploitation of resources; in that regard points out the need for broad all-encompassing ecosystem-based approaches as most likely to be capable of dealing with the multiple challenges facing the Arctic related to climate change, shipping, environmental hazards and contaminants, fisheries and other human activities, along the lines of the EU's Integrated Maritime Policy or Norway's Integrated Management Plan for the Barents Sea and the sea areas of the Lofoten Islands; recommends the Member States to endorse the revised Arctic Council Offshore Oil and Gas Guidelines of 2009;

II. Governance

42. Recognises the institutions and the broad framework of international law and agreements that govern areas of importance to the Arctic such as UNCLOS (including the basic principles of freedom of navigation and innocent passage), the IMO, the OSPAR Convention[9], the North East Atlantic Fisheries Commission (NEAFC), CITES[10] and the Stockholm Convention as well as the existing numerous bilateral agreements and frameworks, in addition to the national regulations in place in the Arctic States; thus concludes that the Arctic region is not to be regarded as a legal vacuum, but as an area with well developed tools for governance; nevertheless points out that, due to the challenges of climate change and increasing economic development, those existing rules need to be further developed, strengthened and implemented by all parties concerned;

---

[9] Convention for the Protection of the Maritime Environment of the North-East Atlantic.

[10] Convention on International Trade in Endangered Species of Wild Fauna and Flora.

43. Emphasises that, although States play a key role in governance in the Arctic, other players – such as international organisations, indigenous and local people and sub-state authorities – also have important roles; points out that it is important to increase trust among those with legitimate interests in the region by taking a participative approach and using dialogue as a way of developing a shared vision for the Arctic;

44. Believes that the impression given by some observers of a so-called scramble for the Arctic does not contribute to fostering a constructive understanding and cooperation in the region; points out that the Arctic States have on several occasions declared their commitment to resolve and in some cases have worked towards resolving possible conflicts of interests according to the principles of international law;

45. Recognises the important role of the AC as the foremost regional forum for cooperation for the whole Arctic region; affirms its commitment not to support any arrangements which exclude any of the Arctic EU Member States, candidate countries or Arctic EEA/ EFTA states; acknowledges the concrete work done in the working groups of the AC with the involvement of the observers and calls on the Commission and EU agencies to continue to actively engage in all relevant working groups whenever possible; favours strengthening the legal and economic base of the AC;

46. Recognises that the challenges facing the Arctic are global and should therefore include all relevant actors;

47. Welcomes the results of major reports which the AC working groups have produced in recent years on Arctic oil and gas, the impacts of warming and emergency response needs;

48. Welcomes the degree of political organisation of indigenous interests in the Sami Parliaments and then Sami Council in Northern Europe and the cooperation among several indigenous organisations on a circumpolar basis and acknowledges the unique role of the AC with regard to the involvement of indigenous people; recognises the rights of the indigenous peoples of the Arctic as set out in the UN Declaration on the Rights of Indigenous Peoples and encourages the Commission to make use of the EIDHR for the benefit of Arctic indigenous people empowerment;

49. Welcomes the broad cooperation on issues such as the protection of the Arctic marine environment (PAME Working Group), not only on a regional level but bilaterally and internationally; interprets in this respect the work done on SAR in the AC as a first step towards mechanisms also to take binding decisions;

50. Welcomes the continuous AC assessment of the scope and structure of its work and is confident that it will continue to broaden the basis for decision-shaping processes to include non-AC actors;

51. Expresses its hope that the AC will further develop its important work and broaden the basis for decision-shaping processes to include other Arctic actors who are upgrading their presence in the Arctic region, and thus involve their knowledge and capacities and take into account their legitimate interests under international law, while at the same time the significantly greater importance of the interests of the Arctic States should be stressed; welcomes the internal procedure within the AC regarding a review of the status of observers and of the possible future scope of the tasks of the AC;

52. Is of the opinion that a strengthened AC should play a leading role in cooperation on the Arctic and would therefore welcome politically and administratively improved capacities of the AC, e.g. the permanent secretariat currently under discussion, more equal sharing of costs, more frequent ministerial meetings and an Annual Arctic Summit on the Highest Level, as proposed by the Foreign Minister of the EU Member State Finland, which is also a Member of the Arctic Council; would further welcome greater involvement of the Parliamentarians of the Arctic to underline the parliamentary dimension and be sure to include relevant non-Arctic players; furthermore insists that continued high-level meetings of an inner exclusive core of States will merely undermine the status and role of the AC as a whole; wishes the AC to maintain its open and inclusive approach and thus to remain open to all stakeholders;

53. Regards the Northern Dimension as a focal point for regional cooperation in Northern Europe; notes that the four partners, namely the EU, Iceland, Norway and the Russian Federation, as well as the Arctic Council, the Barents Euro-Arctic Council, the Council of the Baltic Sea States, the Nordic Council of Ministers, the European Bank for Reconstruction and Devel-

opment (EBRD), the European Investment Bank (EIB), the Nordic Investment Bank (NIB) and the World Bank (IBRD), are participants in the Northern Dimension and that both Canada and the United States hold observer status in the Northern Dimension; stresses the need for close alignment between the Northern Dimension policy and the EU's evolving Arctic policy; notes the Northern Dimension's Arctic Window; highlights the valuable experience of the Northern Dimension partnerships, particularly the new Northern Dimension Partnership on Transport and Logistics and its benefits for cooperation in the Arctic;

54. Confirms its support for permanent observer status for the EU in the AC; recognises that EU Member States are involved in AC work through various international organisations (such as the IMO, OSPAR, NEAFC and the Stockholm Convention) and highlights the need for coherence in all EU policies towards the Arctic; asks the Commission to keep Parliament duly informed about meetings and work in the AC and its Working Groups; stresses meanwhile that the EU and its Member States are already present as members or observers in other international organisations with relevance to the Arctic such as the IMO, OSPAR, NEAFC and the Stockholm Convention and therefore should more coherently focus on the work in these organisations; underlines in this regard in particular the need for coherence in all EU policies towards the Arctic; encourages the AC to also involve civil society and non-governmental organisations as ad-hoc observers;

55. Regards the Barents Euro-Arctic Council (BEAC) as an important hub for cooperation between Denmark, Finland, Norway, Russia, Sweden and the European Commission; notes the work of the BEAC in the fields of health and social issues, education and research, energy, culture and tourism; notes the advisory role of the Working Group of Indigenous Peoples (WGIP) within the BEAC;

III. Conclusions and requests

56. Requests the Commission to develop the existing Inter-Service Group into a permanent inter-service structure to ensure a co-

herent, coordinated and integrated policy approach across key policy areas relevant to the Arctic, such as the environment, energy, transport and fisheries; recommends assigning the co-lead of this structure to the EEAS and DG MARE, the latter acting as a cross-sectoral coordinator within the Commission; further recommends creating an Arctic unit in the EEAS accordingly;

57. Calls on the Commission, in negotiating bilateral agreements, to take account of the fact that the sensitive Arctic ecosystem must be protected, the interests of the Arctic population, including its indigenous population groups, must be safeguarded and the natural resources of the Arctic must be used sustainably, and calls on the Commission to be guided by these principles in relation to all activities;

58. Underlines the fact that the EU and its Member States are main contributors to Arctic-relevant research, regional cooperation and the development of technology relevant to the region and beyond, and requests the Commission to examine the possibilities of developing circumpolar co-funding and co-programming initiatives to enable smoother and more effective cooperation between experts from the countries involved; requests the EU to promote cooperation activities with the USA, Canada, Norway, Iceland, Greenland and Russia in the field of multidisciplinary Arctic research, thereby establishing coordinated funding mechanisms; further requests the Commission to create a means to work directly with Arctic Member States, indigenous organisations and Arctic research institutes in order to help inform the EU about relevant issues, important research topics and matters that concern those living and working in the Arctic to help establish future research activities;

59. Is of the opinion that the EU should develop further its capacities and calls on the Commission to explore and report on the establishment as well as on the continuation of EU activities in the Arctic such as a circumpolar joint multilateral research funding programme providing for easier and less bureaucratic cooperation and joint projects of the research community; requests the Commission to explore as a key priority the establishment of an EU Arctic Information Centre as a joint, networked undertaking, taking into account suitable proposals; notes the proposal by the University of Lapland in this respect; considers that such a centre needs to be capable both of organ-

ising permanent EU outreach to the major actors relevant to the Arctic and of channelling Arctic information and services towards the EU's Institutions and stakeholders;

60. Emphasises that, in order to objectively determine the nature and rate of the changes occurring in the natural environment of the Arctic, it is vital that international teams of scientists be given full access to carry out research in this particularly sensitive area of our planet; points out that the EU is stepping up its presence and involvement, particularly in the European sector of the Arctic, by building joint infrastructure for research and increasing the number of research programmes carried out in the Arctic; supports in particular research teams made up of scientists from many different fields and representing all the countries involved; welcomes the often good and open cooperation in research and takes the view that this research should be open, which would be in the interests of, and make it available for use by, the international community as a whole;

61. Emphasises the contribution of the EU's European Territorial Cooperation objective, as a clear European added value, in particular the cross-border cooperation programmes of Kolartic and Karelia as well as the CBC Baltic Sea Basin programme, which includes the Barents region; requests the Commission to explore how a suitably enhanced Northern Periphery Programme could have a similar impact on an Arctic Strategy in the next programming period;

62. Asks the Commission to support efforts to quickly and efficiently realise the SIOS and EMSO observatories as unique contributions to better understanding and protecting the Arctic environment;

63. Requests the Commission to put forward proposals as to how the Galileo Project or projects like Global Monitoring for Environment and Security that could have an impact on the Arctic could be developed to enable safer and faster navigation in Arctic waters, thus investing in the safety and accessibility of the North-East Passage in particular, to contribute to better predictability of ice movements, better mapping of the Arctic seabed and an understanding of the main geodynamic processes in the area, which are of major importance for the geodynamics of the Earth and for the water cycle in polar regions and in order to enhance our knowledge of unique ecosystems;

64. Calls for all governments in the Arctic region, especially that of Russia, to adopt and endorse the United Nations Declaration on the Rights of Indigenous Peoples adopted by the General Assembly on 13 September 2007;

65. Urges Member States to ratify all the key agreements regarding the rights of indigenous peoples, such as ILO Convention 169;

66. Requests the EU and its Member States to propose, as part of the ongoing IMO work on a mandatory Polar Code for shipping, that soot emissions and heavy fuel oil be regulated specifically; in the event that such negotiations do not bear fruits, requests the Commission to put forward proposals on rules for vessels calling at EU ports subsequent to, or prior to, journeys through Arctic waters, with a view to imposing a strict regime limiting soot emissions and the use and carriage of heavy fuel oil;

o

o o

67. Instructs its President to forward this resolution to the Council, the Commission, the Vice-President/High Representative of the Union for Foreign Affairs and Security Policy, the governments and parliaments of the Member States and the governments and parliaments of the Arctic region states.

Explanatory Statement

I. Introduction

The Arctic Region is attracting more and more attention, due to the effects of climate change, the main trigger of developments. Its effects are of a larger scale than in other regions of the world. At the same time, those changes are affecting other regions of the world, through rising sea levels on the one and consequences for adjacent regional climates on the other hand.

Thus Europe does not only bear a certain responsibility, being one of the main contributors to pollution and green house gas emissions, but also has a particular interest in the Arctic, since it will have to deal with the consequences of the changes taking place there from environmental and climate change issues to the geopolitics of shipping routes and security of supply of resources.

II. Why the EU needs a sustainable and coherent Arctic Policy

Three of the EU Member States are also Members of the Arctic Council, while Iceland applied for membership in the EU. In addition Norway and Iceland are interlinked closely with EU policies trough the EEA agreement, and the EU also has a Partnership agreement with Greenland, which is not part of the EU.

In spite of not having an Arctic coast line, the EU already is an Arctic player in a number of relevant fields[11]. Some of the competences of the EU to regulate issues concerning the Arctic are shared or complementary, some like fisheries are exclusive. It is worth noting that the Lisbon Treaty changed the internal procedures of the EU towards a stronger involvement of the European Parliament as co-legislator. Taking a closer look, the Arctic will be of major importance for a number of reasons.

Climate change is the main driver of change in the Arctic as elsewhere. It is commonly agreed, that the Arctic is a region that is affected earlier and more heavily by climate change and pollution originating in the in-

---

[11] For a comprehensive overview of legal competences of the EU with regard to the Arctic and for a detailed sectoral assessment: Timo Koivurova et al., 'EU Competences affecting the Arctic', study commissioned by the EP.

dustrialised or developing parts of the world. This question needs to be dealt with on a global level, since its causes lie outside the Arctic and in turn will also affect the whole globe.

The EU is already a frontrunner in research and in environmental and climate change policies in the international context and will continue to be so. Notwithstanding the fight against climate change, the EU must acknowledge the need to adapt to the unavoidable changes as well as have a rational assessment of the risks, threats, challenges and opportunities those changes entail.

A growing world population will demand the sustainable and responsible management of the resources available and needed. This will be true with regard to living resources like fish that will contribute to feeding the world population, but also with respect to non living resources like gas and oil or minerals. In that respect the perception of the Arctic as a pristine and untouched place that solely needs to be preserved is not correct. The Arctic, unlike the Antarctic is inhabited and has a tradition of making use of its resources. In particular the indigenous people of the north have a long history of sustainable use of those resources and explicitly reject the idea of 'living in a museum' but rather express their will to develop.

Since the rise of new economies is resulting in an increasing need for resources, energy and minerals, the EU has a natural interest in ensuring security of supply of resources and energy needed for the population and industries in Europe.

Some partners in the Arctic are already today major contributors when it comes to the supply of energy, raw materials and also fish for Europe. The great variety of resources, the potential for renewable energy produced by wind or waves and the invaluable diversity of the Arctic biosphere can only be developed and protected in a holistic and sustainable ecosystem-based-approach as sketched in the EU's Integrated Maritime Policy or in the integrated management plans for example by Norway in the Barents Sea.

Since it is estimated that about a fifth of the remaining hydrocarbon resources are to be found in the Arctic these resources might be of particular importance to the EU until the goal of a low carbon economy will be achieved. In particular natural gas or LNG has the lowest CO_2 emission of all traditional energy forms and could provide a major part in the energy mix and thus function as a bridge into a low carbon economy.[12]

As a main consumer of those products, Europe should make clear that it supports only those activities that are conducted with the highest environmental, safety and administration standards available and hence foster cooperation in a way that best practices can easily be applied elsewhere.

The principle of an eco-system-based management could ensure that the aspects and interests included in the administration of a certain region where activities like, fishing, shipping, exploitation of geological resources and other activities overlap are balanced with the interest to preserve and protect the eco-system.

Another major point of interest for the EU and its Member States is the development of new world trade routes. Businesses have already begun to explore the new possibilities. Last summer the German shipping company Beluga tested the economical possibilities by sending two container ships from Asia to Europe. Developing the northern sea routes would make trade between Europe, Asia and North America faster, thus saving energy, emissions and costs, but also safer, avoiding the pirate ridden seas and included economical risks when using traditional sea routes.[13]

Vital for the development of these sea routes will be the predictability both in terms of safety and marine shipping and in legal and political terms. Even though conditions will remain harsh in the Arctic, the improvement of navigation and shipping technologies would enable a better use of this route. These investments in mapping, sea ice observation, communication and search and rescue structures and alike will determine to what extent this route can be used in the coming decades. The interest of nations like China, South Korea and Singapore highlight the increased importance that is attributed to these sea routes.

The EU, its Member States and European businesses should be actively involved in cooperating in the development of those sea routes not only because they will be of major importance to European businesses, but in particular as the EU is in the unique position also to offer some of the tools needed to develop this route, as illustrated by the better cover-

---

12 For an overview of Arctic resources assessment see Valur Ingimundarson 'The geopolitics of Arctic Natural Resources', study commissioned by the EP.

13 For an analysis see Moe/ Oystein 'Opening of new Arctic Shipping Routes', study commissioned by the EP.

age and reliability which the Galileo System could provide compared to the existing GPS Systems.

Recalling the above mentioned legitimate interests and position of the EU in terms of funding of research, shipping, and consumer power and taking into account the EU's economical importance, Europe has a lot to offer with regard to the protection and the sustainable development of the Arctic region.

III. The Position of the EU in the Arctic Region - Geopolitics and Governance

It is important to understand and asses the actual political and legal situation in the Arctic before making any suggestions on how risks and challenges can best be dealt with.

The Arctic region is not a legal or political vacuum as assumed by some observers. Furthermore it is in principle a sea surrounded by states with far reaching EEZ's on which their legal regime applies. There is a large number of bilateral and some multilateral agreements on national and on regional level. In addition, a number of international treaties, organisations and agreements regulate sectors of relevance to the Arctic.

Having regard to those facts, it becomes clear that the idea of an Arctic Treaty, modelled along the Treaty for the continent of Antarctica, thus land not sea, uninhabited and unclaimed compared to inhabited and state controlled in the Arctic, is not only not promoted by the peoples and states in the Arctic, but also wouldn't be an appropriate way to deal with the challenges in the Arctic.

To start from scratch, neglecting the already existing legal framework, also for protection and preservation, and to engage in a decade long UN Process with the unclear prospect of getting a somewhat international agreement on the Arctic, would result in not dealing with the practical and pressing issues in the Arctic.

All the states have submitted to follow Public International Law in the settlement of disputes and furthermore have or will submit their respective claims for the prolonging of the continental shelf zone to the relevant UN Commission. Thus looking at the legal map of the Arctic it becomes obvious, that almost all the area is or will be within the EEZ of one of the parties. Only very small areas are subject to overlapping claims and as stated above, all parties declared their will to settle disputes according to international law. The delimitation agreement between Russia and Norway concluded 15 September 2010 is insofar very illustrative. This large area can be deleted from the map of overlapping claims.

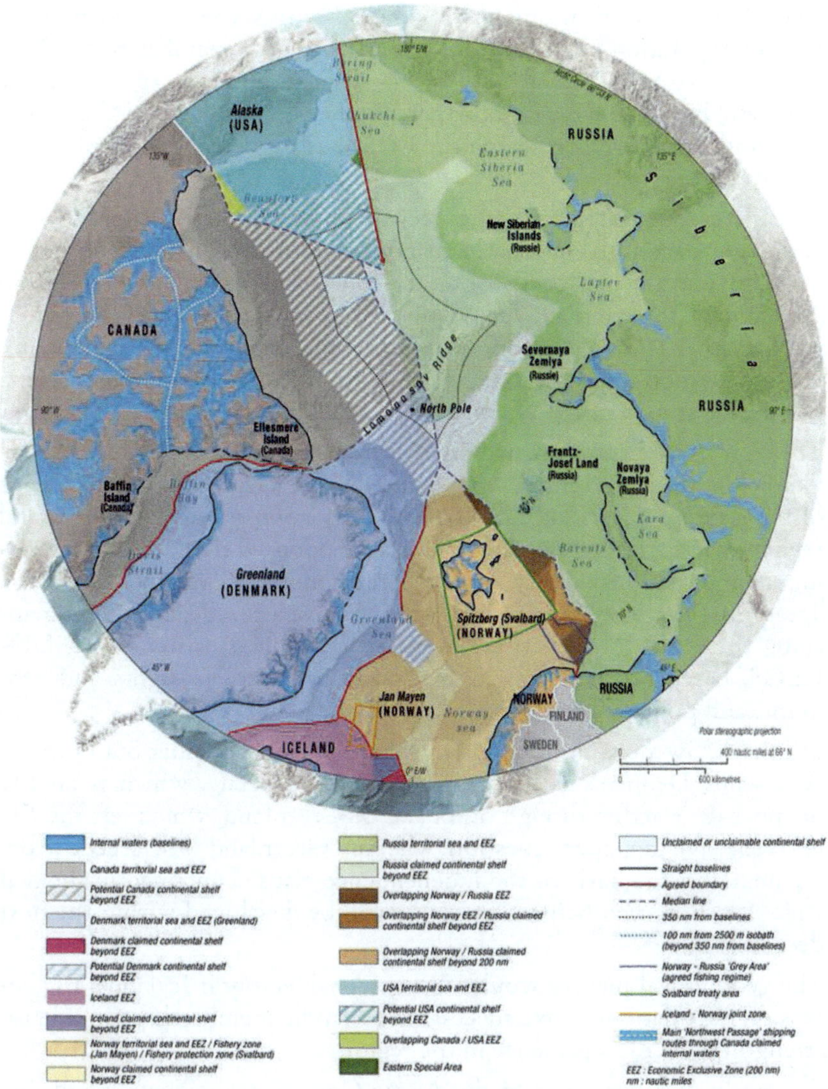

The Legal Situation in the Arctic Ocean Map

Having a look at the map one will get the impression that most of the Arctic 'belongs' to the littoral states. Nevertheless under UNCLOS all states have certain rights such as the right 'to free and innocent passage' in these waters. In addition, other international rules apply.

The EU is involved in one way or the other in several fora of international cooperation in the region, in particular as a member to the Barents-Euro-Arctic-Council and as an ad hoc observer to the Arctic Council. Together with its Northern Dimension policy and the extensive funding of Arctic research the EU is already a recognized player in the Arctic.

The increased strategic importance of the Arctic has been driven by various transnational and national processes: such as debates on global warming and the prospects for an ice-free Arctic in the summer within 20-30 years, the control over Arctic oil and gas deposits and the potential for other commercial opportunities opened by new sea routes; not least by symbolic political acts, such as the Russian decision to put a flag on the seabed of the North Pole in 2007.

The eight Arctic states, which are the permanent members of the Arctic Council – the central international and intergovernmental organizations of the region – view UNCLOS as the only comprehensive multilateral regime that applies to the Arctic and have opposed the idea of concluding an international treaty on the Arctic modelled on the Antarctica Treaty of 1959. The Arctic states want to have a privileged role in managing the region, which they interpret as being consistent with UNCLOS, based on their geographic location, sovereign rights and economic and political interests.

Denmark, Sweden, and Finland are the three EU Member States in the AC, while Denmark is the only Arctic Ocean state, which is an EU member. It is acting though on behalf of Greenland, which left the EU in 1985. It is an open question whether Greenland will secede from Denmark on the basis of the independence clause contained in the Self Rule Act, if its rich natural resources will be developed within the next decades.

The geopolitical picture would change considerably if Iceland's EU accession negotiations prove successful. Icelandic membership would also strengthen the EU's presence in the region.

Currently member states of the Arctic Council are reviewing and discussing the status and rights of observers as well as how the Council should continue to work. With the case of the task force on 'Search and Rescue' the AC for the first time will establish and adopt binding rules, thus taking a step from a pure decision shaping to a decision taking body as some have commented. If that would be the case, the EU would need to asses the situation and make sure that its interests and those of its Member States, in particular on issues such as shipping and fisheries are duly represented and its rights under international agreements are taken into account.

Having said this, and recalling the contribution of the EU and its Member States already today in research, funding, its impact trough EU legislation on environment, climate, fisheries and others as well as the possibilities for cooperation in the future on issues such as the development of mapping and maritime safety, economic development and alike, it can be concluded that the EU has a lot to contribute to the sustainable development of the Arctic, a region that will be of major importance to a world adapting to climate change, facing growing population and scarcity of resources.

Result of Final Vote in Committee

Date adopted	9.12.2010
Result of final vote	+: 46 –: 0 0: 7
Members present for the final vote	Gabriele Albertini, Arnaud Danjean, Michael Gahler, Marietta Giannakou, Ana Gomes, Andrzej Grzyb, Takis Hadjigeorgiou, Richard Howitt, Anneli Jäätteenmäki, Ioannis Kasoulides, Nicole Kiil-Nielsen, Maria Eleni Koppa, Andrey Kovatchev, Wolfgang Kreissl-Dörfler, Eduard Kukan, Alexander Graf Lambsdorff, Vytautas Landsbergis, Krzysztof Lisek, Sabine Lösing, Ulrike Lunacek, Mario Mauro, Kyriakos Mavronikolas, Alexander Mirsky, María Muñiz De Urquiza, Norica Nicolai, Raimon Obiols, Ria Oomen-Ruijten, Pier Antonio Panzeri, Ioan Mircea Paşcu, Vincent Peillon, Alojz Peterle, Bernd Posselt, Hans-Gert Pöttering, Cristian Dan Preda, Fiorello Provera, Nikolaos Salavrakos, Jacek Saryusz-Wolski, Werner Schulz, Charles Tannock, Inese Vaidere, Graham Watson
Substitute(s) present for the final vote	Laima Liucija Andrikienė, Elena Băsescu, Hélène Flautre, Lorenzo Fontana, Kinga Gál, Liisa Jaakonsaari, Elisabeth Jeggle, Metin Kazak, Konrad Szymański, Indrek Tarand, Traian Ungureanu, Janusz Władysław Zemke

Annex 4
Programme

Auswärtiges Amt

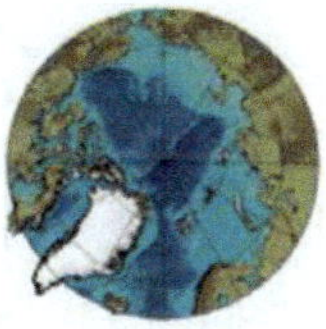

Arctic Science, International Law
and Climate Change

– Legal Aspects of Marine Science
in the Arctic Ocean –
17-18 March 2011

Organized by

The Federal Foreign Office Germany
in cooperation with the
Ministry of Foreign Affairs of Finland

Thursday, 17 March 2011

2.15 p.m.	registration
3.00 p.m.	**Opening: Setting the Scene: Welcome and Keynote Addresses**
	Welcome by the Chair, Susanne Wasum-Rainer
	Guido Westerwelle, Foreign Minister, Germany
	Maria Damanaki, Commissioner, EU
	Jaakko Laajava, Under-Secretary of State, Finland
3.45 p.m.	Short coffee break
4.00 p.m.	**Block 1: Ongoing and future Arctic marine research**

Speakers:

Lars-Otto Reiersen *"Arctic – the Sentinel for Environmental Processes and Effects"*

Michael Gahler *"More relevant research for the EU Arctic Policy"*

Peter Lemke *"The Role of the Arctic in the Global Climate System the Arctic Ocean"*

Keith Alverson *"Ongoing and future Arctic marine research"*

Igor Ashik *"Recent Russian marine research activities in the Arctic Ocean"*

Moderator: **Karin Lochte**

5.30	Coffee break
6.00 p.m.	**Block 2: Marine research in the Arctic Ocean in practice**

Speakers:

Uwe Nixdorf *"Arctic research in practice: on sea, on land, in the air"*

Larry Mayer *"Marine Research in the Arctic: The Perspective of a U.S. Practitioner"*

Bernard Coakley *"Environmental Permitting Constraints on Arctic Marine Scientific Research"*

Alexander Studenetsky *"Marine Scientific Research in waters under jurisdiction of the Russian Federation . Russian Legislation"*

Klemetti Näkkäläjärvi *"Climate change and traditional knowledge"*

Moderator: **Jaakko Laajava**

7.45 p.m.	**Transfer (bus shuttle) to buffet reception** at the Finnish Embassy, Rauchstraße 1, 10787 Berlin
10.00 p.m.	**Transfer (bus shuttle)** from Finnish Embassy **to Potsdamer Platz** and **Hotel Arcotel John F** at the Federal Foreign Ministry

Friday, 18 March 2011

9.00 a.m.	**Block 3: The legal framework**

Speakers:

Christian Marcussen *"Extended continental shelf issues in the Arctic Ocean: A modern land "grab" or an example of cooperation between the Arctic coastal states?"*

Ted L. McDorman *"Setting the Stage: The Continental Shelf and Marine Science in the Arctic Ocean"*

Judge Zhiguo Gao *"Legal Issues of Marine scientific Research in the Arctic: A Chinese Perspective"*

Marie Jacobsson *"Time for regime change? Some views on the international regulation of scientific research"*

Paul Berkman *"Balancing National Interests and Common Interests in the Arctic Ocean"*

Moderator: **Susanne Wasum-Rainer**

10.30 a.m.	Coffee break
11.00 p.m.	**Block 4: Enlarged continental shelves and the freedom of marine science**

Speakers:

Michael Lodge *"The International Seabed Authority and the Arctic"*

Alex Oude Elferink *"Will the establishment of the outer limits of the continental shelf affect the regime for marine scientific research in the Arctic?"*

Betsy Baker *"Principles for Marine Scientific Research Access in the Arctic Ocean"*

Judge Vladimir Golitsyn *"Climate change, marine science and delineation of continental shelf"*

Bjørn Kunoy *„Spatial Limitations in the Establishment of the Outer Limits"*

Moderator: **Judge Rüdiger Wolfrum**

12.45 am.	Lunch break
2.00 p.m.	**Block 5: Particularities and challenges**

Speakers:

Mikko Niini *"The technology drivers towards increased industrial activity in the Arctic Ocean"*

Jean-Claude Gascard *"From DAMOCLES (Developing Arctic Modelling and Observing Capabilities for Long-term Environmental Studies) to ACCESS (Arctic Climate Change impacts on Economic Sectors and Society)"*

Hajo Eicken *"Beyond the 4th International Polar Year as a testbed of internationally coordinated, collaborative Arctic marine research: Lessons learned from sea-ice research initiatives"*

Udloriak Hanson *"Science and research in the Arctic Ocean from an Inuit perspective"*

Sergey Priamikov *"Peculiarities of procedures of application on Marine Scientific Research in EEZ of Arctic ocean coastal states"*

Moderator: **Paula Kankaanpää**

3.30 p.m.	Coffee break
3.45 p.m.	Presentation of results and conclusions by the Chair

Chairs and Speakers

Guido Westerwelle, Federal Minister for Foreign Affairs, Berlin

Maria Damanaki, EU Commissioner for Maritime Affairs and Fisheries, Brussels

Jaakko Laajava, Under-Secretary of State, Ministry for Foreign Affairs, Helsinki

Susanne Wasum-Rainer, Federal Foreign Office, Berlin

Lars-Otto Reiersen, Arctic Monitoring and Assessment Programme of the Arctic Council, Oslo

Michael Gahler, European Parliament, Brussels

Peter Lemke, Alfred Wegener Institute for Polar and Marine Research, Bremerhaven

Keith Alverson, Intergovernmental Oceanographic Commission of UNESCO, Paris

Alexander Studenetsky, Ministry for Science and Education, Moscow

Karin Lochte, Alfred Wegener Institute for Polar and Marine Research, Bremerhaven

Uwe Nixdorf, Alfred Wegener Institute for Polar and Marine Research, Bremerhaven

Larry Mayer, Centre for Coastal and Ocean Mapping/NOAA-UNH Joint Hydrographic Centre, University of New Hampshire

Bernard Coakley, Geophysical Institute of the University of Alaska, Fairbanks

Klemetti Näkkäläjärvi, Sami Parlament, Helsinki

Igor Ashik, Arctic and Antarctic Research Institute, St. Petersburg

Christian Marcussen, Geological Survey of Denmark and Greenland, Copenhagen

Ted L. McDorman, Faculty of Law, University of Victoria

Judge Zhiguo Gao, International Tribunal for the Law of the Sea, Hamburg

Marie Jacobsson, Ministry of Foreign Affairs, Stockholm

Paul Berkman, Scott Polar Research Institute, Cambridge

Michael Lodge, International Seabed Authority, Kingston

Alex Oude Elferink, Netherlands Institute for the Law of the Sea, School of Law, University of Utrecht

Betsy Baker, Vermont Law School, South Royalton

Judge Vladimir Golitsyn, International Tribunal for Law of the Sea, Hamburg

Bjørn Kunoy, Ministry of Foreign Affairs of the Faroes, Tórshavn

Judge Rüdiger Wolfrum, International Tribunal for the Law of the Sea, Hamburg

Mikko Niini, Aker Arctic Technology, Helsinki

Jean-Claude Gascard, Université Pierre et Marie Curie, Paris

Hajo Eicken, Geophysical Institute of the University of Alaska, Fairbanks

Udloriak Hanson, Inuit Tapiriit Kanatami, Ottawa

Sergey Priamikov, Arctic and Antarctic Research Institute, St. Petersburg

Paula Kankaanpää, Arctic Center, University of Lapland, Rovaniemi

The conference takes place at the

Federal Foreign Office,

Unterwasserstr. 10, 10117 Berlin

(Europasaal)

Homepage: www.arktis.diplo.de

With the support of the

Alfred Wegener Institute for Polar and Marine Research, Bremerhaven
Arctic and Antarctic Research Institute, St. Petersburg
Arctic Centre, University of Lapland, Rovaniemi
International Arctic Research Centre, Fairbanks
Max Planck Institute for Comparative Public Law and
International Law, Heidelberg

Annex 5
List of Participants

Name/First Name	Title/Function/Organisation	Country
Alverson, Keith (Speaker)	Director of the Global Ocean Observing System, International Oceanographic Commission of UNESCO, Paris	United States of America
Amling, Dagmar	Federal Foreign Office, Berlin	Germany
Ashik, Igor (Speaker)	Dr., Head Oceanology Department, Arctic and Antarctic Research Institute (AARI), St. Petersburg	Russian Federation
Auffret, Yves	EU-Commission, Brussels	European Union
Aust, Helmut	Dr., Federal Foreign Office, Berlin	Germany
Baker, Betsy (Speaker)	Associate Professor, Vermont Law School, South Royalton	United States of America
Beckmann, Jens	Federal Navy, Bremerhaven	Germany
Belyaev, Oleg	2nd Secretary, Embassy of the Russian Federation	Russian Federation
Berkman, Paul (Speaker)	Head of the Arctic Ocean Geopolitics Programme, Scott Polar Research Institute, University of Cambridge, Cambridge	United Kingdom
Berzinš, Jānis	2nd Secretary, Embassy of Latvia	Latvia
Biebow, Nicole	Dr., Alfred Wegener Institute for Polar and Marine Research, Bremerhaven	Germany

Bloom, Evan T.	Director, Office of Ocean and Polar Affairs, Department of State, Washington D.C.	United States of America
Buckow, Johannes	Federal Foreign Office, Berlin	Germany
Burhenne, Wolfgang E.	Dr., Secretary General, Inter Parliamentary Working Center, Bonn	Germany
Carson, Robert	Assistant Deputy Minister, Department of Executive and Intergovernmental Affairs, Government of Nunavut, Ottawa	Canada
Catoir, Friedrich	Legal Expert, Advisor, European Science Foundation, Strasbourg	Germany
Coakley, Bernard (Speaker)	Dr., Associate Professor of Geology and Geophysics, Geophysical Institute of the University of Alaska, Fairbanks	United States of America
Cousineau, Patrice	Director, Oceans and Environmental Law Division, Department of Foreign Affairs and International Trade, Ottawa	Canada
Damanaki, Maria (Keynote Speaker)	Commissioner for European Maritime Affairs and Fisheries, Brussels	European Union
Damaske, Detlef	Dr., Federal Institute for Geosciences and Natural Resources (BGR), Hanover	Germany
Damm, Sebastian Christoph	Federal Foreign Office, Berlin	Germany
Debusmann, Martin Eduard	Scott Polar Research Institute, University of Cambridge, Cambridge	United Kingdom

Desch, Eberhard	Federal Ministry of Justice, Berlin	Germany
Didukh, Volodymyr	1st Secretary, Embassy of Ukraine	Ukraine
Dijk van, Harmen	1st Secretary, Embassy of The Netherlands	Netherlands
Dolata-Kreutzkamp, Petra	Dr., King's College London, London	United Kingdom
Drewes, Oliver	EU-Commission, Brussels	European Union
Dummermuth, Angelika	Dr., Alfred Wegener Institute for Polar and Marine Research, Bremerhaven	Germany
Eder, Michele Longo	Commissioner, United States Arctic Research Commission, Newport	United States of America
Egerton, Paul	Dr., Director, European Polar Board, European Science Foundation, Strasbourg	France
Eggen, Kjell Kristian	Deputy Director General, Ministry of Foreign Affairs of Norway, Oslo	Norway
Eicken, Hajo (Speaker)	Dr., Associate Professor, Geophysical Institute of the University of Alaska, Fairbanks; Chair of the Steering Committee for the US Interagency Study of Environmental Arctic Change (SEARCH), Fairbanks	United States of America
Eigemann, Eugen	Commander, Federal Ministry of Defence, Bonn	Germany
Fetz, Thomas	1st Secretary, Embassy of Canada	Canada
Flasche, Jan	Federal Foreign Office, Berlin	Germany

Friess, Bernhard	Director, Directorate-General for Maritime Affairs (DG MARE), Brussels	European Union
Fritz, Jan-Stefan	Dr., German Marine Research Consortium, Brussels	Germany
Fuchs, Harald	Dr., Client Solution Executive, iDEAL IMT Germany IBM Global Services, Berlin	Germany
Fujita, Shinya	1st Secretary, Embassy of Japan	Japan
Gaedicke, Christoph	Dr., Federal Institute for Geosciences and Natural Resources (BGR), Hanover	Germany
Gahler, Michael (Speaker)	Member of the European Parliament, Strasbourg	European Union
Gao, Zhiguo (Speaker)	Judge, International Tribunal for the Law of the Sea, Hamburg	China
Garcia, José Luis	Counsellor, Embassy of Spain	Spain
Gascard, Jean-Claude (Speaker)	Université Pierre et Marie Curie (UPMC), Paris	France
Gerdes, Rüdiger	Prof. Dr., Climate Sciences/Sea Ice Physics, Alfred Wegener Institute for Polar and Marine Research, Bremerhaven	Germany
Gluchowska-Wójcicka, Maja	Polish Naval Academy, Gdansk	Poland
Golitsyn, Vladimir (Speaker)	Judge, International Tribunal for the Law of the Sea, Hamburg	Russian Federation
Gounaris, Emmanuel	Dr.; Expert, Minister-Counsellor, Ministry of Foreign Affairs of Greece, Athens	Greece

Gradinariu, Laura	General Department for Legal Affairs, Ministry of Foreign Affairs of Romania, Bucharest	Romania
Haftendorn, Helga	Prof. Dr., FU Berlin	Germany
Hain, Stefan	Dr., Alfred Wegener Institute for Polar and Marine Research, Bremerhaven	Germany
Hakkarainen, Petri	2nd Secretary, Embassy of Finland	Finland
Halinen, Hannu (Moderator)	Ambassador, Arctic Affairs, Ministry for Foreign Affairs of Finland	Finland
Hanson, Udloriak (Speaker)	Senior Policy Liaison for Nunavut Tunngavik Inc., Ottawa	Canada
Hemmerling, Mario	Federal Foreign Office, Berlin	Germany
Henin, Thibaud	Ecologic Institute, Berlin	Canada
Herata, Heike	Dr., Federal Environment Agency, Dessau	Germany
Herget, Sabine	Research Assistant, Head of the office of Franz Thönnes, Member of Parliament, Berlin	Germany
Herzig, P.	Prof. Dr., Director, Leibniz Institute of Marine Sciences (FM-GEOMAR), Kiel	Germany
Hilbert, Jacqueline	Federal Environment Agency, Dessau	Germany
Høgseth, Thomas Bruusgaard	2nd Secretary, Embassy of Norway	Norway

Holfort, Jürgen	Dr., Federal Maritime and Hydrographic Agency, Rostock	Germany
Hosseus, Daniel	Association of German Shipowners, Hamburg	Germany
Hubberten, Hans-Wolfgang	Prof. Dr., Alfred Wegener Institute for Polar and Marine Research, Bremerhaven	Germany
Hubert, Anna-Maria	Bremen Graduate School for Marine Sciences (GLOMAR), Bremen	United Kingdom
Hubert, Casey	Dr., Newcastle University, Newcastle	United Kingdom
Humrich, Christoph	Prof. Dr., Institute for Intercultural and International Studies, University of Bremen	Germany
Hunt, Steve	Assistant Desk Officer, Maritime Policy Unit, Legal Adviser, UK Foreign and Commonwealth Office, London	United Kingdom
Ingenfeld, Eva	University of Heidelberg, Dept. of Geography, Heidelberg	Germany
Ivanov, Vladimir	International Arctic Research Centre (IARC), Fairbanks	United States of America
Jaekel, Anja	Federal Foreign Office, Berlin	Germany
Juntunen, Suvi	Saami Parliament, Helsinki	Finland
Kaiser, Daniel	Intern, Office of Franz Thönnes, Member of Parliament, Berlin	Germany
Kankaanpää, Paula (Moderator)	Dr., Arctic Center, University of Lapland, Rovaniemi	Finland

Kato, Kikuko	Director, Ministry of Foreign Affairs, Tokyo	Japan
Kato, Masahiro	3rd Secretary, Embassy of Japan	Japan
Kaufmann, Sven G.	Higher Regional Court, Berlin	Germany
Kauker, Frank	Dr., Alfred Wegener Institute for Polar and Marine Research, Bremerhaven	Germany
Keil, Kathrin	Graduate School for Transnational Studies (BTS), FU Berlin	Germany
Kiku, Dmitry V.	1st Secretary, Embassy of the Russian Federation	Russian Federation
Klinke, Robert	Federal Foreign Office, Berlin	Germany
Kischitzki, Josephin	Embassy of the United States of America	United States of America
Kloke, Bernd	Federal Ministry of Economics and Technology, Berlin	Germany
König, Doris	Prof. Dr., Bucerius Law School, Hamburg	Germany
Kolodziej, Agnes	Desk Officer, Embassy of Canada	Canada
Korup, Christina Hjort	Embassy of Denmark	Denmark
Kozubovskaya-Pellé, Anastasiya	Dr., Legal Project Manager, European Science Foundation, Polar Research Icebreaker Consortium, Strasbourg	France
Kraemer, R. Andreas	Ecologic Institute, Berlin	Germany
Kristjánsdóttir, Ólöf Hrefna (18.03.)	1st Secretary, Embassy of Iceland	Iceland

Kunoy, Bjorn (Speaker)	Legal Adviser, Ministry of Foreign Affairs of the Faroes, Tórshaven	Faroes
Laajava, Jaakko (Keynote Speaker)	Under-Secretary of State for Foreign and Security Policy, Finnish Ministry of Foreign Affairs, Helsinki	Finland
Leiholt, Casper	1st Secretary, Embassy of Denmark	Denmark
Lembke-Jene, Lester	Science & Technology Manager, European Research Icebreaker Consortium – AURORA BOREALIS, Alfred Wegener Institute for Polar and Marine Research, Bremerhaven	Germany
Lemke, Peter (Speaker)	Prof., Alfred Wegener Institute for Polar and Marine Research, Bremerhaven	Germany
Lindemann, Christian	Federal Ministry for the Environment, Nature Conservation and Nuclear Safety (Strategic Aspects of International Cooperation, Regional Conventions, International Law), Berlin	Germany
Lindström, Guy	Deputy Director, International Department Parliament of Finland, Helsinki	Finland
Lochte, Karin (Moderator)	Prof. Dr., Director Alfred Wegener Institute for Polar and Marine Research, Bremerhaven	Germany
Lodge, Michael (Speaker)	Legal Counsel, International Seabed Authority, Kingston	Jamaica
Makens, Brett	2nd Secretary, -Embassy of the United States of America	United States of America

Makepeace, Jonathan	Senior Desk Officer, Europe and Eurasia Bureau (GUB), Foreign Affairs and International Trade, Ottawa	Canada
Marcussen, Christian (Speaker)	Senior Advisor, Geological Survey of Denmark and Greenland (GEUS), Copenhagen	Denmark
Mayer, Larry (Speaker)	Professor, Center for Coastal and Ocean Mapping, Joint Hydrographic Institute, Durham	United States of America
Mayet, Laurent	Ambassador, Ministry of Foreign Affairs, Paris	France
McDorman, Ted L. (Speaker)	Faculty of Law, University of Victoria, Victoria	Canada
Medert, Henning	CNC – Communications & Network Consulting AG, Berlin	Germany
Menn, Iris	Greenpeace e.V., Hamburg	Germany
Misztal, Andrzej	Deputy Head of Law Division, Ministry of Foreign Affairs, Warsaw	Poland
Monsma, David	Executive Director, Aspen Institute, Washington D.C.	United States of America
Näkkäläjärvi, Klemetti (Speaker)	President, Sami Parliament, Helsinki	Finland
Neumann, Antje	German Institute for International and Security Affairs, Berlin	Germany
Niini, Mikko (Speaker)	Aker Arctic Technology, Helsinki	Finland
Nixdorf, Uwe (Speaker)	Dr., Alfred Wegner Institut for Polar und Marine Research, Bremerhaven	Germany

Notz, Dirk	Dr., Max Planck Institute for Meteorology, Hamburg	Germany
Oude Elferink, Alex G. (Speaker)	Dr., Netherlands Institute for the Law of the Sea (NILOS) – School of Law, Utrecht University, Utrecht	Netherlands
Panait, Valentin	1st Secretary, Embassy of Romania	Romania
Parson, Lindsay	Dr., UNCLOS Group, National Oceanography Centre, Southampton	United Kingdom
Peinert, Rolf	Dr., Managing Director, German Marine Research Consortium, Berlin	Germany
Petschke, Mathias	Head of the Representation of the European Commission in Berlin	European Union
Plouffe, Joël	University of Quebec at Montreal, Center for Geopolitical Studies, Montreal	Canada
Poulsen-Hansen, Per	Ambassador of Denmark	Denmark
Priamikov, Sergey (Speaker)	Dr., Arctic and Antarctic Research Institute (AARI), St. Petersburg	Russian Federation
Prokosch, Peter	UNEP/GRID-Arendal, Arendal	Norway
Przygodzka-Markiewicz, Marta	Lawyer, Szczecin	Poland
Pushkareva, Elvira	Prof. Dr., Max-Planck Institute for Comparative Public Law and International Law; Russian Academy of Sciences, Heidelberg	Russian Federation
Rachold, Volker	Dr., Executive Secretary, International Arctic Science Committee (IASC), Potsdam	Germany

Rangreji, Luther	Dr., Senior Legal Officer, Ministry of External Affairs, New Delhi	India
Reichert, Christian	Dr., Federal Institute for Geosciences and Natural Resources (BGR), Hanover	Germany
Reiersen, Lars-Otto (Speaker)	Dr., Executive Secretary, Arctic Monitoring and Assessment Programme (AMAP), Oslo	Norway
Reinemann, Joachim	Embassy of the Republic of Korea	Korea
Reppe, Silvia	Federal Ministry for the Environment, Berlin	Germany
Rogers, Roland J.	Advisor Marine Law and Environment, National Oceanography Centre, Southampton	United Kingdom
Romanyshyn, Iulian	German Institute for International and Security Affairs, Berlin	Germany
Rünger, Detlev	Ambassador of the Federal Republic of Germany in Oslo	Germany
Savva, Vera	Legal Department of the Council for the Study of Productive Resources, Russian Academy of Sciences, Moscow State Institute of International Relations (University) of the Ministry for Foreign Affairs of Russia, Moscow	Russian Federation
Sachs, Torsten	Dr., Helmholtz Centre Potsdam, GFZ German Research Centre for Geosciences, Potsdam	Germany

Schauer, Ursula	Alfred Wegener Institute for Polar and Marine Research, Bremerhaven	Germany
Schilar, Hannelene	Federal Foreign Office, Berlin	Germany
Schmidt, Johanna	Political Department, Representation of the European Commission in Berlin.	European Union
Schröder, Marcus	Dr., Federal Ministry for the Environment, Nature Conservation and Nuclear Safety, Munich	Germany
Schröder-Hinrichs, Jens-Uwe	World Maritime University, Malmö	Sweden
Schulte, Guido	Nordic Yards Holding GmbH, Wismar	Germany
Schwarz, Joachim	Dr., German Association for Marine Technologies, Hamburg	Germany
Schweisfurth, Theodor	Professor, European University Viadrina, Frankfurt/Oder	Germany
Segura, Serge	Ministry of Foreign Affairs, Legal Dept., Paris	France
Stangnes, Dag	Minister-Counsellor, Embassy of Norway	Norway
Stanley, Margaret	Dr., Embassy of Ireland	Ireland
Steinbrenner, Kathrin	Dr., Federal Foreign Office, Berlin	Germany
Studenetsky, Alexander (Speaker)	Ministry of Education and Science of the Russian Federation, Moscow	Russian Federation
Studenetsky, Alexander (Speaker)	Ministry of Education and Science of the Russian Federation, Moscow	Russian Federation

Symonides, Janusz	Prof., University of Warsaw	Poland
Takei, Yoshinobu	Netherlands Institute for the Law of the Sea, Utrecht University, Utrecht	Netherlands
Tamburelli, Gianfranco	Senior Arctic Official, National Research Council, Roma	Italy
Terstegen, Thomas	Federal Foreign Ministry, Berlin	Germany
Thönnes, Franz	Member of Parliament and Former Parliamentary State Secretary	Germany
Tiroch, Katrin	Research Fellow, Max Planck Institute for Comparative Public Law and International Law, Heidelberg	Austria
Tolvanen, Tapio	1st Secretary, Ministry of Foreign Affairs of Finland	Finland
Tomuschat, Christian	Prof. Dr. Dr., Humboldt University, Berlin	Germany
Tran, Quoc Trung	1st Secretary, Embassy of Vietnam	Vietnam
Trapnell, Jennifer	Yukon Government, Intergovernmental Relations, Executive Council Office, Whitehorse	Canada
Trudeau, Harley	Senior Representative, Yukon Government	Canada
Vöneky, Silja	Prof. Dr., Director of the Institute for the Theory of the States and Philosophy of Law, International Law, University of Freiburg i.Br.	Germany
Wagner, Jakob Ritter von	Federal Foreign Ministry, Berlin	Germany
Wagner, Jan Pit	IT Solutions	Germany

Wasum-Rainer, Susanne (Moderator)	Dr., Director General for Legal Affairs, Federal Foreign Office, Berlin	Germany
Weber, Steffen	EU Arctic Forum European Parliament	European Union
Weiss, Mathieu J.	Counsellor, Embassy of France	France
Wendelberger, Klaus	Federal Foreign Office, Berlin	Germany
Westerwelle, Guido (Keynote Speaker)	Federal Foreign Minister, Berlin	Germany
Winkelmann, Ingo	Dr., Head of Division, Ambassador, Federal Foreign Office, Berlin	Germany
Wojcicki, Konrad	Private Researcher, Gdynia	Poland
Wolff, Julia	University Hamburg	Germany
Wolfrum, Rüdiger (Moderator)	Prof. Dr., Judge, International Tribunal for the Law of the Sea, Hamburg	Germany
Wolski, Jakub T.	Ambassador, Ministry of Foreign Affairs, Warsaw	Poland
Zachariadis, Nadine	Federal Foreign Office, Berlin	Germany
Zackenfels, Stefan	Member of the Berlin City Parliament	Germany
Zysk, Katarzyna	Dr., Norwegian Institute for Defence Studies, Oslo	Norway

Max-Planck-Institut für ausländisches öffentliches Recht und Völkerrecht

Beiträge zum ausländischen öffentlichen Recht und Völkerrecht

Hrsg.: A. von Bogdandy, R. Wolfrum

Bde. 27–59 erschienen im Carl Heymanns Verlag KG Köln, Berlin (Bestellung an: Max-Planck-Institut für Völkerrecht, Im Neuenheimer Feld 535, 69120 Heidelberg); ab Band 60 im Springer-Verlag Berlin, Heidelberg, New York, London, Paris, Tokyo, Hong Kong, Barcelona

---

235 Susanne *Wasum-Rainer*, Ingo *Winkelmann*, Katrin *Tiroch* (eds.): **Arctic Science, International Law and Climate Change.** 2012. XIX, 374 Seiten. Geb. € 84,95 zzgl. landesüblicher MwSt.

234 Mirja A. *Trilsch*: **Die Justiziabilität wirtschaftlicher, sozialer und kultureller Rechte im innerstaatlichen Recht.** 2012. XIX, 559 Seiten. Geb. € 99,95

232 Sandra *Stahl*: **Schutzpflichten im Völkerrecht - Ansatz einer Dogmatik.** 2012. XXX, 505 Seiten. Geb. € 94,95

231 Thomas *Kleinlein*: **Konstitutionalisierung im Völkerrecht.** 2012. XLII, 940 Seiten. Geb. € 149,95

230 Roland *Otto*: **Targeted Killings and International Law.** 2012. XVIII, 661 Seiten. Geb. € 109,95 zzgl. landesüblicher MwSt.

229 Nele *Matz-Lück*, Mathias *Hong* (Hrsg.): **Grundrechte und Grundfreiheiten im Mehrebenensystem - Konkurrenzen und Interferenzen.** 2012. VIII, 394 Seiten. Geb. € 89,95

228 Matthias *Ruffert*, Sebastian *Steinecke*: **The Global Administrative Law of Science**, 2011. IX, 140 Seiten. Geb. € 59,95 zzgl. landesüblicher MwSt.

227 Sebastian *Pritzkow*: **Das völkerrechtliche Verhältnis zwischen der EU und Russland im Energiesektor.** 2011. XXIV, 304 Seiten. Geb. € 79,95

226 Sarah *Wolf*: **Unterseeische Rohrleitungen und Meeresumweltschutz.** 2011. XXIII, 442 Seiten. Geb. € 94,95

225 Clemens *Feinäugle*: **Hoheitsgewalt im Völkerrecht.** 2011. XXVI, 418 Seiten. Geb. € 89,95

224 David *Barthel*: **Die neue Sicherheits- und Verteidigungsarchitektur der Afrikanischen Union.** 2011. XXV, 443 Seiten. Geb. € 94,95

223 Tilmann *Altwicker*: **Menschenrechtlicher Gleichheitsschutz.** 2011. XXX, 549 Seiten. Geb. € 99,95

222 Stephan *Bitter*: **Die Sanktion im Recht der Europäischen Union.** 2011. XV, 351 Seiten. Geb. € 84,95

221 Holger *Hestermeyer*, Nele *Matz-Lück*, Anja *Seibert-Fohr*, Silja *Vöneky* (eds.): **Law of the Sea in Dialogue.** 2011. XII, 189 Seiten. Geb. € 69,95 zzgl. landesüblicher MwSt.

220 Jan *Scheffler*: **Die Europäische Union als rechtlich-institutioneller Akteur im System der Vereinten Nationen.** 2011. XXXV, 918 Seiten. Geb. € 149,95

219 Mehrdad *Payandeh*: **Internationales Gemeinschaftsrecht.** 2010. XXXV, 629 Seiten. Geb. € 99,95

218 Jakob *Pichon*: **Internationaler Strafgerichtshof und Sicherheitsrat der Vereinten Nationen.** 2011. XXVI, 399 Seiten. Geb. € 89,95

217 Michael *Duchstein*: **Das internationale Benchmarkingverfahren und seine Bedeutung für den gewerblichen Rechtsschutz.** 2010. XXVI, 528 Seiten. Geb. € 99,95

216 Tobias *Darge*: **Kriegsverbrechen im nationalen und internationalen Recht.** 2010. XXXV, 499 Seiten. Geb. € 94,95

215 Markus *Benzing*: **Das Beweisrecht vor internationalen Gerichten und Schiedsgerichten in zwischenstaatlichen Streitigkeiten.** 2010. L, 846 Seiten. Geb. € 139,95

214 Urs *Saxer*: **Die internationale Steuerung der Selbstbestimmung und der Staatsentstehung.** 2010. XLII, 1140 Seiten. Geb. € 169,95

213 Rüdiger *Wolfrum*, Chie *Kojima* (eds.): **Solidarity: A Structural Principle of International Law.** 2010. XIII, 238 Seiten. Geb. € 69,95 zzgl. landesüblicher MwSt.

212 Ramin S. *Moschtaghi*: **Die menschenrechtliche Situation sunnitischer Kurden in der Islamischen Republik Iran.** 2010. XXIII, 451 Seiten. Geb. € 94,95

211 Georg *Nolte* (ed.): **Peace through International Law. The Role of the International Law Commission.** 2009. IX, 195 Seiten. Geb. € 64,95 zzgl. landesüblicher MwSt.

210 Armin *von Bogdandy*, Rüdiger *Wolfrum*, Jochen *von Bernstorff*, Philipp *Dann*, Matthias *Goldmann* (eds.): **The Exercise of Public Authority by International Institutions.** 2010. XIII, 1005 Seiten. Geb. € 149,95 zzgl. landesüblicher MwSt.

209 Norman *Weiß*: **Kompetenzlehre internationaler Organisationen.** 2009. XVIII, 540 Seiten. Geb. € 99,95

208 Michael *Rötting*: **Das verfassungsrechtliche Beitrittsverfahren zur Europäischen Union.** 2009. XIV, 317 Seiten. Geb. € 79,95

207 Björn *Ahl*: **Die Anwendung völkerrechtlicher Verträge in China.** 2009. XIX, 419 Seiten. Geb. € 89,95

206 Mahulena *Hofmann*: **Von der Transformation zur Kooperationsoffenheit?** 2009. XIX, 585 Seiten. Geb. € 99,95

205 Rüdiger *Wolfrum*, Ulrike *Deutsch* (eds.): **The European Court of Human Rights Overwhelmed by Applications: Problems and Possible Solutions.** 2009. VIII, 128 Seiten. Geb. € 59,95 zzgl. landesüblicher MwSt.

204 Niels *Petersen*: **Demokratie als teleologisches Prinzip.** 2009. XXVII, 280 Seiten. Geb. € 79,95

203 Christiane *Kamardi*: **Die Ausformung einer Prozessordnung sui generis durch das ICTY unter Berücksichtigung des Fair-Trial-Prinzips.** 2009. XVI, 424 Seiten. Geb. € 89,95

202 Leonie F. *Guder*: **The Administration of Debt Relief by the International Financial Institutions.** 2009. XVIII, 355 Seiten. Geb. € 84,95 zzgl. landesüblicher MwSt.

201 Silja *Vöneky*, Cornelia *Hagedorn*, Miriam *Clados*, Jelena *von Achenbach*: **Legitimation ethischer Entscheidungen im Recht.** 2009. VIII, 351 Seiten. Geb. € 84,95

200 Anja Katarina *Weilert*: **Grundlagen und Grenzen des Folterverbotes in verschiedenen Rechtskreisen.** 2009. XXX, 474 Seiten. Geb. € 94,95

199 Suzette V. *Suarez*: **The Outer Limits of the Continental Shelf.** 2008. XVIII, 276 Seiten. Geb. € 79,95 zzgl. landesüblicher MwSt.

198 Felix *Hanschmann*: **Der Begriff der Homogenität in der Verfassungslehre und Europarechtswissenschaft.** 2008. XIII, 370 Seiten. Geb. € 84,95

197 Angela *Paul*: **Kritische Analyse und Reformvorschlag zu Art. II Genozidkonvention.** 2008. XVI, 379 Seiten. Geb. € 84,95

196 Hans Fabian *Kiderlen*: **Von Triest nach Osttimor.** 2008. XXVI, 526 Seiten. Geb. € 94,95

195 Heiko *Sauer*: **Jurisdiktionskonflikte in Mehrebenensystemen.** 2008. XXXVIII, 605 Seiten. Geb. € 99,95

194 Rüdiger *Wolfrum*, Volker *Röben* (eds.): **Legitimacy in International Law.** 2008. VI, 420 Seiten. Geb. € 84,95 zzgl. landesüblicher MwSt.

193 Doris *König*, Peter-Tobias *Stoll*, Volker *Röben*, Nele *Matz-Lück* (eds.): **International Law Today: New Challenges and the Need for Reform?** 2008. VIII, 260 Seiten. Geb. € 69,95 zzgl. landesüblicher MwSt.

192 Ingo *Niemann*: **Geistiges Eigentum in konkurrierenden völkerrechtlichen Vertragsordnungen.** 2008. XXV, 463 Seiten. Geb. € 94,95

191 Nicola *Wenzel*: **Das Spannungsverhältnis zwischen Gruppenschutz und Individualschutz im Völkerrecht.** 2008. XXXI, 646 Seiten. Geb. € 99,95

190 Winfried *Brugger*, Michael *Karayanni* (eds.): **Religion in the Public Sphere: A Comparative Analysis of German, Israeli, American and International Law.** 2007. XVI, 467 Seiten. Geb. € 89,95 zzgl. landesüblicher MwSt.

189 Eyal *Benvenisti*, Chaim *Gans*, Sari *Hanafi* (eds.): **Israel and the Palestinian Refugees.** 2007. VIII, 502 Seiten. Geb. € 94,95 zzgl. landesüblicher MwSt.

188 Eibe *Riedel*, Rüdiger *Wolfrum* (eds.): **Recent Trends in German and European Constitutional Law.** 2006. VII, 289 Seiten. Geb. € 74,95 zzgl. landesüblicher MwSt.

187 Marcel *Kau*: **United States Supreme Court und Bundesverfassungsgericht.** 2007. XXV, 538 Seiten. Geb. € 99,95 zzgl. landesüblicher MwSt.

186 Philipp *Dann*, Michał *Rynkowski* (eds.): **The Unity of the European Constitution.** 2006. IX, 394 Seiten. Geb. € 79,95 zzgl. landesüblicher MwSt.

GPSR Compliance

The European Union's (EU) General Product Safety Regulation (GPSR) is a set of rules that requires consumer products to be safe and our obligations to ensure this.

If you have any concerns about our products, you can contact us on ProductSafety@springernature.com

In case Publisher is established outside the EU, the EU authorized representative is:

Springer Nature Customer Service Center GmbH
Europaplatz 3
69115 Heidelberg, Germany

Batch number: 10682073

Printed by Printforce, the Netherlands